Analysis and Application of Rare Earth Materials

edited by
ODD B. MICHELSEN

NATO ADVANCED STUDY INSTITUTE

Kjeller, Norway, 23rd - 29th August 1972
organized by
The Netherlands-Norwegian Reactor School
Institutt for Atomenergi, Kjeller, Norway

UNIVERSITETSFORLAGET 1973
Oslo - Bergen - Tromsø

© Universitetsforlaget 1973

ISBN 82-00-04780-6

Printed in Norway by
Foto-Trykk, Oslo

PREFACE

Since World War II remarkable progress has been made in the chemistry and physics of the rare earth (RE) elements. In its wake followed as a natural consequence a similar development in the analytical techniques used for their determination. The improved methods of separation, which have made the RE elements available in purities comparable to those of most other elements of the Periodic Table, and their extended use both in research and for industrial purposes, have set RE analysts a difficult task, in view of the increasingly rigorous requirements of versatility, specificity, sensitivity, and accuracy.

Research on RE elements has been going on in Norway since the first nuclear reactor was started up at Kjeller in 1951, particularly into the separation chemistry of these elements. Initially, when the work was mainly associated with the separation of fission products, little emphasis was placed on the analytical chemistry of the RE elements. The situation changed when interest became focused on the separation of macro-quantities of these elements. Although a wealth of information was available from publications in the field, it soon became clear that RE analysis was a difficult art with problems of its own, which sometimes had to be individually dealt with by the laboratories involved. It was felt that a broad discussion of the state of RE analysis was timely, particularly since the subject rarely came up at the ordinary meetings on RE elements.

A grant from the NATO Scientific Affairs Division enabled a NATO Advanced Study Institute to be set up. It was arranged at the Netherlands-Norwegian Reactor School, Institutt for Atomenergi, Kjeller, Norway, from August 23 to August 29, 1972. The members of the organizing committee were:

J. Alstad, University of Oslo
E. Andersen, Institutt for Atomenergi (chairman)

M. Bonnevie-Svendsen, Institutt for Atomenergi

B. Gaudernack, Institutt for Atomenergi

O.B. Michelsen, Institutt for Atomenergi

The committee considered that it might serve a useful purpose if, in addition to RE analysis, a section on the application of RE elements, which is also a somewhat neglected area at RE conferences, was included. The subject title of the Study Institute thus became: "Analysis and Application of Rare Earth Materials".

This book contains the 23 lectures given at the Institute and also a final chapter on the application of luminescent materials, which is a presentation of the main features of a panel discussion prepared for these proceedings by Dr. Tecotzkv, the chairman of the discussion. It is hoped that the book will provide a representative picture of the present state of RE analysis and application and thus be of interest to all those concerned with these fields.

The style and the quality of the language of the articles in this book are in essence as submitted by the individual authors.

The organizing committee would like to express its gratitude to the NATO Scientific Affairs Division for the grant that made the Study Institute possible and to the management of the Institutt for Atomenergi for placing their facilities at its disposal. Thanks are also due to Mrs. G. Jarrett, who did a great deal of the practical work connected with the meeting, and to Mrs. J. Tveten for her careful retyping of the manuscripts.

The Editor

C O N T E N T S

Page

K.A. Gschneidner, Jr.,

General Review of Scientific and Technological Aspects
of the Utilization of RE Materials   1

K.A. Gschneidner, Jr.,

Fundamental Properties of RE Elements and Their
Implications on the Analytical Chemistry of RE Materials   9

H.A. Das and J. van Ooyen,

Techniques for the Mutual Separation of the Lanthanides
in Analytical Chemistry   25

E. Herrmann,

Determination of Rare Earth Impurities in Pure Rare
Earths by Means of Extraction Chromatography   39

F. Molnár,

Anion Exchange Concentration of Rare Earth Impurities
in Rare Earth Materials for Analytical Purposes   55

E.L. DeKalb,

Emission Spectroscopy for Rare Earth Analyses, and for
the Determination of Impurities in Rare Earth Materials   61

V.A. Fassel,

Flame and Plasma Atomic Absorption, Emission and
Fluorescence Spectroscopy of the Rare Earth Elements   71

M. Bonnevie-Svendsen and A. Follo,

X-Ray Fluorescence Analysis of Rare Earth Elements   87

V.A. Fassel,

Trace Level Rare Earth Determinations by X-Ray Excited
Optical Fluorescence (XEOF) Spectroscopy   109

S. Larach,

Cathode-Ray Excited Emission (CREE) Spectroscopic
Analysis of Trace Rare Earths   123

Page

J. Haaland and D.E. Stijfhoorn,

Trace Elements Characterization in Y_2O_3 by Mass
Spectroscopy and Isotopic Dilution 153

E. Steinnes,

Rare Earth Determination by Neutron Activation Analysis 165

L.A. Haskin and R.L. Korotev,

Determination of Rare Earths in Geological Samples
and Raw Materials 183

G. Blasse,

Theoretical Basis for the Use of Rare Earths
in Optical and Luminescent Materials 213

P.N. Yocom,

Application of Rare Earths in Optical
Materials and Devices 229

J.E. Mathers,

Production of Rare Earth Red Phosphors for Colour
Television and Lighting Applications 241

M. Steinberg,

The Surface Activity of Solid Rare Earth Oxides 263

K.A. Gschneidner, Jr.,

Miscellaneous Uses of RE Materials
(Ceramics, Glasses, Nuclear Applications, etc.) 271

G. Blasse,

Theoretical Basis for the Use of
Rare Earths in Magnetic Materials 281

C.S. Brown,

Production and Application of Rare Earth Ferrimagnetic
Garnets. Part I: Application in Microwave Components 297

C.S. Brown,

Production and Application of Rare Earth Ferrimagnetic
Garnets. Part II: Application in Bubble Domain Stores 315

H. Stäblein,

Permanent Magnets Based on RE Materials 331

  Page

K.A. Gschneidner, Jr.,

    The Production, Quality Control, and Application
    of Pure RE Metals and Alloys   351

Panel Discussion

    Rare Earth Applications in Luminescence
    (Chairman: M. Tecotzky)   359

List of Participants   373

GENERAL REVIEW OF SCIENTIFIC AND TECHNOLOGICAL ASPECTS

OF THE UTILIZATION OF RE MATERIALS

by

Karl A. Gschneidner, Jr.

Ames Laboratory USAEC and Department of Metallurgy

Iowa State University, Ames, Iowa 50010, U.S.A.

ABSTRACT

The major uses and some of the more promising potential applications are examined from a scientific and technological view. This has led to the subdivision of these applications into four groups: those uses based on (1) the similar chemical nature of the rare earths, (2) the 4f electrons, (3) valence states different than +3, and (4) nuclear properties. In general, the mixed rare earths are used primarily in (1) and the separated individual elements in (2), (3), and (4), although some notable exceptions are found. The scientific basis for the various applications is briefly discussed.

1. INTRODUCTION

Since the term "rare earths" represents several different possible groupings of the elements scandium, yttrium and the lanthanides La through Lu, it is best to define this term as used in this and my subsequent papers as referring to yttrium and the 15 lanthanide elements, excluding scandium. Although this definition is different than that agreed upon by the International Union of Pure and Applied Chemistry (which also includes scandium), there are several practical reasons for omitting scandium. These are the minute commercial utilization of this element and its much different chemical behaviour from that of the other Group IIIA elements.

The diverse uses of the rare earth elements can be divided into three major categories and one minor one. The first major category is the one in which uses are based primarily on the similar chemical nature due to the essentially identical outer electronic configuration of three valence electrons (Table 1).

TABLE 1

Electronic Configurations of the Rare Earths[a]

| Element | Solid or Liquid State (Real Life) Configurations | | | | Ground State (Gaseous Configuration) No. Electrons | | |
| | Normal | | Unusual | | | | |
	No. 4f Electrons	No. Valence Electrons	No. 4f Electrons	No. Valence Electrons	4f	5d	6s
La	0	3	---	---	0	1	2
Ce	1	3	0	4	1	1	2
Pr	2	3	1	4	3	--	2
Nd	3	3	--	--	4	--	2
Pm	4	3	--	--	5	--	2
Sm	5	3	6	2	6	--	2
Eu	6	3	7	2	7	--	2
Gd	7	3	--	--	7	1	2
Tb	8	3	7	4	9	--	2
Dy	9	3	--	--	10	--	2
Ho	10	3	--	--	11	--	2
Er	11	3	--	--	12	--	2
Tm	12	3	--	--	13	--	2
Yb	13	3	14	2	14	--	2
Lu	14	3	--	--	14	1	2
Y	0	3	--	--	0	1[b]	2[c]

[a] See footnote on p.6.

[b] 4d electron

[c] 5s electrons

The second major category is the one in which uses are based primarily on the 4f electrons or the lack of them. The third major category is based on either a non-trivalent (an "unusual" valence) state and/or a variable valence state. The minor category is based on the nuclear rather than electronic nature of these elements.

The rare earths are used in the mixed state in the first category, but as individual elements in the other three categories, with one or two notable exceptions. Each of these four categories are discussed in more detail below.

2. USES BASED ON SIMILARITY OF CHEMICAL BEHAVIOURS

In general the chemical (and metallurgical) behaviour of the rare earth elements are quite similar to one another, primarily because their outer electronic configurations consist of three valence electrons, and their minor differences arise primarily because of the slight variation in size due to the lanthanide contraction. Because of this minute variation in the chemical properties there is little need to separate them - the small change (improvement) in the property is not worth the cost of separating one element from another. In some instances there may be a large enough difference between the chemical behaviours of the lights vs. that of the heavies, where a crude separation into two groups is economically justified. Today, however, such applications are minor ones.

Since the rare earths are used as a group, their cost per unit weight is small, and thus make large-volume applications possible. They find use as catalysts - primarily in zeolite molecular sieve catalysts for cracking oil, mischmetal and rare earth silicide additives to ductile iron and steels, mischmetal-iron lighter flints, mixed oxide polishing compounds, and more recently as liners for military shells.

The exact chemical role played by the rare earths in petroleum cracking catalysts is not known, but the addition of 1 to 5% mixed rare earths to the $Na_2O \cdot Al_2O_3 \cdot nSiO_2 \cdot xH_2O$ zeolites does increase the catalytic efficiency by a factor of three, permitting a three-fold reduction in capital investment to produce a given amount of refined petroleum. First introduced in 1961 the lifetime of the rare earth cracking catalyst has exceeded that of most cracking catalysts, which is about five years.

The use of mixed rare earth metals is based primarily on their reactive nature with respect to the non-metallic elements, primarily oxygen, nitrogen and their respective congeners. The rare earths are added to steel to control the sulphur and thus, improve workability and some of the steels' mechanical properties. The use of mischmetal in lighter flints and artillery shell liners is based on the pyrophoric nature of the rare earths in general. But it is primarily the pyrophoric nature of cerium in the mixture, which accounts for the use of mischmetal in these applications. The largest metallurgical application is the addition of rare earths to ductile iron to spheroidize the graphite. Although the mischmetal is more expensive than magnesium, it has replaced much of the magnesium used in this application because the rare earths are not volatile (magnesium is) and thus, the quality of the final product is more readily controlled.

The mixed rare earth oxides are also used to polish glass, but in general the effectiveness of the polishing mixture depends on the amount of cerium dioxide present, the more cerium, the more effective the polishing mixture. The polishing mechanism, whether chemical or physical, or both, is not known.

3. USES BASED ON THE 4f ELECTRON (OR ABSENCE THEREOF)

Since the 4f electron(s) is (are) at least indirectly involved in the application of interest, the individual rare earth element, usually in high purity (99% or higher) is required. For most of these elements it is necessary to use ion exchange and/or liquid-liquid extraction methods to obtain an element of the required purity. Cerium and europium are exceptions, since they can readily be purified by chemical means by making use of their dual valence states (+4 and +3 for Ce and +3 and +2 for Eu). The high cost of separating the individual rare earth elements accounts for the significant increase in cost of the individual element over the mixed rare earths[*], and therefore, the volume of individual rare earths consumed is small. However, in terms of money the sales of the individual rare earths account for about as large a share of the gross expenditure of currency as the mixed rare earths.

---

[*] The increase varies from a low of 7 times the cost of the mixed rare earth oxides for CeO_2 up to 6000 times for Lu_2O_3.

The principal and/or most promising uses of the individual rare earths are as hosts and activators in phosphors, magnetic and electronic materials, optical devices (fibre optics, lasers, etc.), alloy additives, carbon arcs and basic and applied research.

The energy levels associated with the 4f electrons and transitions within the 4f levels and between other electronic levels account for their utilization as activators in phosphors, lasers, and carbon arcs. Today, europium is the most important rare earth activator, being used in colour television as the red phosphor, and as a "colour correcting" phosphor in high pressure mercury vapour lamps. Cerium, samarium and terbium are also important activators. Most of the 4f-containing rare earths can lase, but only neodymium is of any commercial importance. The addition of rare earths to carbon arc cores increases the arc intensity by a factor of ten and the resultant light quality is nearly identical to sunlight if the mixed rare earth fluorides are used. We are quite fortunate in that the natural mixture of rare earths gives the proper light quality, and that this mixture of rare earths, each with its own electronic energy level scheme, is needed to produce this balanced white sunlight. This is the only major use of rare earths which depends greatly on the 4f electron(s) and in which they are used as a mixture.

The lack of 4f electrons in lanthanum and yttrium and the half-filled 4f level in gadolinium accounts for their use as host materials for the various rare earth phosphors. The most important hosts are Y_2O_3, Gd_2O_3, R_2O_2S (R = La, Y, Gd) and YVO_4. The absence of a 4f electron also accounts for the use of yttrium aluminium garnet (YAG) as a laser host for neodymium and as a substitute for decorative diamonds.

The magnetic properties due to the 4f electrons and their interaction with the magnetic electrons of the first transition-metal group element in ferrites, garnets and the cobalt(iron)-rich intermetallic compounds account for the use of rare earths in microwave and bubble devices, and the new family of rare earth-cobalt permanent magnets.

Basic and applied research account for a significant share of the total sales of the individual rare earths. Of course, the future growth and diversification of the rare earth industry depend on results, advances and discoveries made in this area. Although some of the research is chemical in nature and should rightly be placed in the first category, most of the work deals with the 4f related properties and so it is all grouped here.

The final application included in this group is the utilization of La, Ce and Y additions of about 0.1 to 0.2% to cobalt-, nickel- and chromium-based superalloys to improve their corrosion and oxidation resistance in various environments - including high temperatures and salt water. It is difficult to understand the use of the individual elements in this application, since the improvement appears to depend more on the chemical nature of the rare earths, and thus perhaps should be moved to the first category. I venture the guess that the commercially available mischmetal, when the early studies were made, contained sufficiently high impurities (probably Mg, Si, O, Pb, Ca, Al) such that little or no improvements were noted, but that the individual metals available, which were much purer with respect to these impurities, worked. I believe that a high purity mischmetal (with respect to non-rare earth impurities) would work equally well if substituted for La and Ce, and an Y-heavy lanthanide mixed metal for Y. In either event, the mixture should be significantly cheaper.

4. USES BASED ON UNUSUAL VALENCE STATES[*]

The use of rare earths in products in which the elements are in their divalent or tetravalent states would be included in this category, as well as those uses which would involve oxidation-reduction reactions between the normal trivalent state and the unusual divalent or tetravalent state. The restrictions imposed in this category, naturally limits the applications to those rare earths which exhibit dual valence states, i.e., Ce, Eu and Yb, and perhaps Pr, Sm and Tb. In actual practice only cerium and to a very limited extent, europium, are used in their unusual valence states.

The use of CeO_2 as a glass polishing agent is one important commercial application. As noted earlier, CeO_2 may be used in combination with the normal trivalent rare earth oxides, R_2O_3. However, CeO_2 is the most effective compound in the mixture, and accounts for the higher polishing speed,

---

[*] The term "unusual" is used in the sense that it is unusual from the point of view of the rare earths as a whole, rather than from a particular element's point of view.

longer life and superior finishes as compared to cheaper polishing agents, such as iron rouge.

The other major use of CeO_2, which just became commercially viable in 1970, is the addition of this compound to decolorize flint glass used in the manufacture of bottles and jars. This application depends on the oxidation-reduction reaction of cerium to keep the iron impurities in the Fe^{+3} state, since the Fe^{+2} impurity imparts a bluish colour to the glass.

The use of divalent europium as an activator in a number of phosphors is a minor application in this category. The host materials are generally alkaline earth-base salts such as $M_3(PO_4)_2$, MS and $M_2P_2O_7$ where M is an alkaline earth, usually strontium or barium.

5. USES BASED ON NUCLEAR PROPERTIES

Several of the rare earth elements have high neutron capture cross sections and thus are used as control rod materials and burnable poisons. The most likely candidates are Sm, Eu, Gd, Dy and Er. Other rare earths which have very low cross sections, such as cerium and yttrium may be used as diluents in nuclear fuels.

The rare earth elements are utilized in the nuclear field as individual elements. However, extremely high purity, such as required for phosphors, is generally not necessary.

FUNDAMENTAL PROPERTIES OF RE ELEMENTS AND THEIR IMPLICATIONS ON THE ANALYTICAL CHEMISTRY OF RE MATERIALS

by

Karl A. Gschneidner, Jr.

Ames Laboratory USAEC and Department of Metallurgy

Iowa State University, Ames, Iowa 50010, U.S.A.

ABSTRACT

The electronic configurations of the metallic state, the ground state (gaseous atoms) and the trivalent ions of the rare earth elements are given. The relationship between these configurations and the chemical and physical properties of the metals and some of the important compounds is discussed. A separate section is devoted to the spectra of the rare earth elements. Some comments are made regarding the influence some of these properties have on various analytical techniques.

1. INTRODUCTION

In this paper I will try to briefly review some of the important physical and chemical properties of the rare earth metals and some of their compounds. Where appropriate, a comment or two will be made relative to the analytical methods used to analyze rare earth materials.

Although many properties of the rare earths vary slightly as one proceeds from one lanthanide to its neighbour, significant differences exist if one compares the properties of the end members, e.g., the sizes of lanthanum and lutetium differ by 7.5%. Larger differences may also occur because of differences in valence states, e.g., metallic europium and ytterbium are divalent, and the other lanthanides are trivalent, and this valence difference accounts for a difference of more than a billion in the vapour pressures of ytterbium and lanthanum at $1000^{\circ}C$. These properties, however, may be considered to be primarily dependent on the outer valence electron configurations.

Examination of properties directly related to the 4f electron, such as their spectral properties, finds neighbouring lanthanides have large differences. Naturally, one would expect nuclear properties to vary widely for these elements.

2. ELECTRONIC CONFIGURATIONS

Most of the properties, other than the nuclear properties, can be logically treated from an understanding of the electronic configurations of the rare earths. The ground state configurations were given in my first talk, table 1. These can be summarized as being $4f^{n+1}6s^2$ for the majority of the lanthanides, the exceptions being La, Ce, Gd and Lu which have $4f^n5d6s^2$.

The ground state configurations are generally not important with regard to the main emphasis on this Conference, except for the vaporization process (which affects the heat of sublimation and boiling point) in flame and plasma atomic absorption or emission spectroscopy. More important are the trivalent ion configurations, which are also given in table 1 of my previous talk, as the "normal" configuration assuming the valence electrons have been removed from the atom to give the tripositive ion. The "unusual" configurations, also given in the same table, account for a second valence state for Ce, Pr and Tb (tetravalent) and for Sm, Eu and Yb (divalent).

A third electronic configuration which is important in understanding the behaviour of these elements is that of the pure metals, see Table 1 of this paper. When this table is examined, it is seen that the normal electronic configuration is $4f^n$ plus 3 valence electrons.* There are, however, two exceptions, Eu and Yb, which have $4f^{n+1}$ plus 2 valence electrons configurations. As one would expect, this difference in the number of valence electrons has a significant effect on the physical and chemical properties of Eu and Yb as compared to those of the normal trivalent rare earth metals. But not as readily realized is that this difference also has a notable effect on the thermodynamic energies of formation of the trivalent compounds formed by Eu and Yb.

---

* The nature of these valence band electrons (the amount of s, p, d, f character) has been recently discussed by Gschneidner [2], but since this is not essential to this paper the reference is cited only for completeness.

TABLE 1

Electronic Configuration of the Rare Earth Metals at 298°K and 1 atm Pressure

Metal	No. 4f Electrons	No. of Valence (Band) Electrons	Metal	No. 4f Electrons	No. of Valence (Band) Electrons
La	0	3	Tb	8	3
Ce	1*	3*	Dy	9	3
Pr	2	3	Ho	10	3
Nd	3	3	Er	11	3
Pm	4	3	Tm	12	3
Sm	5	3	Yb	14	2
Eu	7	2	Lu	14	3
Gd	7	3	Y	--	3

* Approximate, Gschneidner and Smoluchowski give 0.94 4f and 3.06 valence electrons [1].

3. PHYSICAL PROPERTIES (OTHER THAN SPECTRAL PROPERTIES)

3.1 Metals

The metallic radii of the rare earth metals are shown on the bottom of figure 1. The anomalies of Eu and Yb are quite evident, and are due to their divalent nature. The minor anomaly of Ce is due to a valence slightly greater than 3, [1], see footnote to table 1. The rare earth metals are quite large compared to most of the elements in the periodic table - they range from about the 65th to the 90th percentile on the basis of increasing metallic radius. The rare earths are quite electropositive (low electronegativities - see fig. 1) exceeded only by the alkali and alkaline-earth metals. These two factors play a considerable role in their chemical and metallurgical behaviour, e.g., the rare earths form stable compounds with the halides, chalcogenides and pnictides because of their electronegativity, and because of their size, the rare earth metals are essentially insoluble in the solid state of most of the major industrial metals.

The melting points of the lanthanides are shown in figure 2, and a variation of a factor of two is noted in the values for the trivalent lanthanides.

FIGURE 1

Metallic radii and electronegativities of the lanthanide
metals. The values for yttrium
are almost identically the same
as those of gadolinium. The per-
centages listed in the figure
are the relative change of the
property on proceeding from
La to Lu.

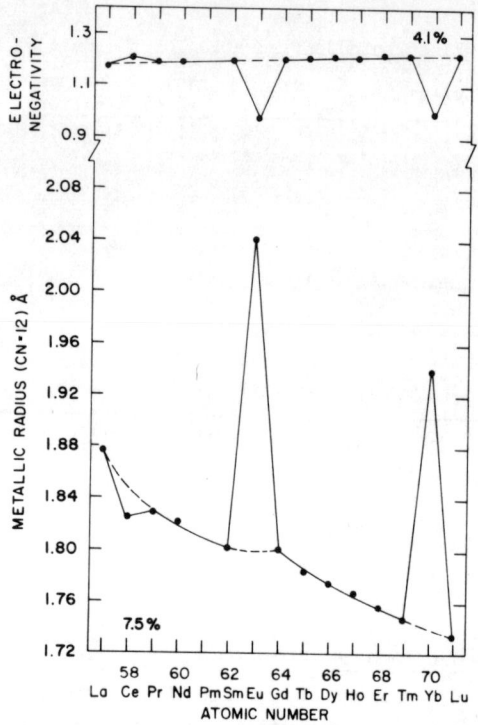

FIGURE 2

The melting points of the
lanthanide metals. The value
for yttrium is 1520°C. The
melting points have been taken
from several sources which
were summarized in [2].

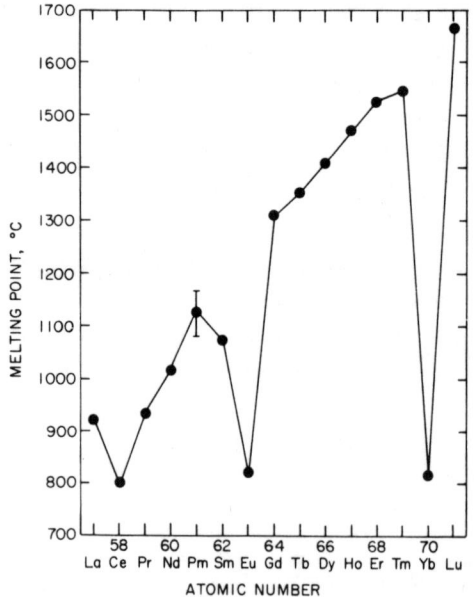

This large variation according to several authors is due to 4f participation in the bonding (see ref. 2 and other references quoted therein). The anomalies of Eu and Yb are quite evident.

The boiling points and the heats of sublimation of the lanthanides are given in figure 3. As one can see there is no apparent systematic variation in these two properties as compared to those seen in the first two figures.

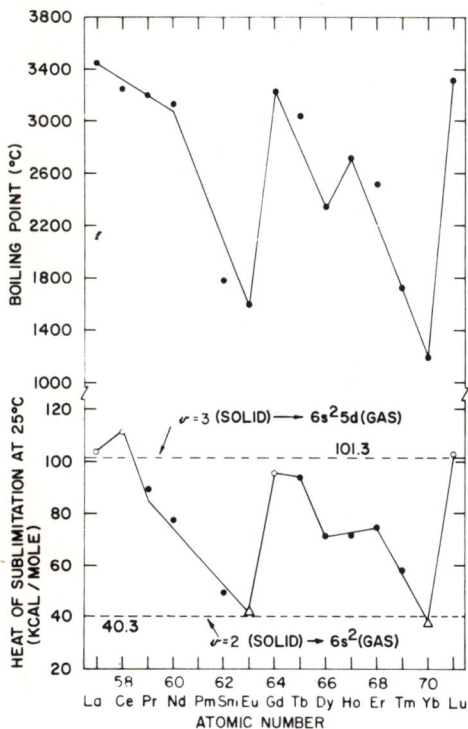

FIGURE 3

The boiling points and heats of sublimation at 25°C (298°K). The values for yttrium are 3338°C and 101.5 kcal/mole, respectively. The letter v means number of valence electrons. These values were taken from Hultgren et al. [3], who, on the basis of new thermodynamic data on the gaseous species, revised the values originally reported by Habermann and Daane [4].

One does note that there is a correlation in variation of the boiling point and the sublimation energy, as would be expected from Trouton's rule. The nature of the variation in these two curves is due in part to a change in the electronic configurations during the vaporization process for more than half of these elements (see tables 1 in this paper and in the previous paper). This has recently been discussed in more detail by Gschneidner [2].

The high vapour pressures of Eu and Yb have a strong influence on the methods one can use to measure some of the non-metallic impurities in these metals, especially oxygen and hydrogen. For hydrogen analysis the Pt bath vacuum fusion procedure normally used for the trivalent rare earth metals will not work, but fusion in a lower melting tin bath at $850^{\circ}C$ yields reliable hydrogen values for Eu and Yb. For oxygen analysis, vacuum fusion in Pt-Sn(C) or Pt(C) cannot be used for these two volatile metals because the distilled metals getter the evolved gas. The best method used to date is neutron activation.

The high vapour pressures of Eu and Yb also influence the results obtained for these elements in matrices of the other rare earth elements using dc arc excited emission spectroscopy. The oxides of these two elements are decomposed by the carbon from the graphite electrode allowing the volatile metallic species to boil-off in the early states of the arcing process. This leads to a large variation in the concentration of Eu or Yb in the arc as a function of time, t. The relative concentration of these two elements is large for small values of t and it becomes smaller as t increases until the sample is depleted before the arcing process is completed. Since the dc arc is an uncontrolled excitation method, coupled with the boil-off problem, the precision of determining Eu and Yb by emission spectroscopy is significantly less than that obtained for the other rare earth elements.

3.2 Compounds

The more important compounds from the viewpoint of this Conference are the oxides and halides. One of the basic parameters governing the formation of these compounds and the crystal structures formed by them is the ionic radius, figure 4. In this case the smooth variation in the property due to the lanthanide contraction is quite evident. Furthermore, the absence of any anomalies is also notable. As will be shown shortly, this type of behaviour of a property of a rare earth compound series, even if there is no valence change, is exceptional.

The melting points of the sesquioxides, R_2O_3, the trifluorides, RF_3, and the trichlorides, RCl_3, are shown in figures 5 - 7, respectively.

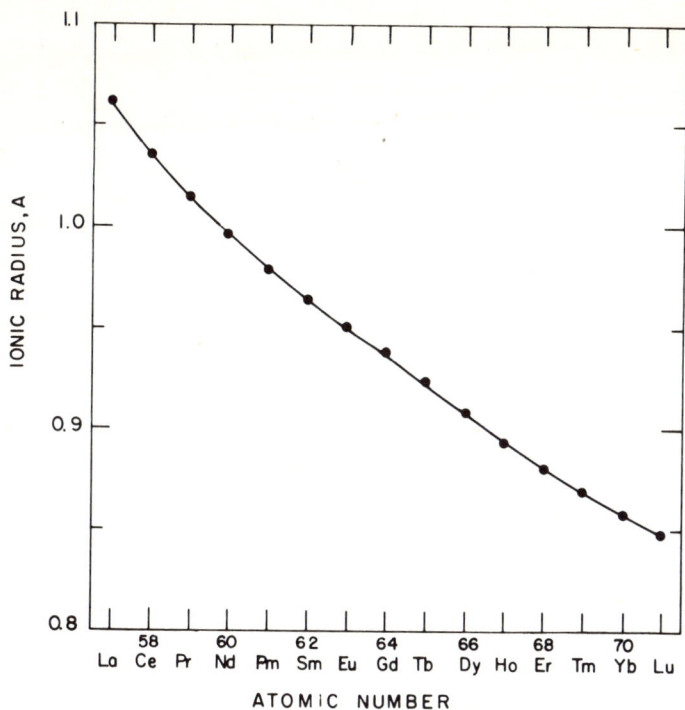

FIGURE 4

The ionic radii for R^{+3} for a coordination number of 6 based on the $O^=$ radius equal 1.380A [5]. The yttrium ionic radius is 0.893A [6].

The melting points of the oxides (fig. 5) show a slow increase (within the experimental precision they may be the same) for the lights and heavies with a sharp increase at Eu-Gd and Yb-Lu. The significance of these steps is not understood. The low melting point for Ce_2O_3 may be associated with the fact that CeO_2 is the more stable oxide and that Ce_2O_3 may disproportionate to $Ce+CeO_2$ in the absence of oxygen, or oxidize to CeO_2 in the presence of air. The rare earth oxides are one of the most refractory compounds formed by the rare earth elements; however, some of the rare earth borides, carbides,

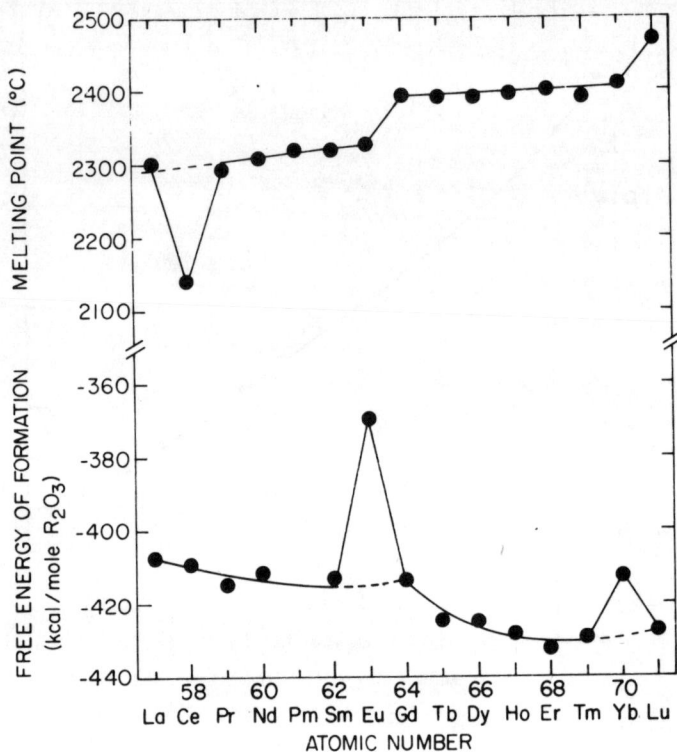

FIGURE 5

The melting points and free energies of formation at $298^{\circ}K$ of the rare earth sesquioxides. Most of the melting points are taken from Foex [7], the remaining are taken from several sources - Ce_2O_3 [8], Pm_2O_3 [9], Tm_2O_3 and Lu_2O_3 [8] and Y_2O_3 ($2376^{\circ}C$) [10]. The free energies of formation are taken from Holley et al. [11]. For Y_2O_3 the ΔG_f^{298} = -434.3 kcal/mole Y_2O_3 [11].

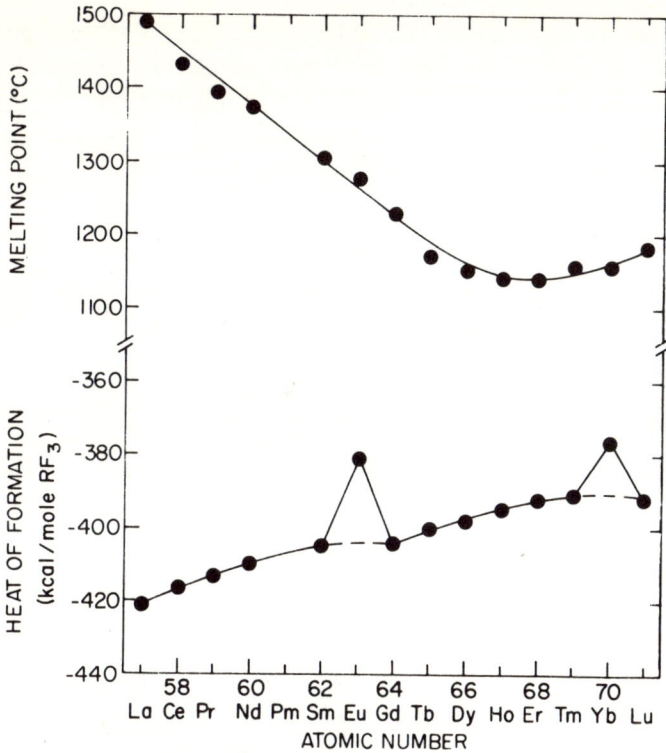

FIGURE 6

The melting points and the estimated heats of formation at 298°K
of the rare earth trifluorides. The data are taken from Thoma [12].
The melting point of YF_3 is 1155°C [13] and the heat of formation
is -410.8 kcal/mole YF_3 [14].

nitrides and phosphides do melt higher than the oxides.

The melting points of the fluorides (figure 6) and the chlorides
(figure 7) decrease with increasing atomic number until a minimum is reached
in the middle of the series, and then the melting points begin to increase.
Interestingly, the bromides and iodides show a similar minimum, except it

occurs closer to the beginning of the lanthanide series - the position of
the minimum (in terms of the lanthanide atomic number) is inversely proportional to the atomic number of the halide. The melting points of the fluorides are intermediate (as compared to the high values for the oxides) and
those of the chlorides (and also the bromides and iodides) are considered
to be low.

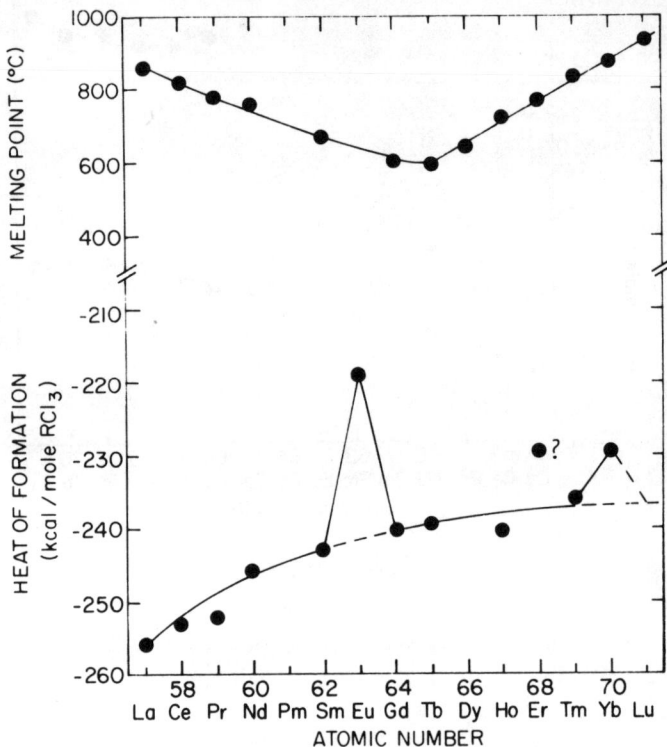

FIGURE 7

The melting points and the heats of formation at 298°K of the rare earth trichlorides. The melting points are taken from Brown [15] and the heats of
formation from the results summarized by Gschneidner [16]. The melting point
of YCl_3 is 709°C [15] and the heat of formation is -239.0 kcal/mole YCl_3 [14].

4. CHEMICAL PROPERTIES
4.1 Metals

The rare earth metals react readily with oxygen and moisture. The lights, La, Ce, Pr and Nd, and divalent Eu react at atmospheric conditions and must be handled carefully to avoid the pick-up of oxygen and in the case of Eu also hydrogen. This is especially important when analyzing the metals for these two impurities. Of the four trivalent metals listed above, La and Ce are the most reactive; the Pr and Nd reactivities are significantly less and decrease with increasing atomic number. The remaining rare earth metals do not oxidize readily at atmospheric conditions, but over a 6-month to one-year period of time only Sm and Yb tarnish due to the formation of an oxide film. The colour change is from a bright metallic silver colour to a dull blackish grey. The purity of the metal has a great deal to do with its reactivity with air; the purer the metal the less the reactivity. The difference in reactivity may in part be due to the fact that the first four lanthanides form a hexagonal oxide, which flakes off, while the other lanthanides form a monoclinic (Sm, Eu and Gd) or a cubic modification (the remaining lanthanides plus Y) which forms a very thin protective layer which prevents further oxidation.

Europium reacts with atmospheric moisture to form $Eu(OH)_2 \cdot H_2O$ [17], rather than the monoclinic Eu_2O_3. Thus one needs to prevent this surface reaction from occurring when determining both oxygen and hydrogen in Eu metal. The reactivity of the rare earths is also quite useful in other analytical techniques involving the metals, since they are readily dissolved in aqueous solution by common mineral acids, except HF, or converted to the stable R_2O_3 oxide, except for Ce, Pr and Tb which form CeO_2, Pr_6O_{11} and Tb_4O_7 by heating in air at $800 - 1000^\circ C$. The former is important in wet chemistry and spectrophotometric methods and the latter, especially for the light lanthanides, in preparing samples for spectroscopic analysis in which the oxide is the preferred chemical species. At one time all the rare earth metals could be easily converted to the oxide by ignition in air, but as the metals became purer, it became more difficult to oxidize them completely, and thus today, they are first dissolved in acid before converting to the oxide.

4.2 Compounds

The free energy of formation of the rare earth sesquioxides is shown
in figure 5 and the heats of formation for the trifluoride (estimated values)
and trichlorides are shown in figures 6 and 7, respectively. The notable
feature in all three of these graphs is the fact that the free energies or
heats of formation for Eu and Yb compounds are significantly less negative
than those of the remaining lanthanide compounds, even though Eu and Yb are
trivalent ions in these compounds. As has been pointed out several years
ago [18] this difference is due to the fact that both Eu and Yb are divalent
in the standard state (metallic Eu and Yb), and a promotion energy of 23
kcal/g at. for Eu and 9 kcal/g at. for Yb is required to form the trivalent
state for these two elements. Thus the anomaly in the metallic state accounts
for an anomaly in a property of a compound series in which all the lanthanides are in the trivalent state.

The magnitudes of these energies of formation, on a per g-at.-of-metal
basis, are among the highest for all of the elements in the periodic table
which form compounds with the respective anion. Again this reflects the
strong reactivity of the rare earth metals with the non-metallic elements.

The rare earth metals form essentially insoluble hydroxides, fluorides
and oxalates in aqueous solution. For analytical purposes the oxalate is
used to quantitatively precipitate the rare earths individually or as a
group, but in general neither the hydroxide nor the fluoride is used for
precipitating these elements.

The light lanthanide sesquioxides (A-form) slowly pick up moisture and
carbon dioxide, and thus should be used as freshly ignited oxide for analytical chemistry purposes. This absorption may also account for the fact that
the metals of these oxides are so readily oxidized because the oxide reacts
with air, which breaks down the coherent protective coating.

5. SPECTRAL PROPERTIES

The preponderant analytical methods to be discussed at this Conference
are based upon the inner electron energy levels and transitions between them.
The complexity of the spectra of most of the individual rare earth elements

has been known for many years. But even today there is a great deal of effort being expended in most of the technologically advanced countries to catalog, index and identify the energy levels of the rare earth spectra which still remain unsolved. Fortunately, for those of us who rely on precise analytical results in our researches and those who make these analytical measurements, a complete knowledge and understanding of these spectra are not necessary to obtain qualitative and quantitative chemical information about our samples.

One can obtain an idea about the complexity of the spectra of a rare earth element, especially one with a partially filled 4f level, from an examination of figure 8, which shows the energy levels of the trivalent rare earth ions. It must be remembered that the spectra consist of transitions between energy levels and thus there are many more lines in a spectrum than there are energy levels. For example, Dieke and Crosswhite [19] note that for Sm^{+3}, if one just takes into account the four lowest energy levels ($4f^5$, $4f^45d$, $4f^46s$ and $4f^46p$), there is a total of 1994 levels (198, 977, 208 and 611, respectively) which gives rise to 306, 604 allowed transitions.* The situation becomes even more complex for the new few lanthanide ions up to and including Ho^{+3}.

Although the basic electronic energy levels shown in figure 8 will not change, the energy levels can be shifted by as much as 100 cm^{-1} by the environment in which the R^{+3} ion is placed. The shift will depend on the electronic charge of neighbouring ions or atoms, and their symmetry about the particular rare earth ion. For some techniques, this will have some effect, but in others it can be completely ignored.

None of the spectroscopic methods makes use of the complete spectra of a given rare earth. Some techniques, such as spectrophotometric analysis, will rely on unique absorption bands (a series of closely spaced electronic transitions which are not readily resolved) of a given element. Others, such as emission spectroscopy, generally make use of a large number of lines (electronic transitions) which, in general, have been resolved. And, finally, some methods, such as x-ray excited optical fluorescence and cathode-ray excited emission, make use of one or only a few of the electronic transitions.

---

* Assuming parity and J-section rules are obeyed along with the $\Delta\ell = \pm 1$ rule for electrons that change orbits.

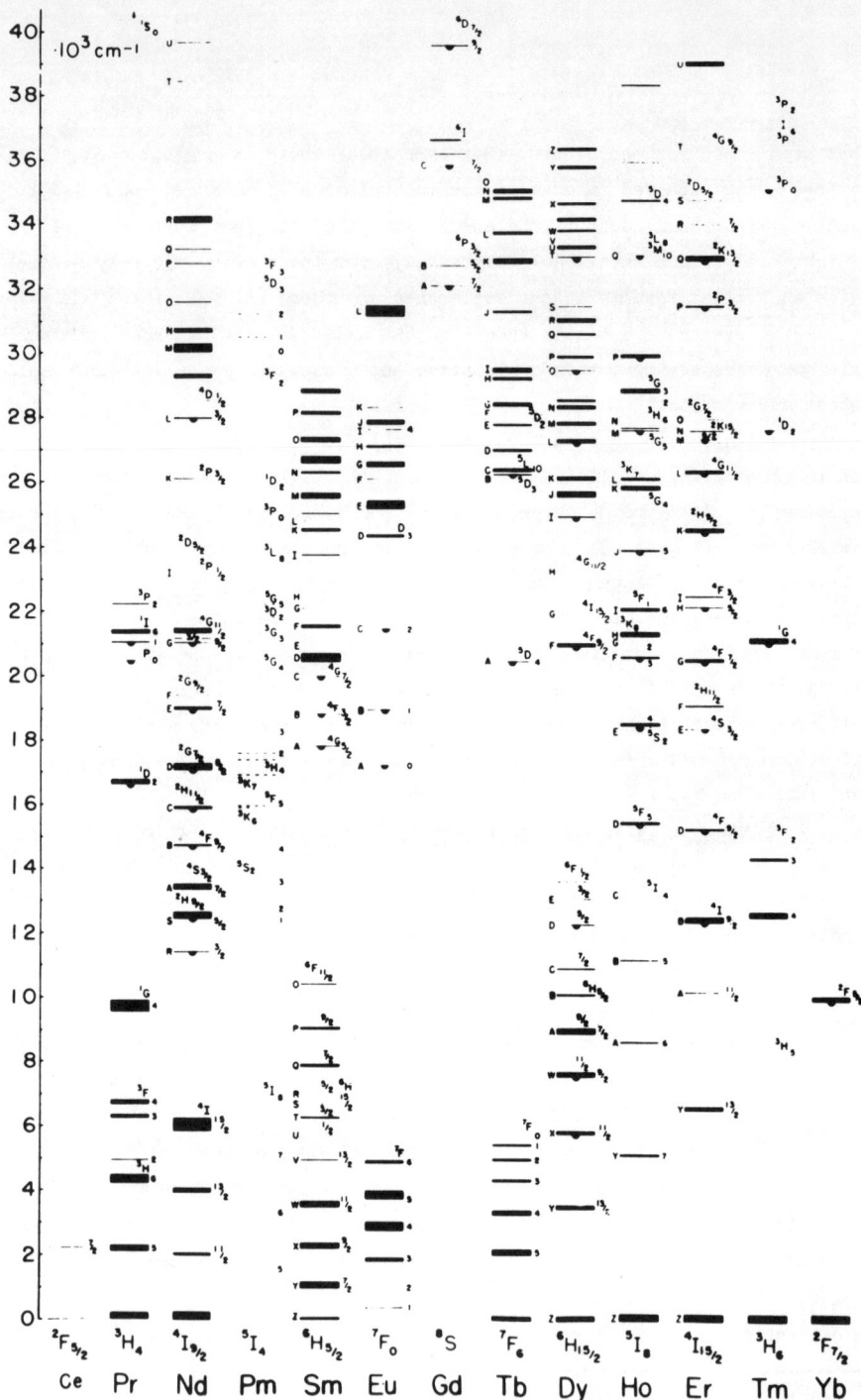

FIGURE 8

The energy levels of the trivalent lanthanide ions
Ce[+3] through Yb[+3] as given by Dieke and Crosswhite [19].

The trivalent rare earth ions of La, Gd and Y do not have any low-lying energy levels and thus there are no electronic transitions which give rise to spectral lines in the visible and UV regions. This permits one to use the materials based on these three ions as hosts. Furthermore, rare earth impurities in La-, Gd- and Y-based materials are generally easily detected with no interferences by using spectral methods.

I am sure that the many experts in the various fields will give in their papers much more detail, pointing out some of the salient features and the limitations of their analytical methods. Perhaps in their talks additional information involving the relationship between the fundamental properties of the rare earths and the analytical method will become evident.

References
[1] K.A. Gschneidner, Jr. and R. Smoluchowski, J. Less-Common Metals 5, 374 (1963).
[2] K.A. Gschneidner, Jr., J. Less-Common Metals 25, 405 (1971).
[3] R. Hultgren, R.L. Orr, P.D. Anderson and K.K. Kelley, Selected Values of Thermodynamic Properties of Metals and Alloys, Wiley, New York (1963), and supplements thereof, issued irregularly by the authors, Department of Mineral Technology, College of Engineering, University of California, Berkeley.
[4] C.E. Habermann and A.H. Daane, J. Chem. Phys. 41, 2818 (1964).
[5] D.H. Templeton and C.H. Dauben, J. Am. Chem. Soc. 76, 5237 (1954).
[6] Calculated using the method of Templeton and Dauben [5] from the lattice constant of Y_2O_3.
[7] M. Foex, Rev. Int. Hautes Temp. Refract. 3, 309 (1966).
[8] O.A. Mordovin, N.I. Timofeeva and L.N. Drozdova, Izv. Akad. Nauk SSSR, Neorg. Mater. 3, 187 (1967); Eng. transl., Inorg. Materials 3, 159 (1967).
[9] T.D. Chikalla, C.E. McNeilly, J.L. Rasmussen and J.L. Bates, U.S. Atomic Energy Comm. Rept., BNWL-SA-3818 (Sept. 1971).
[10] T. Noguchi, M. Mizuno and T. Yamada, Bull. Chem. Soc. Japan 43, 2614 (1970).

[11] C.E. Holley, Jr., E.J. Huber, Jr. and F.D. Baker, p.343 in <u>Prog. Sci.
Tech. Rare Earths</u>, Vol. 3, L. Eyring, ed., Pergamon Press, New York
(1968).

[12] R.E. Thoma, p. 90 in <u>Prog. Sci. Tech. Rare Earths</u>, Vol. 2, L. Eyring,
ed., Pergamon Press, New York (1966).

[13] F.H. Spedding and D.C. Henderson, J. Chem. Phys. $\underline{54}$, 2476 (1971).

[14] D.D. Wagman, W.H. Evans, V.B. Parker, I. Halow, S.M. Bailey, R.H.
Schumm and K.L. Churney, NBS Tech. Note 270-5 (March 1971), U.S.
Government Printing Office, Washington, D.C.

[15] D. Brown, <u>Halides of the Lanthanides and Actinides</u>, Wiley-Interscience,
London (1968).

[16] K.A. Gschneidner, Jr., J. Less-Common Metals $\underline{17}$, 1 (1969).

[17] F.H. Spedding, J.J. Hanak and A.H. Daane, Trans. Met. Soc. AIME $\underline{212}$,
379 (1958).

[18] K.A. Gschneidner, Jr., J. Less-Common Metals $\underline{17}$, 13 (1969).

[19] G.H. Dieke and H.M. Crosswhite, Appl. Optics $\underline{2}$, 675 (1963).

TECHNIQUES FOR THE MUTUAL SEPARATION OF THE LANTHANIDES IN ANALYTICAL CHEMISTRY

by

H.A. Das and J. van Ooyen

Reactor Centrum Nederland, Petten (NH), The Netherlands

1. INTRODUCTION

"Wet" methods of analysis are not popular nowadays. For some applications, however, chemical separation remains necessary.

First, there are the elements which can be detected barely, or not at all by purely instrumental methods. The sample should be separated completely with a high or quantitative yield. Then there are the separations for determination by mass spectrometry. The aim is here to obtain a very pure sample; the yield may be low. Finally the preparation of samples for reactor physical measurements can be mentioned. The analytical chemist is requested to submit a sample of considerable size which at the same time meets the most rigorous standards of purity.

Three techniques have been shown to be of particular interest in the mutual separation of the lanthanides: Ion-exchange, reversed phase extraction chromatography and high-voltage electrophoresis. The procedures based upon cation-exchange can be divided in separation at normal and at high pressure.

Reversed-phase chromatography is of special interest to the analysis of fission products and for burn-up determinations.

High-voltage electrophoresis on paper strips is applied in the activation analysis of silicate samples. The main advantage of this technique is its inherent speed and elegance.

As a rule the isolation of the group of the lanthanides (and yttrium) is performed by either (co)precipitation or liquid-liquid extraction. The application of these techniques to the mutual separation of rare earths in industrial processes lies outside the scope of this discussion. Spectro-

photometry is a convenient method for the measurement of the total amount
of lanthanides present in a liquid sample. It is often used as a means to
follow a separation process.

2. SPECTROPHOTOMETRY

Two methods for spectrophotometric analysis are available. One utilizes
the band absorption spectra of the coloured ions. The other method is based
upon the absorption spectra of coloured complexes with dyes. The absorption
by the lanthanide ions can be enhanced by the formation of inorganic complexes.

The application of spectrophotometry to the determination of the lanthanides usually implies two things:

a) Separation of the group of rare earths from other, possibly
   interfering, elements.
b) The determination of the total amount of the combined lanthanides.

The isolation of the group of the rare earths can be performed in various ways. The most common techniques are (co)precipitation and liquid-liquid
extraction. In spectrophotometric determinations these methods are most frequently applied. However, ion-exchange can be used too.

Korkisch [2.1] has given an excellent review of both liquid-liquid extraction and precipitation methods. The most important extractants are
tri-n-butyl phosphate (TBP) and bis-(2-ethylhexyl)-o-phosphoric acid (HDEHP).
The trivalent rare earth ions behave nearly identical, although the extraction coefficient increases from La to Lu. The extractability of tetravalent
Ce is much greater. Oxidation with $KBrO_3$ and extraction from HNO_3 solutions
up to 10 N result in the isolation of Ce. Liquid-liquid extraction or precipitation offers an analytical possibility to separate the other rare earths.

In general the spectrophotometric methods are not very well suited for
the determination of the individual rare earths in the presence of each other.
Sandell and Vickery [2.2] have given a survey of the possibilities of applying spectrophotometry to the determination of the lanthanide ions and of the

inorganic complexes.

The absorption bands chosen by various workers and the interferences by other lanthanides are discussed by Vickery. The limits of detection are varying widely. Beer's law is obeyed up to concentrations of $\simeq$ 10 mg/ml. It is possible to determine Ce by its very strong absorption band in the ultra-violet at 2536 Å [2.3]. Here $\geq$ 5 µg/ml may be determined. By use of selected index peaks, that may either be located in the ultra-violet, the visible, or the near infrared portion of the spectrum, spectrophotometric methods have been recommended for all the rare earths, except La, Pm, and Lu [2.4]. Very few absorption peaks are free from interference if the sample contains all lanthanides.

A high sensitivity ($\geq$ 2 µg/ml) can be obtained for Ce, when it is measured in a 3 M K_2CO_3 solution at pH = 11 - 12. A survey of the determinations based on this technique can be found in ref.[2.5].

The complexes of the lanthanides with EDTA enable a reasonably sensitive determination ($\geq$ 20 µg/ml) [2.5].

The formation of coloured complexes with dyes offers a much more sensitive method of determination.

The alizarinsulphonate method [2.6] is useful down to concentrations of 1 µg/ml. The absorbance at 5500 Å is measured. The pH-value should be carefully controlled by the change of phenol red to yellow. Yttrium interferes.

The other complex-forming dyes, mentioned in older literature, are briefly discussed by Sandell [2.2]. The most interesting is aurintricarboxylic acid [2.7], which enables the determination of $\simeq$ 10 µg/ml of the lanthanides. Useful methods were described by Korkisch [2.8] and Frits et al. [2.9]. The lower limit of the methods is $\simeq$ 0.5 ppm. It is necessary to mention here the possibilities of fluorescence. The elements Ce, Eu, Sm, Tb, Dy, Tm, and Ho have been determined in borax, sodium phosphate or calcium tungstate beads by excitation with an iron arc [2.10].

By reduction to the divalent state in a lattice of CaF_2 or NaCl, very sensitive (down to 1 ng) determinations of Sm, Eu, Tm, and Yb by ultra-violet fluorescence are possible [2.10]

For the control of separation procedures the spectrophotometric determination with an organic complexant is fast and sensitive.

3. MUTUAL SEPARATION OF THE LANTHANIDES BY ION-EXCHANGE

3.1 Cation-exchange

The technique which is mostly applied for the mutual separation of the lanthanides is cation-exchange. The method is based upon the steady decrease in size with atomic number and the corresponding increase in binding energy of ligands.

The presence of organic complexing agents is compulsory, since the affinities of the various lanthanides for the resin are very similar. The first organic complexants which proved successful were citrate, malate, glycolate, lactate and EDTA [3.1]. The obtained separation factors increased in this order. The highest values ($\simeq 1.5$) were obtained with 0.02 M EDTA solutions at pH = 3.6.

In 1956, Choppin et al. announced [3.2] a new eluant for the separation of the actinide elements, α-hydroxy isobutyric acid (AHIB), which also was applied to the lanthanides in the same year.

The separation of trace amounts was performed on a Dowex-50 X-12 400 mesh column of 50 x 2 mm at a flow rate of 1 ml cm^{-2}min^{-1} and a temperature of 87°C. The separation factors, given in table 3.I were found to decrease only $\simeq 10\%$ when the temperature was lowered to 25°C. The reagent was found superior to EDTA with respect to the solubility of its lanthanide salts.

A great average separation factor at a constant flow of the eluent is impractical. Under constant operating conditions, the elution method gives rise to a characteristic pattern wherein the elution bands tend to become broader and relatively farther away from each other. The conditions cannot be optimized for each part of the elution sequence. The idea of a stepwise or gradual change of the elution conditions thus suggests itself automatically. It was put forward by Nervik, Freiling and others [3.3] following the ideas of Tiselius et al. Stewart [3.4] in 1956 applied a gradual decrease in temperature from 95°C to 25°C and made a comparison with the results obtained at constant temperatures of 25 and 95°C.

Variation of the concentration was applied by a Russian group [3.5]. Trace amounts of the lanthanides were separated on a fine fraction ($\simeq 15$ μ) of a cation-exchange resin by gradient elution with a neutral solution of ammonium lactate, gradually increasing from 0.19 to 0.50 M at 90°C. The flow rate was one drop per minute. The obtained separation factors are roughly equal to those quoted in the mentioned earlier publications, but here the necessary elution volume for neodymium is only four times higher than that for lutetium.

TABLE 3.I

Reported separation factors for the lanthanides. The numbers in brackets refer to the list of references

Element	[3.2]	[3.3]		[3.13]		
Dy						
	1.8	1.8				
Tb						
	2.0	1.6 }				
Gd			2.1	2.1		1.2^5
	1.4	1.5 }				
Eu				1.3 }		
	1.8	2.0	1.4	1.3^5 }	1.3	
Sm				1.3^5		1.1^5
	1.7	1.6 }	1.3			
Pm						
	2.1	2.1 }		1.2^5	1.1	
Nd						
	1.7 }		1.1			
Pr		1.5				
Ce				1.1^5		
La						

Massart [3.6] gave a quantitative description of the gradients obtained by mixing two AHIB eluant solutions in various sophisticated ways. For AHIB, it is also possible to change the concentration of the active anion by a slight change of the pH-value under usual working conditions, that is at pH $\simeq$ 3 - 4 and at concentrations between 0.1 to 0.4 M.

The separation of the elution peaks is generally very good. The peak-to-valley ratio is of the order of 10^2 - 10^3. The lanthanides are recovered from the eluate by adding HNO_3 till the pH-value is 2, and adsorption on a small Dowex-50 column.

The number of publications on the application of AHIB is considerable. Korkisch gives a survey of the work until 1969 [3.7], together with summaries of the experiments with the other complexing agents. Complete sets of separation factors were published by Zeligmann, Wish et al. and Massart et al. [3.8].

A remarkable application is the isolation of the lanthanides from waste. Trace quantities of fission products were separated by Seyb, Wolfsberg and Wish [3.9] by gradient elution methods. The concentration of the complexing anion is varied by changing the pH-value of the eluant. The mutual separation of lanthanides, obtained as one fraction in the treatment of waste on a technical scale, was performed with complexants other than AHIB, due to the relatively high cost of that reagent [3.10]. In these processes displacement chromatography is used to reduce the column length. The equilibrium theory of this method was given by Helffrich [3.10].

The first larger scale application of elution chromatography with AHIB was the elaborate purification of gram quantities of ^{147}Pm for reactor-physical measurements on fission products [3.11]. A comparative study of various other complexants was published by Wheelwright [3.12].

In conclusion one can say that cation-exchange using dilute solutions of AHIB at pH-values of 3 - 4 is the most powerful method for the mutual separation of the lanthanides on an analytical scale.

3.2 Anion-exchange

Lanthanides are not adsorbed on strongly basic resins from nitric acid solutions. Only tetravalent cerium is strongly retained. Addition of nitrates enhances the adsorption [3.13].

The separation of the adjacent lanthanides has been performed [3.14] using various concentrations of $LiNO_3$ (3 - 7 M) and NH_4NO_3 (10 M). The application of anion-exchangers to the mutual separation of the lanthanides was also demonstrated by the experiments of Korkisch and Faris [3.13]. They made use of the fact that the absorption of the rare earths - and most other ions - on the anion-exchanger increases considerably when a miscible organic solvent is added to the liquid phase [3.15]. The most convenient elution liquids are 90% - 10% or 95% - 5% mixtures of the lower alcohols with 5 M

HNO$_3$. Reasonably good separation can be obtained by anion-exchange of the
EDTA complexes of the lanthanides [3.16]. The same applies to the use of
AHIB and lactic acid at pH-values of ≃ 4.6 as eluants and Dowex-1 X 10,
200 - 400 mesh [3.17]. The anionic complexes of the type MX_4^- are adsorbed.
Nevertheless, anion exchange is not favoured by analytical chemists for the
separation of the rare earths.

3.3 Application in Activation Analysis

The application of cation-exchange to the determination of the lanthanides in various inorganic samples was described by Falcoff, May and Piccot
[3.18]. The irradiated sample is dissolved in Na_2CO_3 - K_2CO_3 in the presence of La-carrier, and the rare earths are coprecipitated with $Fe(OH)_3$
and then precipitated as the oxalates.

The precipitate is dissolved in some 6 N HCl, the solution is diluted
and led through a column of 35 cm Dowex 50W X 8, 200 - 400 mesh. The rare
earths are separated by gradient elution with 0.5 M AHIB varying in pH from
2.7 to 3.1. The chemical yield was ≃ 95%.

Cation-exchange with NH$_4$-lactate after oxalate precipitation was applied
by Shima and Honda [3.19]. The chemical yield was determined, prior to the
mutual separation, by gravimetry of the added La-carrier. A careful study
of the various precipitation methods for the separation of the lanthanide
group was published by Massart and Hoste [3.20]. Alimarin et al. determined
the rare earths in uranium by adsorption of the elements on a cation-exchanging
resin prior to the irradiation [3.21]. After elution with 5 N HCl, evaporation and irradiation, separation by cation-exchange is applied. AHIB (0.17 M,
pH = 4.65) was the complexing agent. The column (200 x 2 mm) was filled with
a 400 mesh cation-resin. Not all the rare earths were separated: Yb and Lu,
Ho and Dy, Nd-La remained together.

Higuchi et al. performed a separation of the lanthanides into two fractions by cation-exchange, using EDTA as the complexant [3.22]. The heavier
lanthanides (Ho - Lu) ran through. The chemical yield was determined by the
reactivation technique. The works of Desai et al. and of Brunfelt and
Steinnes [3.22] are examples of the application of anion-exchange to the determination of lanthanides in rocks. The elements Ce and Eu are separated by

elution of Eu with 10% HNO_3 - 70% ethylalcohol - 20% water from Dowex-1 X 8, 100 - 200 mesh. Chemical yield was determined by reactivation.

In conclusion we can say that ion-exchange at ordinary pressure and temperature can be used for the separation of trace quantities of the lanthanides, or for the separation of greater amounts of the elements into two groups.

Gradient elution is required if all the lanthanides are to be recovered. For the complete separation of greater amounts, the dimensions of the column become rather large. A reduction of the column volume is possible by applying ion-exchange under higher pressure.

4. SEPARATION OF THE LANTHANIDES BY HIGH-PRESSURE CATION-EXCHANGE
4.1 General

The average radius of the resin beads affects the ion-exchange process in three ways:

a) The velocity of the ion-exchange process depends primarily on the diffusion in the resin of the ions to be adsorbed. The time needed for this step is approximately proportional to the square of the average radius of the resin beads.

b) The pressure drop over the column, necessary to maintain a certain velocity of flow, is inversely proportional to the square of the average bead-radius.

c) The theoretical plate height, Δ, increases considerably with the average bead radius [4.1].

The use of fine resin fractions implies the application of considerable pressure on the resin column. A compromise has to be found between this disadvantage and the resulting better performance of the column. Such a system was first realized for the separation of the lanthanides by Buxton, Campbell and Ketelle in 1966 [4.2].

Usual working pressure was 50 - 60 atm, although the apparatus could go up to $\simeq$ 150 atm. The resin was Dowex 50 X 12 of 15 - 20 μ or 20 - 40 μ grain size. As eluants 0.3 - 0.8 M solutions of AHIB at pH 4.0 to 4.8 were used. The influence of loading, temperature and flow rate on the separation of 50 - 200 mg amounts of Nd and Pr was studied. The dimensions of the resin bed were either 42 x 0.90 cm or 85 x 0.66 cm. The influence of loading was marked but that of temperature was virtually absent. With increasing flow rate the elution bands tended to become broader. To investigate the analytical possibilities of a high pressure system for the separation of other lanthanides, a device was built in which the working pressure is $\simeq$ 90 atm.

4.2 Experimental Equipment

The experimental setup consists of the following parts:

a) A stainless steel column of 1450 mm length, with a diameter of 3/8" and surrounded by a heating jacket of 1" outer diameter. The column ends in a stainless steel "Hoke" in line micron-filter. The amount of dry (Biorad AG 50 W X-12, 6-29 μ) resin needed is $\simeq$ 68 gram. The specific capacity is 5.0 ± 0.1 meq x (gram dry resin)$^{-1}$. The capacity of the column is thus $\simeq$ 0.34 equivalents.

b) A "Beckmann Accu-Flo" pump. The plunger stroke is adjustable.

c) A contact manometer.

d) A pressure relief which opens at 115 atmospheres.

e) A bleed line for the liquid, which can be opened by hand.

f) A glass syphon with an electrical contact. The eluate flows from the syphon through a glass tube to the fraction collector.

g) A "Pleuger Junior F" automatic fraction collector with 240 tubes of 25 ml.

h) A "Tamson" 3 liter thermostat, equipped with a pump.

j) A 2 liter polythene storage vessel for the eluant solution.

k) The various components are connected by 1/8" (0.035" wall thickness) stainless steel tubing. Four "Hoke Flomite" ball valves make part of this tubing.

l) The water is pumped from the thermostat through armoured PVC tubes of 1/2" outside diameter.

m) The apparatus is mounted in a frame made of wood and "dexion" construction elements and surrounded by a shielding of polyester-covered steel fence.

5. SEPARATION BY EXTRACTION AND ELECTROPHORESIS ON PAPER STRIPS

The separation of the lanthanides by paper chromatography in nitric acid media and using a number of organic solvents was investigated by Lederer [5.1], and after him by many others [5.2].

In summing up the results, it can be said that a method which is capable of separating all the lanthanides has not yet been found. Increase of the separation factors by addition of indifferent salts such as $LiNO_3$, was tried by Danon [5.3]. For adjacent lanthanides only the separation of La and Ce could be performed.

Better prospects are offered by extraction chromatography [5.4]. The stationary phase is filter paper, silica gel, kieselguhr or "Corvic", usually loaded with bis-(2-ethylhexyl)-o-phosphoric acid (HDEHP) or tri-n-butylphosphate (TBP). The separation factors obtained with HDEHP are even higher than those found in the separation of the lanthanides by cation-exchange with AHIB as the complexing agent. The distinct advantages of extraction chromatography are its elegance and inherent speed. The technique is applied to the analysis of fission products, to determine the burn-up [5.5].

Electrophoresis on paper strips offers great possibilities for the
mutual separation of the rare earths [5.2]. With high-voltage electrophoresis all the lanthanides can be separated in less than one hour. The usual
field-strength is 50 - 80 V/cm. The same complexing agents as in cation-
exchange are used here; AHIB [5.6] is again the most convenient reagent.

Good examples of the application of high-voltage electrophoresis are
the analysis of "lunar soil" by Herr et al. [5.7], based on the work of
Aizetmüller [5.6], and the work of Kiesl on meteorites [5.8]. Herr uses
0.8 M AHIB as the electrolyte and a 50 cm long acetyl-cellulose foil
(Schleicher and Schüll). The temperature is 2 - 3°C and the current is
1.2 mA. The chemical yield is determined as the recovered amount of the
lanthanum carrier, added at the moment of sample dissolution. Kiesl deter-
mines the recovery by adding a spike of ^{153}Gd ($T_{\frac{1}{2}}$ = 242 d).

References

[2.1] J. Korkisch, Modern methods for the separation of rarer metal ions,
       Pergamon Press, 233-255 (1969).

[2.2] E.B. Sandell, Colorimetric determination of traces of metals,
       Interscience, 744-746 and 748-749 (1959).

[2.3] R.C. Vickery, Analytical Chemistry of the rare earths, Pergamon
       Press, 64-80 (1961).
       C. Gordon et al., Anal. Chem. 29, 1531 (1957).
       D. Hure et al., Anal. Chem. 9, 415 (1937)

[2.4] D.L. Stewart and D. Kato, Anal. Chem. 30, 164 (1958).

[2.5] R.C. Vickery, ibid., 76-77.

[2.6] R.W. Rinehart, Anal. Chem 26, 182 (1954).

[2.7] L. Holleck et al., Z. Anal. Chem. 146, 103 (1955).

[2.8] J. Korkisch et al., Talanta 10, 865 (1963).

[2.9] J.S. Frits et al., Anal. Chem. 30, 1776 (1958).

[2.10] E.B. Sandell, ibid., 746-748.

[3.1] B.H. Ketelle and G.E. Boyd, J. Am. Chem. Soc. 69, 2800 (1947).
       S.G. Thompson et al., J. Am. Chem. Soc. 72, 2798 (1950); J. Am. Chem.

Soc. 76, 3444 (1954).

S.W. Mayer and S.W. Freiling, J. Am. Chem. Soc. 75, 5647 (1953).

F.H. Spedding et al., J. Am. Chem. Soc. 72, 2354 (1950).

R.A. Glass, J. Am. Chem. Soc. 77, 807 (1955).

K. Street et al., J. Am. Chem. Soc. 72, 4832 (1950).

L. Wish et al., J. Am. Chem. Soc. 76, 3444 (1954).

[3.2] G.R. Choppin et al., J. Inorg. Nucl. Chem. 2, 66 (1956).

[3.3] W.E. Nervik, J. Phys. Chem. 59, 690 (1955).

E.C. Freiling and L.R. Bunney, J. Am. Chem. Soc. 76, 1021 (1954).

H.L. Smith and D.C. Hoffman, J. Inorg. Nucl. Chem. 3, 243 (1956).

[3.4] D.C. Stewart, J. Inorg. Nucl. Chem. 4, 131 (1957).

[3.5] B.K. Preobrazhenskii et al., J. Inorg. Chem. USSR II, 5, 1164 (1957).

[3.6] D.L. Massart and W. Bossaert, J. Chromatog. 32, 195 (1968).

[3.7] J. Korkisch, Modern methods for the separation of rarer metal ions, Pergamon Press, 203-207 (1969).

[3.8] M.M. Zeligman, Anal. Chem. 37, 524 (1965).

L. Wish and S.C. Foti, J. Chromatog. 20, 585 (1965).

D.L. Massart and W. Bossaert, J. Chromatog. 32, 195 (1968).

[3.9] K.E. Seyb and G. Hermann, Z. Elektrochemie 64, 1065 (1960).

K. Wolfsberg, Anal. Chem. 34, 518 (1962).

L. Wish and S.C. Foti, J. Chromatog. 20, 585 (1965).

[3.10] F. Spedding et al., US-Patent 2798-789, 9 July 1957.

J. Mickler et al., Monatshefte f. Chemie 97, 287 (1966).

B. Weaver et al., J. Am. Chem. Soc. 75, 3943 (1953).

F. Helfferich and D.B. James, J. Chromatog. 46, 1 (1970).

[3.11] H.A. Das et al., Report RCN-131, Sept. 1970.

[3.12] E.J. Wheelwright, J. Inorg. Nucl. Chem. 31, 3287 (1969).

[3.13] J. Korkisch et al., Talanta 10, 865 (1963); 11, 523 (1963).

J.P. Faris and J.W. Warton, Anal. Chem. 33, 1265 (1961).

J. Korkisch, Modern methods for the separation of rarer metal ions, Pergamon Press, 214-233 (1969).

[3.15] J. Korkisch and G.E. Janauer, Talanta 9, 957 (1962).

[3.16] M. Wald, Bericht 236/V, Polska Akad. Nauk (1961).

[3.17] G.R. Choppin et al., Proc. of the IAEA Conf., 283 (1962).

[3.18] R. Falcoff et al., Bull. Soc. Chim., 3257 (1967).

[3.19] M. Shima and M. Honda, Geochim. Cosmochim. Acta 31, 1995 (1967).

[3.20] D.L. Massart and J. Hoste, Anal. Chim. Acta 42, 7 (1968).

[3.21] I.P. Alimarin et al., J. Radioanal. Chem. 4, 45 (1970).

[3.22] H. Higuchi et al., J. Radioanal. Chem. 5, 207 (1970).

[3.23] H.B. Desai et al., Talanta 11, 1249 (1964).

       A.O. Brunfelt and E. Steinnes, Chem. Geol. 2, 199 (1967).

[4.1]   E. Glueckauf, Report AERE C/R 1356 (1957).

[4.2]   S.R. Buxton and D.O. Campbell, Report ORNL-TM-1876 (1967).

       D.O. Campbell and B.H. Ketelle, Iorg. Nucl. Chem. Letters 5, 533
       (1969); Report ORNL-4272, 99 (1969).

[5.1]   M. Lederer, Progrès récent de la chromatographie Paris, Vol. III,
       (1956); Nature 176, 462 (1955); Anal. Chim. Acta 15, 46 (1956).

[5.2]   See for a review:
       J. Korkisch, Modern methods for the separation of rarer metal ions,
       Pergamon Press, 223-233 (1969).

       C. Test, Anal. Chem. 34, 1556 (1962).

       E. Cerrai and C. Testa, J. Inorg. Nucl. Chem. 25, 1045 (1963).

       T.B. Pierce and R.S. Hobbs, J. Chromatog. 12, 74 (1963).

[5.3]   J. Danon and M.C. Levi, J. Chromatog. 3, 193 (1970).

[5.4]   See for a review:
       M. Lederer, Chromatogr. Rev. 9, 115 (1967), and

       G. Nickless, Advan. Chromatogr. 5, 121 (1968).

       D. Markland and F. Hecht, Microchim. Acta, 970 (1963).

       T.B. Pierce and P.F. Peck, Analyst 89, 662 (1964).

       T.B. Pierce and R.F. Flint, Anal. Chim. Acta 31, 595 (1964).

       A. Daneels et al., J. Chromatogr. 18, 141 (1965).

       T.B. Pierce and R.F. Flint, J. Chromatogr. 24, 141 (1966).

       H. Holzapfel et al., J. Chromatogr. 20, 580 (1965); 24, 153 (1966).

       E. Gagliardi, G. Pocorny, Microchim. Acta, 977 (1966).

       T. Shimizu and R. Ishikura, J. Chromatogr. 56, 95 (1971).

[5.5]   S.F. Marsh, Anal. Chem. 39, 641 (1967).

- 38 -

[5.6] K. Aizetmüller et al., Atomkernenergie 10, 264 (1965);
       Microchim. Acta (6), 1089 (1964).
       K. Bächmann, Radiochim. Acta 4, 124 (1965).
       W. Kraak and G.D. Wals, J. Chromatogr. 20, 197 (1965).
[5.7] W. Herr et al., in Activation analysis in geochemistry and cosmo-
       chemistry, Oslo, 219 (1971).
       W. Kiesl, ibid. 243.

DETERMINATION OF RARE EARTH IMPURITIES IN PURE RARE EARTHS BY MEANS OF EXTRACTION CHROMATOGRAPHY

by

E. Herrmann[*]

Laboratory for Nuclear Problems

The Joint Institute for Nuclear Research, Dubna, U.S.S.R.

ABSTRACT

Determination of rare earth impurities in very pure rare earth materials is frequently difficult without a preconcentration of the tracers. We have investigated the use of extraction chromatography with undiluted di(2-ethyl-hexyl) phosphoric acid (HDEHP) as a stationary phase for the concentration of tracers with lower atomic number relative to that of the macro element. Enrichment factors of about 10^6 and yields of more than 80% were obtained with excellent reproducibility when $A_{macro}/D_{micro} \geq 1.5$. D_{micro} is the partition coefficient of the trace impurity, A_{macro} is the position of the front, normalized to the unit HDEHP-volume on the column, and depends on the loading B (expressed in %) of the column with rare earth ions and on the partition coefficient E of the macro element measured in trace concentration. This dependence can be expressed by the empirical equation:

$$lg\ A = lg\ E - 0.036B - 0.15$$

The influence of macro amounts of a given rare earth element on the partition behaviour of other lanthanides present in trace concentration has also been investigated.

The amount of an individual rare earth in the concentrate was determined by one of the following methods:

optical spectrometry,
neutron activation analysis,

---

[*] Present address: Technische Universität Dresden, 8027 Dresden, DDR.

spectrophotometry with arsenazo III,
complexometric titration with DTPA.

In the last two cases it is necessary to use a preliminary separation of
the components in the concentrate. This was done either by the cation exchange - α-hydroxy-isobutyrate method or by means of extraction chromatography using a HDEHP microcolumn.

The standard deviation of the determinations depends only on the analytical procedure, the preconcentration being of negligible importance in this
respect.

1. INTRODUCTION

Up to the present time the determination of individual lanthanide impurities in very pure rare earth materials has often been difficult without
a preconcentration of the tracers. We have used for this purpose the method
of extraction chromatography with di(2-ethylhexyl) phosphoric acid (HDEHP)
as the stationary phase, which has already been applied by us for the separation of nuclear reaction products from erbium, dysprosium and gadolinium
targets irradiated with 660 MEV protons.

By this method we succeeded in separating e.g. carrier free dysprosium
and lanthanides lighter than dysprosium from erbium targets (2 g) in 40 minutes [1]. A very important feature of this method is that for given conditions as in the case of a given rare earth amount and a given acid concentration of the element, the elution curve is very reproducible. The characteristics of this method have made it possible to work out a method for the analysis [2] of high purity rare earth materials which has been further developed in the National Investigation and Development Institute of the Rarer
Metal Industry (GIREDMET) in Moscow [3,4].

In this paper we will show some of the principles and results of this
preconcentration method.

It is well known that one of the best extractants for the separation of
rare earth elements is di(2-ethylhexyl) phosphoric acid (HDEHP) that gives a
separation factor > 2 for most lanthanide pairs. By an excess of undiluted
HDEHP rare earth ions are extracted from diluted mineral acid solutions

according to eq.(1)

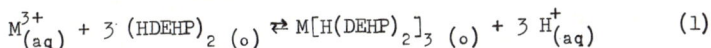

$$M^{3+}_{(aq)} + 3\ (HDEHP)_{2\ (o)} \rightleftarrows M[H(DEHP)_2]_{3\ (o)} + 3\ H^+_{(aq)} \tag{1}$$

It is supposed that it forms a chelate of the structure I [5]

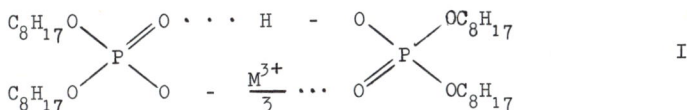

Supposing that no other complexes than complex I is formed and neglecting the activity coefficients, one obtains for the distribution coefficient, D, the following relation:

$$D = \frac{[M[H(DEHP)_2]_3]_{(o)}}{[M^{3+}]_{(aq)}} = K\frac{[(HDEHP)_2]^3_{(o)}}{[H^+]^3_{(aq)}} \tag{2}$$

where K is the equilibrium constant of reaction (1). The value of K increases with the atomic number of the lanthanide element.

Owing to chemical reactions (1), at a given HDEHP starting concentration, the equilibrium concentration of free HDEHP in the organic phase decreases with increasing loading. Therefore, as follows from eq.(2), the distribution coefficient also decreases.

At high concentrations of the rare earth ions in the aqueous phase, the organic phase is able to take up more rare earth ions than stoichiometrically possible according to eq.(1). As a result an undesirable gel may be formed as a third phase. Peppard [6,7] has shown that what actually takes place is the formation of a three-dimensional polymeric complex by substitution of $M^{3+}/3$ for hydrogen in formula I (eq.(3)):

$$M\ H(DEHP)_{2\ 3\ (o)} + M^{3+}_{(aq)} \rightleftarrows 2\ \underline{M(DEHP)_3}_{(o)} + 3\ H^+_{(aq)} \tag{3}$$

The capacity of a given organic phase for rare earth ions, calculated on the basis of a polymer formation of the $M(DEHP)_3$ type, is termed theoretical maximum capacity.

If the organic phase is fixed on a column of inert absorbent, e.g. silanized silica gel, silanized Kisselguhr or organic polymer, and the aqueous phase is passed through this column, we arrive at the extraction chromatographic technique which combines the features of both the extraction - and the ion-exchange type of operation.

We have previously shown that for supporting HDEHP, porous materials with a mean pore diameter ≥ 35 Å are very suitable [8].

The relationship between the distribution coefficient, D, from eq.(2) and the peak location on the elution curve is given by eq.(4):

$$D = \frac{V - V_o}{V_s} \tag{4}$$

(V - volume of eluate to the peak maximum, V_o - free column volume, V_s - volume of the stationary HDEHP phase on the column).

In order to carry out practical separations for the concentration of tracer impurities in rare earth materials, one must have some knowledge of the behaviour of the macro component on the column, as well as of its influence on the behaviour of the tracers.

2. BEHAVIOUR OF THE MACRO COMPONENT

What a given HDEHP column can absorb with respect to rare earth elements is usually somewhat less than the amount that can be calculated on a theoretical basis. This is due to kinetic and thermodynamic reasons.

The effective column capacity increases to some extent with smaller grain size of the supporting material and with higher temperatures. The conclusion may be drawn from this that the pores of the material are plugged by the solid rare earth-HDEHP polymer, which makes the saturation of the extractant with rare earth ions inside the grains become difficult.

Figure 1 illustrates the influence of the acid concentration of the aqueous solution on the column's effective capacity for Gd. It can be seen that a complete loading of the organic phase is not possible even at a pH of 2.

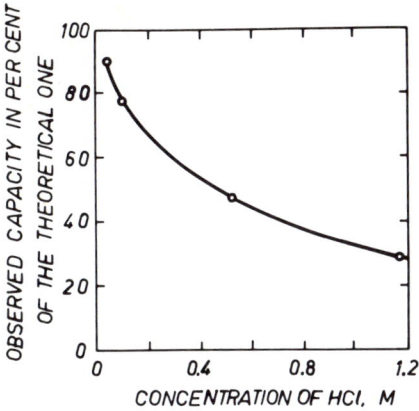

FIGURE 1

Measured capacity for Gd of a column as a function of the HCl concentration. The Gd concentration in the feed solution was kept constant at 40 mg/ml.

The influence of the amount of gadolinium on the shape of the elution curve is shown in figure 2. It is seen that increasing the amount of Gd(III), loaded on the column from a 0.05 M HCl solution, the volume of the eluate up to the peak maximum decreases, the frontal part of the peak becomes sharper and the tailing more pronounced. (The dotted curve refers to carrier-free radioactive gadolinium). The shapes of the peaks can be explained by the decrease of distribution coefficients with increasing loading of the HDEHP phase (see above). This means that $(\partial D/\partial [M]) < 0$, and therefore, larger rare earth amounts move more quickly than smaller ones. This also implies that the front of the peak is self-sharpening and the tail is limited by the moving velocity of the tracers.

For practical separations it is important to know the eluate volume after which the concentration of the macro component exceeds a given value, e.g. 10^{-6} moles/l. This volume we shall call "front position". In order to obtain a value independent of the given column size, the front position, V_A, is normalized, which results in an eq.(5) analogous to eq.(4):

$$A = \frac{V_A - V_o}{V_s} \tag{5}$$

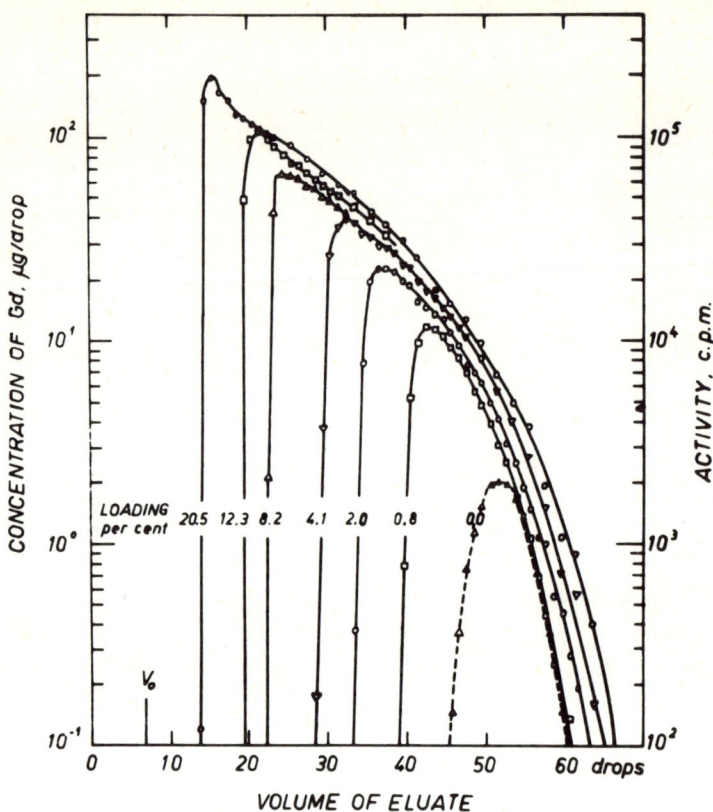

FIGURE 2

Elution curves of gadolinium as a function of the loading. Dotted curve -
carrier-free ^{146,149}Gd. Loading for curves with full line: 0.8%(0.1mg);
2.0%(0.25mg); 4.1%(0.6mg); 8.2%(1.0mg); 12.3%(1.5mg) and 20.5%(2.5mg Gd).
Column: 2.2mm Φ x 68mm; Silica gel KSK No. 2.5 (15μm) with 0.6 ml HDEHP/g
silica gel; Eluent: 0.87 M HCl; Flow rate: 1.3 ml/cm^2·min.

Figure 3 shows the logarithm of the normalized front position, A, for
erbium, gadolinium and yttrium as a function of the column loading, B (in

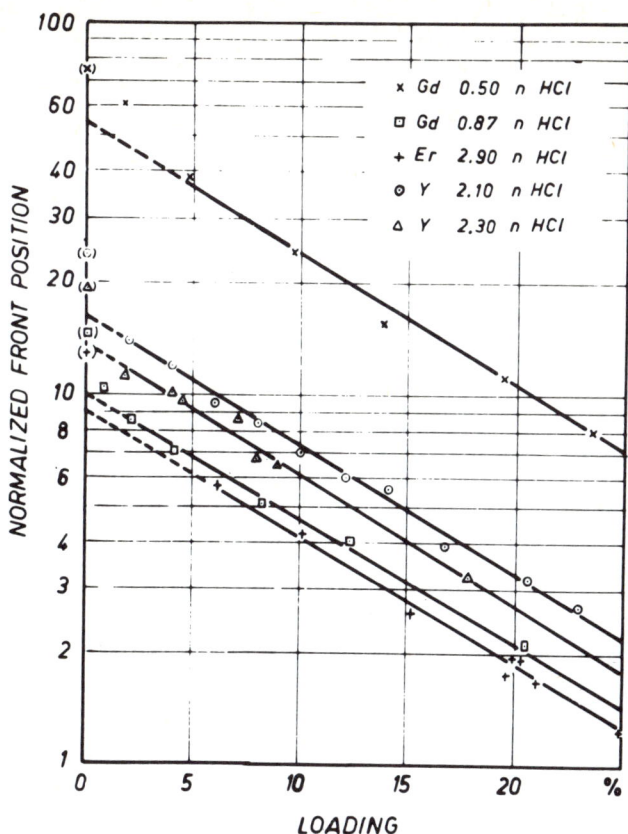

FIGURE 3

Normalized front position A as a function of the loading. (In brackets on the ordinate the position coefficients of the elements in tracer concentrations)

percent of the theoretical maximum capacity). In the range from 3 to 25 per cent of loading one obtains parallel straight lines for all these elements and acid concentrations. After extrapolation to B = 0, in all cases, the straight lines intersect the ordinate at about the same distance from the distribution coefficients, D, measured for tracers of the appropriate elements. So it is possible to write the following empirical equation (6):

$$\log A = \log D - 0.15 - 0.036B \qquad\qquad (6)$$

By means of this equation we can predict the behaviour of macro amounts of other rare earth elements, now studied in detail.

From a comparison of eqs.(2) and (6) it is seen that the front position as well as the distribution coefficient depend inversely on the third power of the acid concentration in the aqueous phase. Experimental evidence for this preconsideration is demonstrated in figure 4, where the slope of the line is - 2.8.

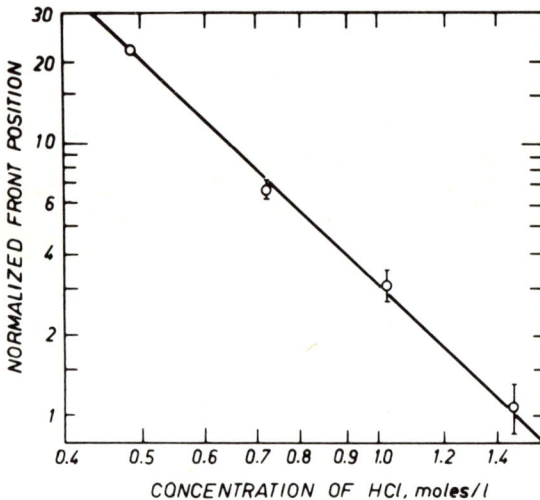

FIGURE 4

Dependency of the normalized front position, A, on the HCl concentration (Gd loading 12.4%).

3. SEPARATION OF TRACERS OF LOWER ATOMIC NUMBER THAN THE MATRIX ELEMENT

The peak position and peak width for a given tracer lanthanide element of lower atomic number than the matrix element are independent of the loading of the column as long as the peaks of these two components are not over-lapping. For practical separations it means that at a value of D_{micro} = 1.5 ± 0.5 one can separate the components if

$$A_{macro} : D_{micro} \geq 1.5 \qquad (7)$$

As an example we will consider the separation of terbium tracer from macro amounts of erbium: With increasing loading of the column with erbium the distance between the two peaks as well as the ratio A_{Er}/D_{Tb} decrease. The higher loading has the effect of enhanced tailing on the tracer peak. However, it is seen from figure 5 that the distribution coefficient of terbium and also the HETP remain constant up to a loading of about 21%. At a higher

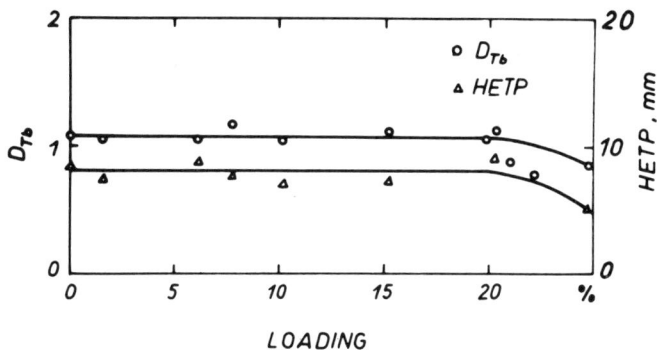

FIGURE 5

The distribution coefficient, D, and HETP (height equivalent to a theoretical plate) for Tb as functions of the loading of the column with Er. Column: 25mm ϕ x 480 mm; Silica gel: "Eisenach" 0.08 - 0.1mm with 0.6 ml HDEHP/g silica gel; Eluent: 3.0 M HCl; Temperature: 40°C. Flow rate: 0.8 ml/cm^2·min.

loading of the column with erbium the peaks overlap, and, as a result of displacement effects, the terbium peak appears earlier. Finally, at a very high loading, the two elements are eluted in one peak, and only in the front of this peak a small enrichment of the tracer can be obtained.

The main cause for the tailing of the tracer peak is the inclusion of the terbium into the solid erbium - HDEHP - polymer by sorption on the column from dilute acid solutions [4]. It is possible to decrease this effect if sorption is carried out with solutions of higher acidity. However, the danger is that a part of the macro amount breaks through the column.

The fixation of the micro component in the polymer $M(DEHP)_3$ makes it impossible to isolate the tracers from the matrix in good yields if the loading of the column is more than about 50% of the maximum column capacity, even in the cases of $A_{macro} : D_{micro} \gg 1.5$.

Figure 6 shows the separation of traces of terbium and dysprosium from 2 g of erbium at a loading of the column of 21%. In all the experiments, the

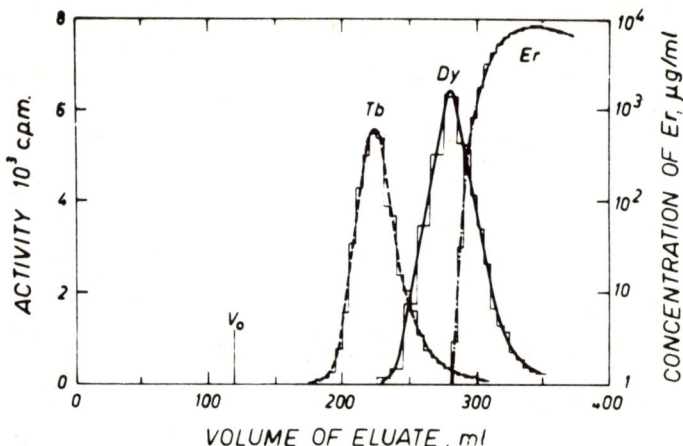

FIGURE 6

Separation of traces of Tb and Dy from 2 grams of Er. Column: 26mm ϕ x 410mm. Silica gel: 100 g, 0.08 - 0.1 mm, with 60 ml HDEHP; Eluent: 2.5 M HCl; Temperature: 40°C; Flow rate: 0.8 ml/cm^2·min.

erbium appears in the eluate after passing through the column of 284 ± 3 ml
of 2.5 M HCl. In the erbium-free liquid there are 43 ± 3% of dysprosium and
78 ± 7% of terbium. The rare earths lighter than terbium are eluted with a
yield of about 80% or more. The enrichment factor of the tracers is about
10^6. The separation process is finished after elution with less than three
free column volumes.

Figure 7 illustrates a more difficult problem, namely the separation of
Tb, Dy and Ho from 1 g of Y. The separation factor, $\beta_{Y/Ho}$, for traces of
these elements in our system is only 1.8. Therefore, we can utilize only
4.4% of the column capacity, and the separation is finished only after passing through the column of about seven free column volumes of 2.5 M HCl. The
time needed for the separation in this case is about 3 hours.

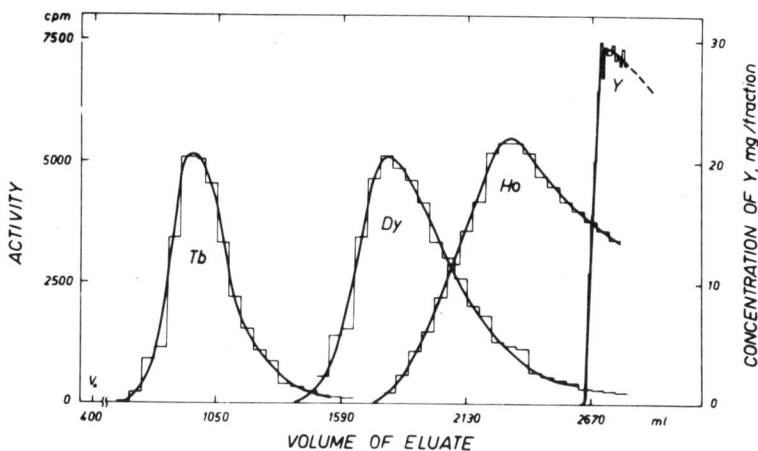

FIGURE 7

Separation of traces of Tb, Dy and Ho from 1 gram of Y. Column: 50mm Φ x
450mm; Silica gel: 325 g, 0.06 - 0.07 mm, with 256 ml HDEHP; Eluent: 2.5 M
HCl; Temperature: 40°C; Flow rate: 0.8 ml/cm^2·min.

4. SEPARATION OF LANTHANIDE TRACERS OF HIGHER ATOMIC NUMBER THAN THE MATRIX ELEMENT

In the separation of tracers of higher atomic number than the matrix element the loading effect on the behaviour of the micro component is significantly smaller than in the case discussed in the preceding section. Since the tracer has the greater distribution coefficient, it is sorbed on top of the column in spite of the formation of the rare earth - HDEHP - polymer, by displacing the macro component in the complex. During the elution, the macro component migrates ahead of the tracer, and therefore, has only a small influence on the shape of the tracer peak. The loading in such separations may be quite high, but the enrichment factors, because of the tailing effect, are about two orders of magnitude smaller.

5. DETERMINATION OF THE TRACER IMPURITIES

After the extraction chromatographic separations, the tracer impurities are obtained in a relatively large volume, amounting up to about 3 l. For the concentration of this solution to about 15 ml, parallel to the elution process, a circulation-evaporator of quartz, shown in figure 8, was successfully used. The evaporation speed adjusted by means of the current input, was equal to the elution speed.

The effluent contains about 1 - 2 mg/l of silicic acid. It is coprecipitated with the rare earth tracers by neutralization of the solution with ammonia. The desorption of the rare earths can be carried out with 1 - 2 ml of 1 M HCl. The yields of this concentration method are more than 95%.

The further processing of the concentrate depends on the analytical method chosen. The methods used are summarized below:

Method	Sensitivity	Standard deviation	Remarks	Ref.
a	$2 \cdot 10^{-5} - 1 \cdot 10^{-3}$	19 %	different tracers different matrices	[3.4]
b	$1 \cdot 10^{-8} - 3 \cdot 10^{-6}$		La,Ce,Pr,Sm,Er,Gd, Tb in Y	[2,3]
c	$2 \cdot 10^{-5} - 4 \cdot 10^{-5}$	11 %	La···Dy in Er	[2]
d	$5 \cdot 10^{-6}$	8%		[9]

a - optical spectral analysis; b - neutron activation analysis;
c - spectrophotometry with arsenazo III; d - complexometric titration with DTPA.

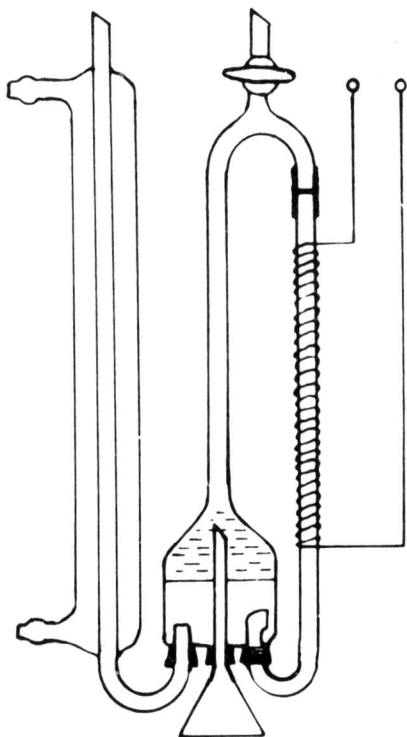

FIGURE 8

Circulation evaporator

The standard deviation of the determination depends almost only on the analytical procedure, not at all on the preconcentration.

In the last three cases, the concentrate of the tracer impurities were resolved into fractions of different elements. This was done either by the cation-exchange - alpha-hydroxyisobutyrate method or by means of extraction chromatography using a HDEHP microcolumn.

After elution from the HDEHP columns, it is especially favourable to apply the complexometric microtitration method with DTPA and arsenazo III as an indicator. This method is very simple, using a silanized 0.1 ml micropipette equippped with a special top as shown in figure 9.

The eluate is collected on a small teflon disk and
then evaporated nearly to dryness. The residue is
dissolved in one drop of sodium acetate buffer solution containing 0.007% of arsenazo III. For
lighter rare earth elements the pH is 4 - 4.5, for
the heavier ones it is 3. Then the pipette is
dipped into the mixture and 10^{-4} M DTPA solution
is added until the colour changes to pink.

6. CONCLUSIONS

The extraction chromatographic preconcentration method outlined above has been used until now
for the following elements: Y, La, Ce, Nd, Sm, Eu,
Gd, Ho and Er. It is difficult to carry out separations in the region of praseodymium or to determine ytterbium and lutetium. In the first case the
separation factors for the neighbouring rare earths
are close to unity, in the second one the distribution coefficient of the elements are very high, and
therefore large volumes of eluent are needed for
their elution. The columns can be used repeatedly
and are capable of doing hundreds of separation
cycles.

The method of extraction chromatography is
useful for the control of rare earth products and
is applied in the rare earth industry.

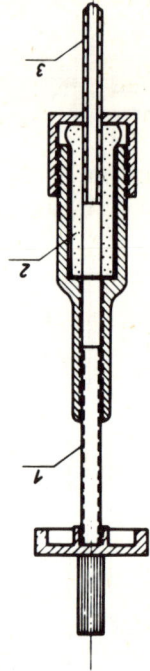

FIGURE 9

Pipetting device.
1-fine-threaded screw
piston; 2-vacuum tubing
3-pipette.

References

[1] E. Herrmann, H. Grosse-Ruyken, N.A. Lebedew and W.A. Chalkin, Radio-
    chimija 6, 756 (1964) (Russ.).

[2] J. Bosholm, H. Grosse-Ruyken, N.A. Lebedew, E. Herrmann and W.A. Chalkin,
    Abbrv. Engl. translation of scientific papers presented in russian of the
    XX-th International IUPAC-Congress, Moscow 1965, P/E 93.

[3] M.G. Semskowa, N.A. Lebedew, Sch.G. Melamed, O.F. Saunkin, G.W. Suchow, W.A. Chalkin, E. Herrmann and G.I. Schmanenkowa, Sawodskaja Laboratorija 33, 667 (1967) (Russ.).

[4] G.I. Schmanenkowa, M.G. Semskowa, M.G. Melamed. G.P. Pleschakowa and G.W. Suchow, Sawodskaja Laboratorija 35, 897 (1969) (Russ.).

[5] D.F. Peppard, G.W. Mason, W.I. Driscoll and R.I. Sironen, J. Inorg. Nucl. Chem. 7, 276 (1958).

[6] D.F. Peppard in The Rare Earths, F.H. Spedding and A.H. Daane, ed., New York - London 1961, pp. 38-54.

[7] D.F. Peppard, G.W. Mason and W.I. Driscoll, J. Inorg. Nucl. Chem. 4, 334 (1957).

[8] E. Herrmann, J. Chromatog. 38, 498 (1968).

[9] E. Herrmann, Dissertation, Dresden 1966.

ANION EXCHANGE CONCENTRATION OF RARE EARTH IMPURITIES IN
RARE EARTH MATERIALS FOR ANALYTICAL PURPOSES

by

F. Molnár

Joint Institute for Nuclear Research, Dubna, USSR.

ABSTRACT

For a total analysis of rare earth impurities in very pure rare earth matrix elements it is necessary to concentrate these impurities before their analysis. For this purpose an anion exchange method, based on the chromatographic separation of the impurities and the matrix element using aqueous methanol solutions of ammonium nitrate as eluents, was developed. It was stated that from the cerium sub-group matrices all rare earth impurities, except the heavier adjacent one, from the yttrium sub-group matrices only cerium sub-group impurities, and - depending on the atomic number of the matrix element - terbium and dysprosium, can be separated. The enrichment factor for this concentration procedure is about 10^6. The concentrated impurities can be processed further as required by the method chosen for their analytical determination.

For the determination of rare earth impurities in high purity rare earth materials, in general, it is necessary to concentrate these impurities before their analysis. For this purpose, first of all, chromatographic methods should be employed.

It was found that the anion exchange method, developed for processing of cerium sub-group targets irradiated with high energy protons [1,2], is applicable for the concentration of rare earth impurities in cerium sub-group rare earth materials [3]. This method is based on the anion exchange chromatography of the rare earth elements carried out with aqueous-methanol

- 56 -

solutions of ammonium nitrate.

The distribution coefficients, K_d, of the light rare earths in these anion exchange systems increase exponentially with increasing methanol concentration. At a fixed value of the methanol content K_d is a linear function of the ammonium nitrate concentration:

$$K_d = a \, [NH_4NO_3]$$

where a is a constant, which, under given experimental conditions (resin, methanol concentration, temperature), is characteristic of the individual rare earth elements. The distribution coefficients show a decreasing trend with increasing temperature, but the separation factors do not change.

The values of a and the corresponding separation factors for the cerium sub-group rare earth elements in the 80% CH_3OH - NH_4NO_3 - IRA 400 system at $20^\circ C$ are summarised in Table 1. It can be seen that the separation factors are high enough for an effective chromatographic resolution of the neighbouring elements. The mean value of the separation factors in the range from lanthanum to gadolinium is 1.81.

TABLE 1

Values of a and separation factors for the light rare earths
in the 80% CH_3OH - NH_4NO_3 - IRA 400 system at $20^\circ C$

Element	Gd	Eu	Sm	Pm	Nd	Pr	Ce	La
a	17.2	26.1	42.0	81.4	156	330	631	1065
Separation factor		1.52	1.61	1.94	1.92	2.12	1.91	1.69

In figure 1 the separation of rare earth impurities (and that of promethium) from a neodymium sample is shown. As can be seen, under the given experimental conditions, about 80% of the praseodymium and 100% of all other rare earth impurities are separated from the neodymium. The impurities heavier than neodymium are eluted between the second and seventeenth fractions, while the lighter ones can be removed from the column, after the elution of the neodymium, with about two free bed volumes of 0.02 M nitric acid solution.

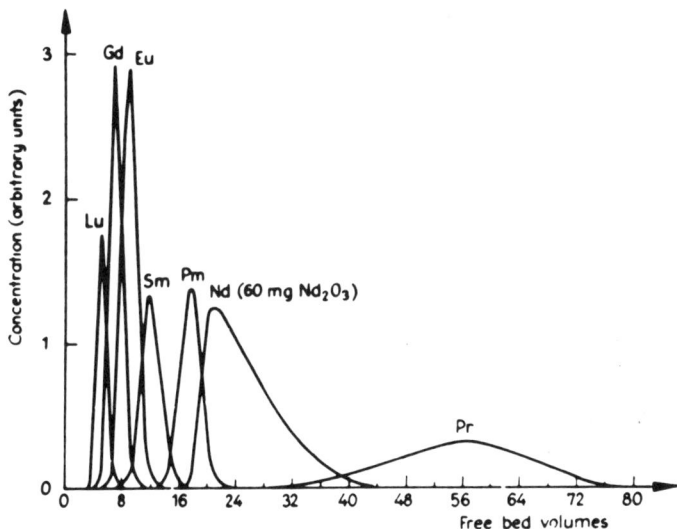

FIGURE 1

Separation of rare earth impurities (and that of promethium) from a neodymium sample. Column: 2 mm x 250 mm Dowex 1X8 (200-400 mesh; nitrate form). Eluent: 80% CH_3OH - 0.01 M HNO_3 - 0.3 M NH_4NO_3. t = $50°C$.

Since in the range of the light rare earths the separation factors between the adjacent elements differ from each other to a slight extent only, the yields for the separation of rare earth impurities from any cerium subgroup sample are similar to those shown in figure 1. Therefore, as a general rule, it can be stated that under the elution conditions in which the concentration maximum of the sample is to be eluted between the twentieth and twenty-fifth free bed volume fractions, all rare earth impurities, except the heavier adjacent one, can be separated partly or entirely from the sample. The concentrate of the impurities heavier than the sample is obtained in

10 - 15 free bed volumes of the eluent in question, while the lighter ones
are concentrated in a small volume of dilute nitric acid solution. The
time needed for the whole concentration procedure is about four hours.
This may be reduced by about 30% if the nitrate concentration of the eluent,
after the beginning of the elution of the sample, is decreased to an appropriate value. It has been observed that this change in the nitrate concentration does not cause any considerable decrease in the separation yields
of the lighter impurities.

Unfortunately, in the range of the yttrium subgroup elements, the values
of the separation factors are close to unity [4]. Therefore, it should be
expected that from the yttrium sub-group samples only cerium sub-group impurities, and - depending on the atomic number of the sample - terbium and
dysprosium can be separated [5]. This is illustrated in figure 2 where the
separation of rare earth impurities from an yttrium sample is shown.

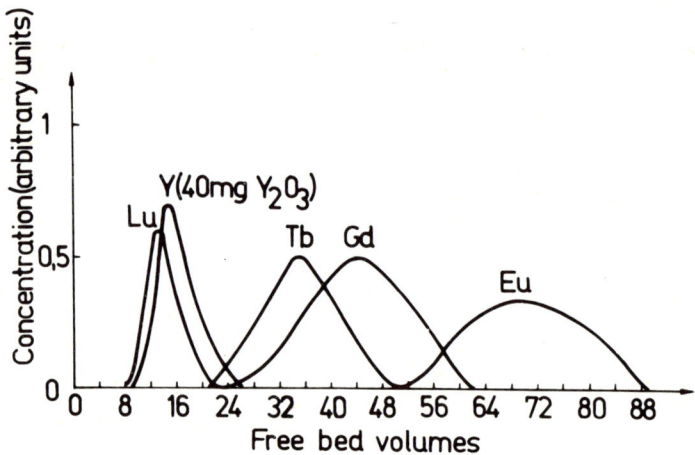

FIGURE 2

Separation of rare earth impurities from an yttrium sample. Column:
6 mm x 250 mm Dowex 1X8 (200-400 mesh; nitrate form). Eluent: 85%
CH_3OH - 0.01 M HNO_3 - 1.5 M NH_4NO_3. t = 50°C.

It is seen that, under the given experimental conditions, about 75% of the terbium, 95% of the gadolinium and 100% of the rare earths lighter than gadolinium are separated from the yttrium. They can be eluted, after the elution of the yttrium, with about two free bed volumes of 0.02 M nitric acid solution. It should be noted that the separation yield of dysprosium has not been evaluated, but its separable amount was found sufficient for an analytical determination. The concentration procedure in the case of an yttrium sample requires about three hours.

Although carrier-free radioactive isotopes and microamount impurities are completely eluted from the column with a small volume of dilute nitric acid solution, in the case of ion exchange of macroamount rare earths the column retains some material, well detectable by activation techniques. This can be washed out only by prolonged rinsing with dilute nitric acid. To overcome this inconvenience, it is necessary to use a separate column for each sample element.

With the anion exchange concentration method outlined above a concentration factor of about 10^6 can easily be achieved. The concentrated impurities may be processed further as required by the method chosen for their analytical determination.

References

[1] F. Molnár, A. Horváth and V.A. Khalkin, J. Chromatog., 26, 215 (1967); 26, 225 (1967).

[2] F. Molnár, Dissertation, Budapest 1968.

[3] F. Molnár and N.A. Lebedev, J. Radioanal. Chem., 2, 91 (1969).

[4] J.P. Faris and J.W. Warton, Anal. Chem., 34, 1077 (1962).

[5] É. Kulus, F. Molnár and E. Szabó, J. Radioanal. Chem., 7, 347 (1971).

EMISSION SPECTROSCOPY FOR RARE EARTH ANALYSES, AND FOR THE DETERMINATION OF IMPURITIES IN RARE EARTH MATERIALS

by

Edward L. DeKalb

Ames Laboratory USAEC, Iowa State University,

Ames, Iowa 50010, U.S.A.

ABSTRACT

Emission spectroscopy is the primary and most powerful analytical tool for the chemical analysis of rare earth materials, and for the determination of rare earth impurities in other materials. However, many rare earth elements have a complex emission spectrum, so that spectral equipment and analytical procedures must be selected with care. Rare earth analyses can be classified into four broad categories: 1) the determination of rare earths in a "pure" rare earth; 2) the determination of other impurities in a rare earth; 3) the determination of rare earths in a mixture of various rare earths; and 4) the determination of rare earth impurities in other materials. Each of these four types of analyses presents the spectroscopist with unique problems which may be solved in several ways. The present state-of-the-art, as measured by detection limits, is reviewed for the first two categories.

1. INTRODUCTION

Emission spectroscopy is the primary analytical tool for the chemical analysis of rare earth (RE) materials and for the determination of RE impurities in other materials. The widespread application of this analytical approach results from the identity of chemical properties of the RE series of elements, so that except in a few cases, the behaviour of the REs in chemical analyses approximates that of a single element.

Rare earth similarities do not extend to their spectra, however, whether in the X-ray, optical, or infrared regions of the spectrum. The spectrum of an RE element is unique, and the most useful property available to the analytical chemist.

The spectra of many of the RE elements are very complex, however, so that the choice of a suitable spectrometer or spectrograph is restricted to those which can provide adequate resolution. Of the RE elements, Ce, Pr and Nd have the most complex spectra, with Sm, Tb, Dy, Ho, Er and Tm only a little better. Y, La, Eu, Gd, Yb and Lu have relatively simple spectra. In general, the lowest detection limits are obtained for elements with a simple spectrum. A complex spectrum has another serious implication: the choice of analytical spectral lines for impurity determinations is likely to be severely restricted because of coincident or unresolved spectral lines from the principle constituent. This often means that the most sensitive analytical lines cannot be used, which also restricts detection limits.

We have found that a reciprocal linear dispersion of 2.5 Å/mm is adequate for most analytical determinations involving the REs. Furthermore, the optical system must be as perfect as possible if analytical determinations involving the complex spectra are to be attempted.

It is convenient to classify RE analyses into four broad categories: 1) the determination of REs in a "pure" RE; 2) the determination of other impurities in an RE; 3) the determination of REs in a mixture of various REs; and 4) the determination of REs in other materials.

2. DETERMINATION OF RARE EARTH IMPURITIES IN A RARE EARTH

The analytical methods developed by Fassel and co-workers[1-5] for the determination of RE impurities in an RE metal or oxide have been adopted by numerous analysts, and minor modifications to the basic approach have been reported by several authors. The more or less pure RE oxide is mixed with an equal amount of graphite, and is subjected to dc arc excitation at ∼ 15 amperes in an undercut graphite electrode. Although the dc arc is an uncontrolled type of excitation system this simple approach provides exceptionally good analyses because of the unique similarity in the physical and chemical properties among many of the REs. In particular, similar vaporization characteristics make it possible to utilize the internal standard principle to its full advantage.

Detection limits as low as a few tenths of a ppm can be obtained if
both the impurity and matrix RE element have relatively simple spectra.
More often, the detection limits are in the range of 10 to 1000 ppm. Precision is unusually good for dc arc excitation, with ± 5% often reported.

Modifications to Fassel's basic method include 1) the use of controlled atmospheres to eliminate interfering cyanogen bands [6-10]; 2) variation in the arcing current, sample electrode dimensions, and the sample
weight; 3) the addition of a buffer such as CsCl [9]; and 4) the use of
an ac arc instead of the dc arc, which is claimed to improve both the detection limits and the precision [11,12].

The use of dissolved REs as the sample has also been reported. Asumi,
Kato and Miyake [13] claimed RE impurity detection limits of 2 to 40 ppm in
complex-spectra RE elements. They used spark excitation to a rotating disc
electrode (RDE) in a controlled atmosphere, and a RE solution containing 60%
methanol.

Separation or preconcentration steps have been tried by several spectroscopists to improve detection limits. Antonov, Drygin and Kalmykov [14]
obtained an enrichment factor of 15 for La, Pr and Nd by extracting the major constituent, Ce, and analyzing the evaporated residues. Melamed, Kostygov and Leshchenko [15] used ion exchange procedures to obtain enrichments
of 15 to 250 for the determination of RE elements in Y, La, Pr and Nd.
Zemskova and co-workers [16] also used a chromatography method to obtain
enrichment factors as high as 100.

For direct spectral determinations without preconcentration, the best
detection limits of which I am aware are tabulated in table 1. Most of
these were taken from refs.[6] and [10].

3. DETERMINATION OF OTHER IMPURITIES IN RARE EARTHS

The spectral determination of non-rare earth impurities in REs is complicated not only by the complex spectra of many of the RE elements but also
by the very different volatilization rates of the impurities to be determined. Simple dc arc methods are usually found to suffer from poor precision

TABLE 1

Detection Limits for RE in RE Determinations

RE Impurity	Detection Limit (ppm)	Matrix	RE Impurity	Detection Limit (ppm)	Matrix
Sc	0.5	Y_2O_3	Gd	0.5	Y_2O_3
Y	2	Sm_2O_3	Tb	5	Y_2O_3
La	3	Y_2O_3	Dy	1	Y_2O_3, Eu_2O_3
Ce	10	Y_2O_3, Sm_2O_3	Ho	5	Y_2O_3
Pr	5	Y_2O_3	Er	0.5	Y_2O_3
Nd	6	Sm_2O_3	Tm	0.5	Y_2O_3
Sm	3	Y_2O_3	Yb	0.01	Y_2O_3
Eu	0.5	Y_2O_3	Lu	0.1	Y_2O_3

if a matrix element spectral line is used for internal standardization, since the impurity and matrix elements will likely be present in the arc column at different times, and will therefore experience different arc fluctuations. Four approaches to the vaporization situation can be taken. Added reference elements with vaporization rates similar to the rates of the impurity elements to be determined can be used for internal standardization.

Vaporization differences can be minimized by using a spark-type of excitation which introduces the sample into the excitation zone by a sputtering process. At the Ames Laboratory, this second approach has been emphasized. Uniarc excitation, which is a combination of a spark and a unidirectional ac arc, has provided reliable analyses with adequate sensitivity for many impurities in most of the REs, and with a precision of better than ± 10%. A briquetted sample of one part RE oxide mixed with nine part pelleting graphite is used. The complex RE matrix spectra do in some cases interfere with the more sensitive spectral lines of impurity elements. For Pr and Nd, which have the most complex spectra, this problem is particularly serious, and alternate analytical approaches must be used if determinations at lower concentration levels are required.

Osumi, Kato and Miyake [13] found the RDE-spark excitation-controlled atmosphere-methanol solution technique to be effective with both non-rare earth and rare earth impurities in neodymium. Apparently the complex neodymium spectrum is suppressed using their excitation scheme.

The preceding techniques sought to avoid vaporization differences. The third approach to vaporization differences, the carrier distillation technique, makes use of these differences to improve detection limits and to suppress the complex matrix spectrum. About 50 to 100 mg of sample oxide is blended with a "carrier" and loaded into a deeply cratered electrode, which is then subjected to dc arc heating and excitation. The volatile impurities distill away from the more refractory matrix, and are swept into the analytical gap with the aid of the carrier. Carriers which have been used with RE samples include Ga_2O_3 [12,17,18], Ga_2O_3 + AgF [19], Ga_2O_3 + LiF [20], and AgCl [21]. Enhancement factors of 100 or more are obtained for those impurities which are more volatile than the matrix under the selected arcing conditions.

The separation of most non-rare earth impurities from a rare earth matrix is not particularly difficult, and this fourth approach has been used by many spectroscopists to improve detection limits or to eliminate the RE spectrum. Detection limits in the fractional ppm range can thus be obtained for a wide variety of impurities.

An interesting example of a separation determination has been developed at the Ames Laboratory for the analysis of residual amounts of Ta in RE metals. Ta is one of the more important impurity possibilities in metal production, at least in metals prepared at the Ames Laboratory, where the metals are cast or deposited in Ta crucibles. Unfortunately, Ta is in the refractory group of impurities, and also has relatively poor spectral detectability. Our method is outlined in figure 1. Ta can be separated from any of the REs and from a great many other materials by this method. Because of its chemical similarity to Ta, Nb is used as a carrier, as an internal control of the separation steps and also as an internal reference element for the spectral analysis. A double separation is sometimes necessary to

reduce RE concentrations to negligible levels. By adjusting the original
sample weight and the amount of Nd added, the concentration range of the
methods can be selected for any Ta level greater than one ppm.

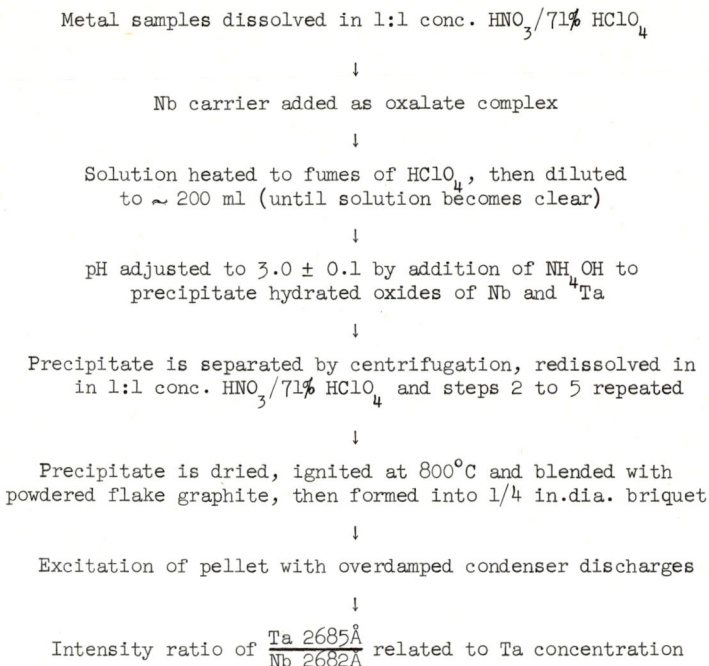

Metal samples dissolved in 1:1 conc. $HNO_3/71\%$ $HClO_4$

↓

Nb carrier added as oxalate complex

↓

Solution heated to fumes of $HClO_4$, then diluted
to ~ 200 ml (until solution becomes clear)

↓

pH adjusted to 3.0 ± 0.1 by addition of NH_4OH to
precipitate hydrated oxides of Nb and Ta

↓

Precipitate is separated by centrifugation, redissolved in
in 1:1 conc. $HNO_3/71\%$ $HClO_4$ and steps 2 to 5 repeated

↓

Precipitate is dried, ignited at 800°C and blended with
powdered flake graphite, then formed into 1/4 in.dia. briquet

↓

Excitation of pellet with overdamped condenser discharges

↓

Intensity ratio of $\frac{Ta\ 2685Å}{Nb\ 2682Å}$ related to Ta concentration

FIGURE 1

FLOW DIAGRAM FOR THE DETERMINATION
OF TRACE TANTALUM IN THE RARE EARTHS

Without prior separation or preconcentration steps, the detection limits shows in Table 2 have been reported for non-rare earth impurities in various rare earth oxides.

TABLE 2

Detection Limits for Determination of Non-Rare Earths in Rare Earths

Detection Limits	Elements
1 ppm or less	Ag, Al, As, B, Ba, Bi, Ca, Cd, Co, Cr, Cu, Fe, K, Li, Mg, Mn, Na, Ni, Pb, Si, Sn, Ti, V, Zn
2 to 10 ppm	Be, Ge, Mo, Sb, Zr
11 to 100 ppm	Hf, Nb, Sr, Ta, Th, W

4. ANALYSIS OF RARE EARTH MIXTURES

This type of spectral analysis is quite similar to the determination of RE impurities in a "pure" RE, except that no single RE element of known concentration is present in the sample to serve as the reference element for internal standardization. The most common approach to this problem, and the one Fassel [22] developed in the late 1940's, is to add to the sample an excess of some element which can be independently determined and is chemically similar to the REs to be determined. Fassel used CeO_2, since its concentration in the sample can be measured through oxidation-reduction reactions. The problem of analyzing a mixture of REs thus becomes one of determining REs in a CeO_2 matrix. The primary disadvantage of this approach is that detection limits are raised by the dilution factor. Determinations in the range from about 1 to 90% can be readily accomplished, which is usually adequate for this particular analytical problem.

We now feel that there are better spectrometric ways of solving this analytical problem, and Dr. Fassel will describe two of them in the following paper.

5. DETERMINATION OF RARE EARTHS IN OTHER MATERIALS

A thorough review of the various approaches to such analyses by emission spectroscopy is beyond the scope of this paper, but some of the basic procedures can be mentioned. Although direct excitation of the sample will provide sufficient sensitivity, precision and accuracy for some samples, separation of the REs as a group ordinarily precedes the analysis. Separation not only allows for preconcentration of the REs so that lower detection limits are possible, but also simplifies the analytical problem when geological samples of widely varying composition are encountered. Nearly every type of separation of REs from a matrix can be found in the literature. Ion exchange and extraction procedures seem to be most popular, with precipitation steps suggested almost as often.

Regardless of the type of separation, one of the most fruitful refinements is to concentrate the REs in an RE carrier. Yttrium is frequently used as the carrier for a number of reasons [23]. The ionic radii of Y^{3+} is about the same as for Tb^{3+} and Dy^{3+}, so that in chemical properties yttrium is right in the middle of the RE series. Its emission spectrum is relatively simple, so that detection limits are not hampered by interfering matrix lines. In addition, any separation scheme devised can be monitored by a radioactive tracer, ^{90}Y, which is a β emitter and is easily obtained from ^{90}Sr.

Other carriers have also been suggested. Lanthanum [24] has been used, but is somewhat less desirable than yttrium since La_2O_3 is quite hygroscopic. Non-rare earths, such as Ca and Al [25] have also been used as carriers, particularly with precipitation separations.

At the Ames Laboratory, emission spectral methods for the analysis of RE impurities in compounds of U, Th, and Zr, and in various other materials such as rock and ores have been developed in the past. However, for most of these analyses, alternate analytical methods appear to be superior. For instance, RE impurities can be determined in U_3O_8, ThO_2 and ZrO_2 at fractional ppm levels without separation or concentration steps, by using X-ray excited optical fluorescence. This analytical technique will be described by Dr. Fassel in a subsequent paper.

In conclusion, emission spectroscopy is certainly one of the most use-

ful analytical methods available for RE analyses of all types, and despite
the recent development of competing analytical methods, it will undoubtedly
continue to be the method of choice for most RE determinations.

References

[1] V.A. Fassel and H.A. Wilhelm, J. Opt. Soc. Am. 38, 518 (1948).

[2] V.A. Fassel, H.D. Cook, L.C. Krotz and P.W. Kehres, Spectrochim.
     Acta, 5, 201 (1952).

[3] V.A. Fassel, B. Quinney, L.C. Krotz and C.F. Lentz, Anal. Chem. 27,
     1010 (1955).

[4] R.N. Kniseley, V.A. Fassel, B.B. Quinney, C. Tremmel, W.A. Gordon
     and W.J. Hayles, Spectrochim. Acta, 12, 332 (1958).

[5] R.N. Kniseley, V.A. Fassel, R.W. Tabeling, B.G. Hurd and B.B. Quinney,
     ibid, 13, 330 (1959).

[6] M. Sato, H. Matsui and T. Matsubara, Bunseki Kagaku, 20, 215 (1971).

[7] S.V. Grampurohit and V.P. Bellary, India At. Energy Comm. Report
     B.A.R.C. 472 (1970).

[8] E.M. Hammaker, G.W. Pope, Y.G. Ishida and W.F. Wagner, Appl. Spectrosc.
     12, 161 (1958).

[9] Y. Osumi, A. Kato and Y. Miyake, Z. Anal. Chem. 255, 264 (1971).

[10] D.L. Nash, Appl. Spectrosc. 22, 101 (1968).

[11] L.E. Zeeb, J.T. Rozsa, D.C. Manning and J. Stone, Report No. APEX-431
     (Del.), National Technical Information Services, U.S. Dept. of Commerce,
     Springfield, VA.

[12] J.T. Rozza and J. Stone, Develop. Appl. Spectry. 1, 187 (1962).

[13] Y. Osumi, A. Kato and Y. Miyake, Z. Anal. Chem. 251, 7 (1970).

[14] A.V. Antonov, A.I. Drygin and Yu.A. Kalmykov, Zavod. Lab. 33, 967 (1967).

[15] Sh.G. Melamed, A.S. Kostygov and T.V. Lishchenko, ibid, 30, 1339 (1964).

[16] M.G. Zemskova, N.A. Lebedev, Sh.G. Melamed, O.F. Saunkin, G.V. Sukhov,
     V.A. Khalkin, E. Kherrmann and G.I. Shmanenkova, ibid, 33, 667 (1967).

[17] R.L. Slyusareva, L.I. Kondrat'eva and Sh.I. Peizulaev, ibid, 31, 557
     (1965).

[18] L.I. Pavlenko, N.V. Laktionova and Yu.S. Sklyarenko, Zh. Anal. Khim.

22, 104 (1967)

[19] Y. Osumi, A. Kato and Y. Miyake, Z. Anal. Chem. 255, 103 (1971).

[20] J.H. Muntz, Spectrochim. Acta, 24B, 207 (1969).

[21] A.B. Whitehead, B.C. Piper and H.H. Heady, Appl. Spectrosc. 20, 107 (1966).

[22] V.A. Fassel, J. Opt. Soc. Am. 39, 187 (1949).

[23] H.J. Hettel and V.A. Fassel, Anal. Chem. 27, 1311 (1955).

[24] A.N. Zaidel, N.I. Kaliteevskii, A.A. Lipovskii, A.N. Razumovskii and P.P. Yakimova, Fiz. Sbornik L'vov. Univ., 1958 No. 4, 37.

[25] C.L. Waring and H. Mela, Jr., Anal. Chem. 25, 432 (1953).

FLAME AND PLASMA ATOMIC ABSORPTION, EMISSION AND
FLUORESCENCE SPECTROSCOPY OF THE RARE EARTH ELEMENTS

by

Velmer A. Fassel, Richard N. Kniseley, and Constance C. Butler

Ames Laboratory USAEC and Department of Chemistry,

Iowa State University, Ames, Iowa 50010, U.S.A.

ABSTRACT

In comparison to the very complex spectra emitted by the rare earth
elements in arc or spark discharges, their corresponding flame emission
spectra are relatively simple. Historically, the first analytical applications of flame emission spectra utilized observations on the rare earth
monoxide band systems, which constitute the principal spectral features in
the spectra emitted by non-reducing flames. Because of the spectral overlap of the band systems, their analytical utility was rather limited. The
observation that fuel-rich O_2-C_2H_2 or N_2O-C_2H_2 flames provided a striking
increase in the atomization efficiency, and hence a greater free-atom population in the flames, greatly broadened the scope of application of flame
spectroscopy. The enhancement factors in the line emission spectra achieved
by simply changing the stoichiometry of either O_2-C_2H_2 or N_2O-C_2H_2 flames
are typically several orders of magnitude.

Flame emission generally exhibits greater powers of detection than
atomic absorption measurements by factors ranging from 2 to 80. As a consequence, flame emission spectroscopic techniques have found a wider application, at least in the Ames Laboratory, to the analysis of rare earth mixtures and to the determination of low concentrations of the rare earth elements.

Although the ultimate potential of electrically generated "flame-like"
plasmas has not been evaluated, the preliminary results so far obtained
suggest that these plasmas may replace combustion flames as excitation sources for the analysis of rare earth mixtures.

1. FLAME ATOMIC ABSORPTION, EMISSION, AND FLUORESCENCE

1.1 Introduction

The conventional arc or spark emission spectra of most of the rare earth elements are very complex, possessing thousands of lines of rather uniform intensity and lacking the characteristically intense lines found in the spectra of other elements. Even under high dispersion the probability of line interference is high and difficulties are frequently encountered in locating interference-free lines. The promise that simpler spectra can be obtained under less energetic excitation conditions than those prevailing in arcs and sparks motivated several early investigators to explore the spectra emitted in lower temperature combustion flames. Lundegårdh [1] first recognized the possibility of using flame spectra for the determination of several elements in this group. In a series of papers published during the period 1929 to 1941, Piccardi [2-10] described the principal features of the oxyhydrogen flame spectra of most of the rare earth elements. Piccardi, as well as Lundegårdh, introduced the rare earths into the flames as aerosols of aqueous solutions. Under these moderate excitation conditions both Lundegårdh and Piccardi observed that most of the spectra were remarkedly devoid of any atomic lines. However, Piccardi noted that most of the rare earth elements have a tendency to form stable diatomic monoxide (MO) molecules in the flame. Since these molecules are relatively stable, their presence in the flame was manifested by the characteristic diatomic band emission spectra. Pinta [11] extended the observations to air-acetylene flames and made a thorough study of the analytical utility of these spectra.

Although Piccardi showed that these band systems could be used to advantage in monitoring rare earth fractionations, the overlap of the individual spectra greatly restricted the general analytical utility of these band systems. The feasibility of applying appropriate interference corrections was studied by Pinta [4], but this approach was practically limited to simple binary or ternary mixtures.

There are several components in these band systems which occur in spectral regions relatively free of interference. The most prominent of these are the LaO bands at 4372, 4418, 7403 and 7910 Å. Although De Albinati [12], Ishida [13] and Menis et al. [14] found that these bands were subject to

various degrees of interference by other cations present in the solution,
Tremmel [15] observed that under his experimental conditions other rare
earths present in mixtures did not exert significant interelement effects
or spectral interferences on the 7910 Å LaO band.

The oxyhydrogen flame spectra of all rare earths except cerium plus
the elements yttrium and scandium have been observed by Rains et al. [16].
They deduced that the LaO band systems described above, the ytterbium atom
line at 3988 Å and the NdO bands at 6606 and 7120 Å should be useful for the
determination of the parent elements in rare earth mixtures. Using air-
acetylene flame excitation, Poluektov and Nikonova [17] employed the LaO
emission at 7940 Å, the YO emission at 5972 to 5988 Å or 6132 to 6165 Å, and
the atom lines of ytterbium and europium at 3988 and 4594 Å, respectively,
for determining the parent element in some rare earth mixtures. Since the
standard addition technique was used, the upper limit of determination was
restricted to the range 10 - 30%.

Several other authors [16-19] have also utilized the rare earth band
spectra for analytical purposes. However, because of the high incidence of
spectral interference among the various rare earth monoxide band systems,
it is unlikely that the analytical utility of the band spectra will be extended significantly beyond the applications discussed above.

1.2 Atomic Line Spectra Emitted in Flames

Although several investigators [11,15,16,25] reported the presence of
a few atomic lines in their flame spectra, the intensities were so weak that,
with the exception of Eu and Yb, the lines possessed little analytical potential. In all of these flame spectra, the predominant features were the monoxide band systems [2-10], indicating that these molecules are stable at
ordinary flame temperatures. Even the relatively high temperatures prevalent
in stoichiometric oxyacetylene flame ($\sim 3200°$K) were insufficient to effectively dissociate these molecules. Collaborative thermodynamic data can be
found in the relatively high dissociation energies of these molecules. The
high stability of these monoxide molecules greatly reduces the free-atom
concentration in the flame and thus the atomic line emission will be weak or
non-existent.

Fassel, Curry and Kniseley [26] found that flame chemistry, rather than flame temperature, was an important consideration in the production of free-atoms of the rare earth elements. Their studies have shown that the increase in the free-atom population is a direct result of the high concentration of carbon-containing species that are prevalent in this and other high temperature hydrocarbon flames [27,28]. The reaction between atomic oxygen and the carbonaceous species is highly exothermic and thus the concentration of atomic oxygen in these flames should be very low [29-32]. This low concentration of atomic oxygen tends to shift the rare earth monoxide dissociation equilibrium in favour of the existence of a relatively high concentration of free atoms in the flame gases. The improvement in free atom production in the fuel-rich hydrocarbon flame is illustrated for Sc in figure 1.

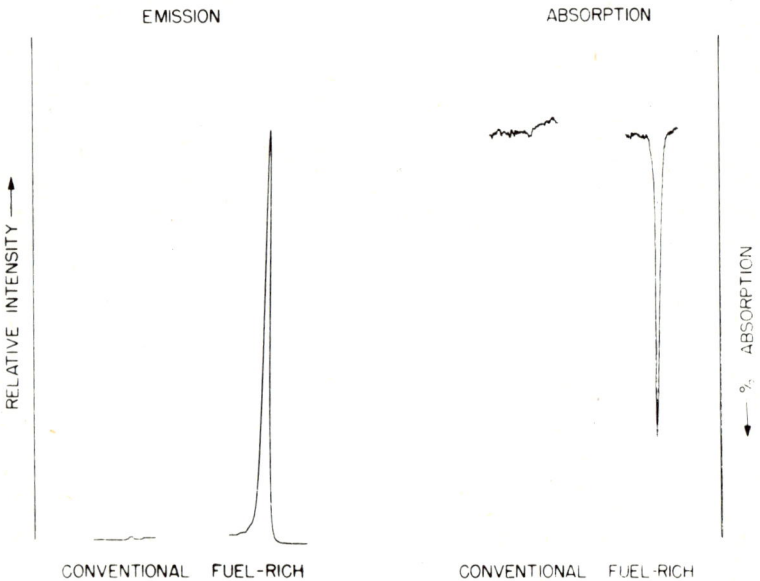

FIGURE 1

Emission and absorption by scandium in a stoichiometric and a fuel-rich nitrous oxide-acetylene flame.

The emission and absorption enhancement factors observed in the fuel-rich
flame vary considerably, with the greatest enhancements observed for those
rare earth elements which have the most stable monoxides [32]. Thus for
both the flame atomic absorption and emission determination of the rare
earths, flame stoichiometry is a critical variable. Very fuel-rich flames
are necessary for the efficient formation of free atoms of those elements
that form stable monoxides (e.g., La and Ce), whereas near stoichiometric
flames produce maximal absorption or emission for Eu and Yb, whose monoxide
dissociation energies are relatively low.

The original work on the flame excitation of the line spectra of the
rare earth elements was done with a total consumption burner (Beckman-type)
that produced a highly turbulent flame. For most of the subsequent investigations in both emission and absorption, premixed O_2-N_2-C_2H_2, [33,34],
N_2O-C_2H_2 [35-40] or O_2-C_2H_2 were employed [41]. For N_2O-C_2H_2 and O_2-C_2H_2,
slot burners of the type normally used in atomic absorption instruments were
found particularly useful for emission observations as well.

The compressed wavelength scales usually employed in published flame
emission spectra convey the impression that the spectra of these elements
are very complex. In reality the spectra are relatively simple, as shown
in figure 2. In this figure, the spectrum of samarium, one of the most complex in the group, is shown in both compressed and expanded form. Even
though the spectra were recorded with a low dispersion 0.5 m grating spectrometer, the simplicity is readily apparent in the lower spectrum.

1.3 Flame Atomic Absorption Spectroscopy (FAAS)

Most of the early observations of the atomic absorption spectra of the
rare earth elements were made on free atoms generated in King-type furnaces
charged with rare earth salts [42-44]. The atomic absorption spectra of
these elements were first studied in flames by Mossotti and Fassel [45,46].
Later observations were made on free atoms formed in turbulent, fuel-rich
oxyacetylene and premixed oxyacetylene or nitrous oxide-acetylene flames
burning on slot burners [38,39,41,48-50].

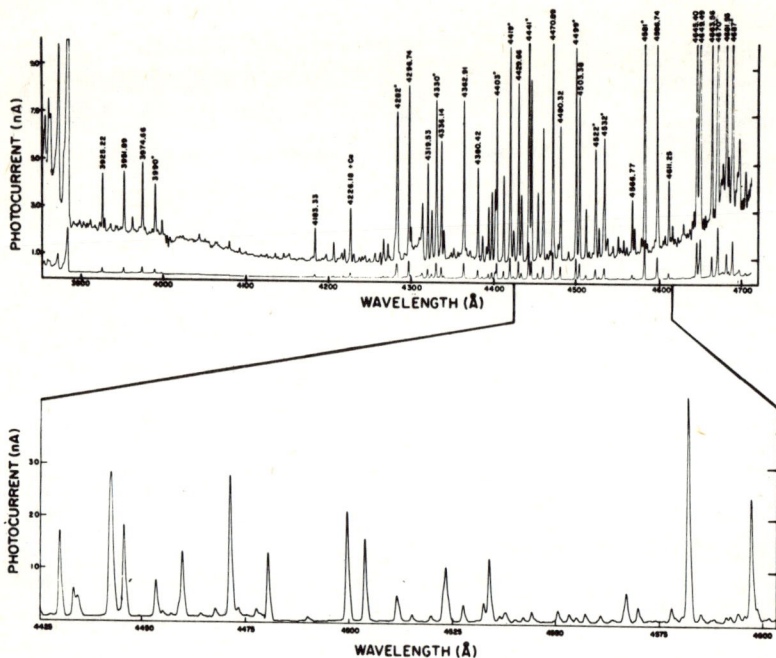

FIGURE 2

Recording of a portion of the flame emission spectrum
of Sm using both a compressed and expanded wavelength scale

1.4 Flame Atomic Fluorescence Spectroscopy (FAFS)

There have been no reports on the application of FAFS to the determination of the rare earths at any concentration. Since high temperature, reducing flames are required for the efficient atomization of the rare earth element, it is not likely that FAFS measurements will surpass the powers of detection or the convenience of measuring the free atoms in flames in simple
thermal emission.

1.5 Detection Limits

Experimentally determined detection limits provide the analyst with realistic numerical data for comparing capabilities of several techniques for determining trace elements and for estimating the lowest concentrations determinable under specific circumstances. The analyst may draw these data from a variety of compilations [33-41,47,48,49-55]. The most recently published comparisons [55] may be viewed as reflecting the present "state of the art". A side-by-side comparison of flame emission spectroscopy (FES) and atomic absorption spectroscopy (AAS) extracted from [55] values is shown in table 1. It is seen that FES detection limits are generally superior to those observed in AAS by factors ranging from 2 up to 80. As a consequence FES techniques have found a far wider application, at least in the Ames Laboratory, to the analysis of rare earth mixtures and to the determination of low concentrations of the rare earth elements.

1.6 Analytical Applications: Flame Atomic Absorption and Emission Spectroscopy

One of the most useful analytical applications of flame atomic spectra is for the analysis of rare earth mixtures, since classical chemical methods are generally not applicable. Flame atomic emission spectra possess several advantages over the other spectra used for the analysis of rare earth mixtures. First, the spectra are striking in their simplicity compared to the arc or spark spectra. It is therefore possible to achieve adequate spectral resolution with small table-model spectrometers, whereas large, high dispersion spectrographs are required when arc or spark spectra are observed. Secondly, all of the rare earths exhibit line spectra of sufficient intensity to possess analytical utility. The group of rare earth elements which evade detection by the spectrophotometric techniques are readily observed in the flame spectra. In addition, there is no evidence of interelement effects, which contrasts sharply with the selective enhancement and absorption effects observed in x-ray fluorescent spectrometric measurements. An early comprehensive assessment of the actual utility of flame emission spectroscopy for the analysis of complex rare earth mixtures was based on the spectra emitted in a highly turbulent flame formed by a "total consumption" Beckman-type

TABLE   1

Comparison of Experimentally Measured Detection Limits[*]

Element	FES	AAS	ICP
Ce	10	---	0.007
Dy	0.05	0.2	0.004
Er	0.04	0.1	0.001
Eu	0.0005	0.04	0.001
Gd	2	---	0.007
Ho	0.02	0.1	0.01
La	0.01[a]	2	0.003
Lu	1	3	0.008
Nd	0.7	2	0.05
Pr	0.07	4	0.06
Sc	0.03	0.1	0.003
Sm	0.2	0.6	0.02
Tb	0.03[a]	2	0.2
Tm	0.02	0.08	0.007
Y	0.03[a]	0.3	0.0002
Yb	0.002	0.02	0.0009

[*] Detection limits are defined as the concentrations ($\mu g/ml$) required to produce a signal level at least twice as great as the standard deviation of the total background noise fluctuations.

[a] Band emission observed.

burner [56]. This method was later adapted for use with the premixed $N_2/O_2/C_2H_2$ and $N_2O-C_2H_2$ [33,34] flames. A summary of the suggested analytical lines for the $N_2/O_2/C_2H_2$ flame can be found in ref.[51]. A typical analytical curve is shown in figure 3. The matrix composition in each of the standards used to prepare this analytical curve was changed. All of the experimental points cluster rather uniformly along a single congruent curve, with no apparent systematic deviations for any of the matrices. This behavi-

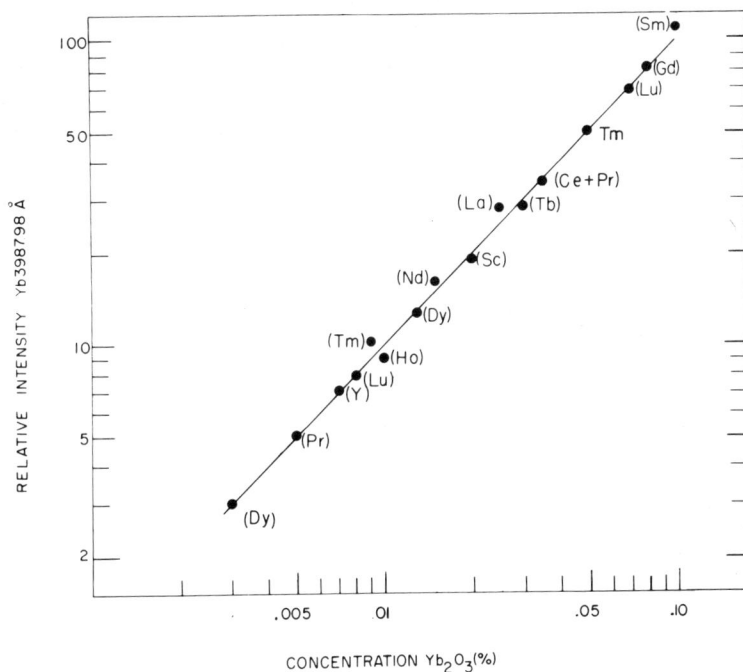

FIGURE 3

Analytical curve for the determination of low concentrations
of ytterbium in rare earth mixtures

our documents the high degree of freedom of this technique from both spectral and interelement interferences. The analytical curves can be extended
all the way to 100%; a typical example is shown in figure 4. The analytical data summarized in tables 2 and 3 reflect the accuracy of flame emission
analyses of rare earth mixtures.

Applications of atomic absorption techniques to rare earth analytical
problems have been described by Anderson [57], Kinnunen and Lindsjö [58,59],
Jaworowski, Weberling and Bracco [60], Chau [61], Kriege and Welcher [62],

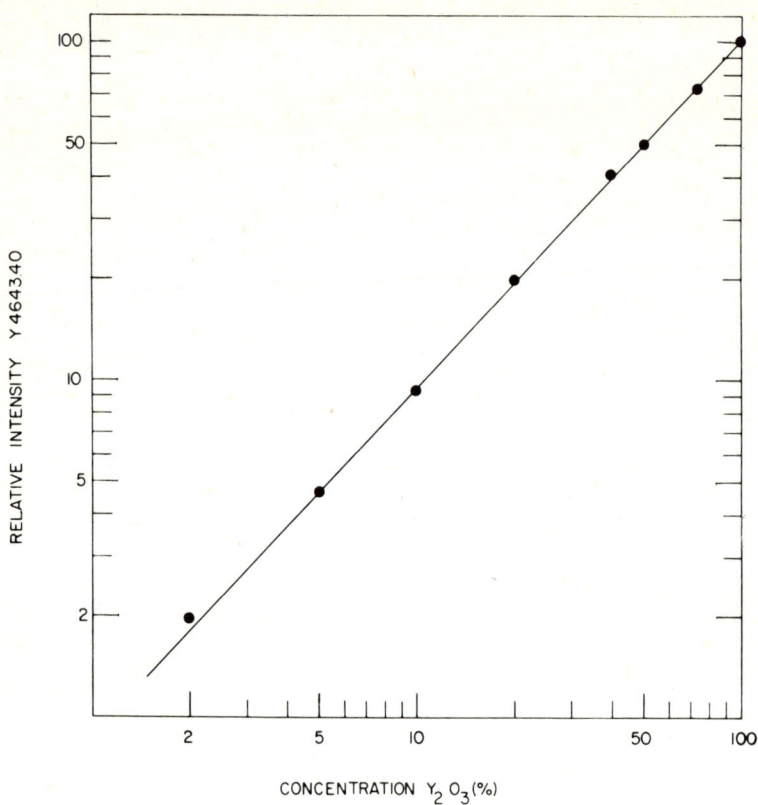

FIGURE 4

Analytical curve for the determination
of yttrium in rare earth mixtures

and Fernandez and Manning [63]. In general, the powers of detection reported were inferior to those found in flame emission, confirming the comparative data shown in table 1. Although "chemical" or "solute vaporization interferences" are recorded more frequently in the published literature for atomic absorption than for emission applications, the degree of

TABLE 2

Comparison of Flame Emission and Spectrophotometric Analyses for Nd and Sm
in Rare Earth Concentrates from Solvent Extraction Experiments

Sample no.	Neodymium		Samarium	
	Flame Emission (wt.%)	Spectrophotometric (wt.%)	Flame Emission (wt.%)	Spectrophotometric (wt.%)
L-1	76.0	76.5	22.5	24.0
L-2	63.0	64.5	34.5	36.5
L-3	56.0	56.0	43.5	44.3
L-4	27.5	27.1	74.5	74.3
L-5	17.0	17.0	84.5	84.9
L-6	12.5	12.6	88.5	88.1

TABLE 3

Results of Flame Emission Analysis of Rare Earth Concentrates
from a Synthetic Xenotime Ore

	% Present	% Found
La_2O_3	3.2	3.6
CeO_2	4.4	4.3
Pr_6O_{11}	.8	1.1
Nd_2O_3	2.8	2.6
Sm_2O_3	1.2	1.1
Eu_2O_3	.10	.10
Gd_2O_3	2.8	2.7
Tb_4O_7	1.3	1.7
Y_2O_3	62	59
Dy_2O_3	7.6	6.9
Ho_2O_3	1.8	1.6
Er_2O_3	5.9	5.7
Tm_2O_3	.80	.73
Yb_2O_3	5.0	5.2
Lu_2O_3	.30	.46
Sc_2O_3	---	---

interference should normally be the same for both techniques under identical experimental conditions.

2. PLASMA ATOMIC EMISSION

"Electrical flames" or electrically generated "flame-like" plasmas possess physical and spectroscopic properties that make them potentially very useful as free atom generators and excitation sources for the determination of the rare earths. The ultimate potential of these plasmas has not been evaluated, but the preliminary result so far obtained in our laboratories are indeed encouraging. For example, the detection limits that have been measured in an electrodeless induction-coupled plasma (ICP) (see table 1) are at least two to three orders of magnitude lower than the lowest values so far reported by flame emission. A schematic representation of this plasma is shown in figure 5. In these plasmas a stream of argon is heated inductively by the eddy currents induced inside the coil space by high frequency currents flowing in the coils. It is possible to "shape" the plasmas in such a way as to form a toroidal, "doughnut-hole" configuration. The sample solution aerosol may be efficiently injected into the plasma through this doughnut hole. The aerosol particles experience a temperature of $\sim 6000^{\circ}K$ as they pass through the plasma. Ordinarily the spectra of the free atoms generated are observed in the tail flame, where the lower temperature ($\sim 4000^{\circ}K$) produces simple spectra relatively free of the Ar continuum and line radiation generated in the core of the plasmas.

The various advantages of an analytical system based on induction-coupled plasma excitation will be discussed in the lecture. One of the distinctive advantages of the system is its analytical range. For example, analytical curves for the determination of Yb and Tm in rare earth mixtures exhibited strict linearity from 0.01 up to 100 wt.% in the mixture.

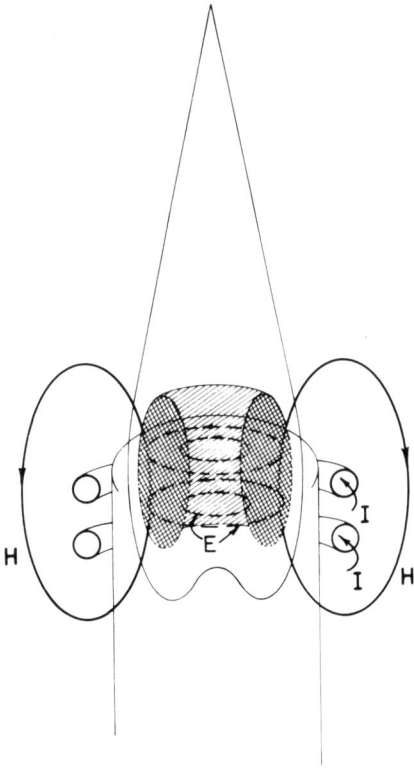

FIGURE 5

Schematic representation of an
induction-coupled plasma showing
the magnetic fields H produced
by current I flowing in the induction coil, and the eddy current
E induced in the plasma

References

[1] H. Lundegårdh, "Die Quantitative Spektralanalyse der Elemente". Part II.
     Fischer, Jena, 1934.

[2] G. Piccardi, Nature 124, 129 (1929).

[3] G. Piccardi, Atti. Accad. Lincei. 14, 578 (1931). C.A. 26, 5031 (1932).

[5] G. Piccardi and A. Sberna, Atti. Accad. Lincei. 15, 309 (1932). C.A. 26,
     5869 (1932).

[6] G. Piccardi, Atti. Accad. Lincei. 15, 577 (1932). C.A. 26, 5869 (1932).

[7] G. Piccardi, Atti. Accad. Lincei., Classe Sci. Fiz. Mat. Nat. 25, 44
     (1937). C.A. 31, 7790 (1937).

[8]   G. Piccardi, Atti. Accad. Lincei., Classe Sci. Fiz. Mat. Nat. 25, 86
      (1937). C.A. 31, 7790 (1937).

[9]   G. Piccardi, Spectrochim. Acta, 1, 249 (1939). C.A. 34, 6541 (1940).

[10] G. Piccardi, Spectrochim. Acta, 1, 533 (1941).

[11] M. Pinta, J. recherches centre natl. recherche sci. Labs. Bellevue,
      (Paris) 21, 260 (1952).

[12] J.F. Possidoni De Albinati, Anales Asoc. quim. arg. 43, 106 (1955).

[13] M. Ishida, J. Chem. Soc. Japan (Pure Chem. Sect.) 76, 60 (1955).

[14] O. Menis, T.C. Rains and J.A. Dean, Anal. Chim. Acta, 19, 179 (1958).

[15] C.G. Tremmel, Masters Degree Thesis, Iowa State University, Ames, Iowa
      (1958).

[16] T.C. Rains, H.P. House and O. Menis, Anal. Chim. Acta, 22, 315 (1960)

[17] N.S. Poluektov and M.P. Nikonova, Ukrain. Khim. Zhur. 25, 217 (1959).

[18] V. Patrovsky, Collection Czech. Chem. Commun. 26, 2445 (1961).

[19] O.A. Shiryaeva and Sh.G. Melamed, Zavodsk. Lab. 30, 183 (1964).

[20] E. Sudo, H. Toto and S. Ikeda, Sci. Rept. Res. Inst. Tohoku Univ. Ser.
      A, 12, 401 (1960).

[21] L.A. Ovchar, S.A. Mischenko and N.S. Poluektov, Zh. Prikl. Spektr. 3,
      306 (1965).

[22] G. Svehla and P.J. Slevin, Talanta 15, 978 (1968).

[23] E. Ruf, Angew. Chem. 73, 338 (1961).

[24] N.S. Poluektov, R.A. Vitkun and L.A. Ovchar, Zh. Anal. Khim. 15,
      264 (1960).

[25] L.A. Ovchar and N.S. Poluektov, Zh. Anal. Khim. 22, 45 (1967).

[26] V.A. Fassel, R.H. Curry and R.N. Kniseley, Spectrochim. Acta, 18,
      1127 (1962).

[27] H.A. Broida and K.E. Shuler, J. Chem. Phys. 37, 933 (1957).

[28] A.G. Gaydon and J.G. Wolfhard, "Flames, Their Structure, Radiations
      and Temperature". (2nd ed). Chapman and Hall, London, 1960.

[29] C.P. Finimore and G.W. Jones, J. Phys. Chem. 62, 178 (1958).

[30] J.O. Rasmuson, "An Experimental and Theoretical Evaluation of the
      Nitrous Oxide-Acetylene Flame as an Atomization Cell for Flame Spectro-
      scopy", Ph.D. Thesis, Iowa State University, Ames, Iowa (1970).

[31] J.E. Chester, R.M. Dagnall and M.R.G. Taylor, Anal.Chima Acta 51, 95
      (1970).

[32] V.A. Fassel, J.O. Rasmuson, R.N. Kniseley and T.G. Cowley, Spectrochim. Acta, 25B, 559 (1970).

[33] R.N. Kniseley, A.P. D'Silva and V.A. Fassel, Anal. Chem. 35, 910 (1963).

[34] A.P. D'Silva, R.N. Kniseley and V.A. Fassel, Anal. Chem. 36, 1287 (1964).

[35] E.E. Pickett and S.R. Koirtyohann, Spectrochim. Acta, 23B, 235 (1968).

[36] R.N. Kniseley, C.C. Butler and V.A. Fassel, Anal. Chem. 41, 1494 (1969).

[37] M.D. Amos, The Element (Aztec Instruments) No. 17 (1967).

[38] G.D. Christian, Anal. Lett. 1, 845 (1968).

[39] G.D. Christian and F.J. Feldman, "Atomic Absorption Spectroscopy", Wiley-Interscience, New York, 1970.

[40] D.N. Hingle, G.F. Kirkbright and T.S. West, Analyst 94, 864 (1969).

[41] J.A. Fiorino, R.N. Kniseley and V.A. Fassel, Spectrochim. Acta, 23B, 413 (1968).

[42] F.W. Paul, Phys. Rev. 49, 156 (1936).

[43] R. Zalubas and M. Wilson, J. Res. Nat. Bur. Std. A69, 59 (1965).

[44] L.F.H. Bovey and W.R.S. Garton, Proc. Phys. Soc. 67A, 291 (1954).

[45] V.A. Fassel and V.G. Mossotti, Anal. Chem. 35, 252 (1963).

[46] V.G. Mossotti and V.A. Fassel, Spectrochim. Acta, 20, 1117 (1964).

[47] R.K. Skogerboe and R.A. Woodriff, Anal. Chem. 35, 1977 (1963).

[48] M.D. Amos and J.B. Willis, Spectrochim. Acta, 22, 1325 (1966).

[49] N. Shifrin and J. Ramirez-Munoz, Appl. Spectros. 23, 358 (1969).

[50] N. Shifrin and J. Ramirez-Munoz, Appl. Spectros. 23, 365 (1969).

[51] R.N. Kniseley, V.A. Fassel and C.C. Butler, "Atomic Emission and Absorption Spectrometry of the Rare Earth Elements" in "Analytical Flame Spectroscopy". R. Mavrodineanu, Ed., Philips' Gloeilampenfabrieken, Eindhoven, Netherlands.

[52] W. Slavin, "Atomic Absorption Spectroscopy", Interscience, New York, 1968 .

[53] V.A. Fassel and D.W. Golightly, Anal. Chem. 39, 466 (1967).

[54] V.A. Fassel, V.G. Mossotti, W.E.L. Grossman and R.N. Kniseley, Spectrochim. Acta, 22, 347 (1966).

[55] G.D. Christian and F.J. Feldman, Appl. Spectros. 25, 660 (1971).

[56] A.P. D'Silva, R.N. Kniseley, V.A. Fassel, R.H. Curry and R.B. Myers, Anal. Chem. 36, 532 (1964).

[57] J.W. Anderson, Paper presented at the 5th Natl. Meeting of Society for Applied Spectroscopy, Chicago, June 1966.

[58] J. Kinnunen and O. Lindsjö, Chem. Analyst, 56, 25 (1967).

[59] J. Kinnunen and O. Lindsjö, Chem. Analyst, 56, 76 (1967).

[60] R.J. Jaworowski, R.P. Weberling and P.J. Bracco, Anal. Chim. Acta, 37, 284 (1967).

[61] Yiu-Kee Chau, Talanta 15, 421 (1968).

[62] O.H. Kriege and G.G. Welcher, Talanta 15, 781 (1968).

[63] F. Fernandez and D.C. Manning, At. Absorption Newslett. 7, 57 (1968).

X-RAY FLUORESCENCE ANALYSIS OF RARE EARTH ELEMENTS

by

M. Bonnevie-Svendsen and A. Follo

Institutt for Atomenergi, Kjeller, Norway.

ABSTRACT

X-ray fluorescence analysis is used for the direct determination of rare earth elements in rock samples, minerals, rare earth concentrates, products and process solutions. Only a minimum of sample pretreatment is required. Solid samples are mostly powdered and measured as briquets. Liquid samples can be measured as received, but for larger series of process samples a filter technique is preferred. The X-ray spectra are much simpler than optical spectra, and experience shows that essentially interference-free first or second order lines can be found for most rare earth elements in practically all types of rare earth mixtures. X-ray fluorescence is a rapid and precise method, but careful precautions are required to avoid errors caused by general or selective absorption - enhancement effects. There is a variety of modes to correct for these effects. The use of internal standards, calibration curves, spike and dilution methods as well as cross-checking with other methods is discussed and illustrated by examples.

1. INTRODUCTION

X-ray fluorescence analysis of X-ray secondary emission spectroscopy is a useful and versatile method for the determination of rare earth (RE) elements in a variety of materials. The method is applicable to the whole concentration range from 100% down to about 0.005%. In favourable cases amounts down to 0.0005% can be detected. Solutions, powders, compact and supported samples can be analysed. Typical applications are analysis of RE in minerals, ores, RE concentrates, products and process solutions. Only a minimum of sample preparation is required and contaminations or losses by chemical sepa-

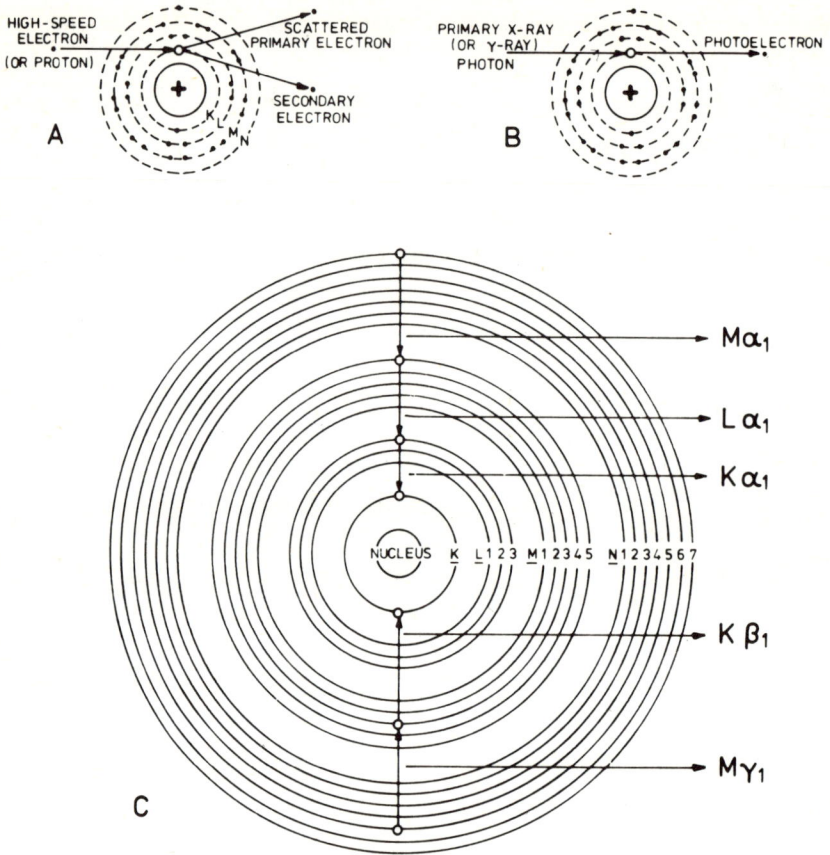

FIGURE 1

Excitation of characteristic X-ray line spectra. A: Primary excitation.
B: Secondary excitation. C: Typical series of electron transitions
that may follow the creation of a K-shell vacancy.

rations are thus minimized. The measurements are rapid and can easily be automated.

The X-ray spectra are much simpler than optical spectra. Since only inner orbitals are involved, the variation of wavelength with atomic number is not periodic but is a uniformly progressive function of Z. Thus the problems encountered in RE analysis are the same as for any other group of consecutive elements.

While X-ray spectrometry with primary excitation has been used for nearly 60 years, the prototype of the first commercial secondary emission spectrometer did not appear until 1948. Since then there has been a rapid development in instrumentation, techniques and applications. The main publications on X-ray fluorescence analysis of RE elements appeared in 1955-61 [1-6]. Later evaluations of sensitivity and applicability are based on these early reports without regard to recent instrumental improvements.

At Institutt for Atomenergi the method has been applied since 1967. It has been used for analysis of RE elements in rock samples, minerals, RE concentrates, products (up to 99.5% pure oxides) and process solutions. Together with isotope dilution mass spectrometry and atomic absorption spectrometry it is the main analytical tool for the control of the RE separation process.

2. PRINCIPLE OF THE METHOD

The characteristic X-ray spectrum of an element is generated when the element is exposed to electrons or primary X-ray photons with sufficient energy to expel an electron from an inner orbital of its atoms as illustrated in figure 1. The created vacancy gives rise to a series of electron transitions with resultant emission of X-ray photons. A typical spectrum of Sm and a schematic plot of the involved electron transitions are shown in figure 2.

The photon energy of a spectral line $(E = h\nu = hc/\lambda)$ is the energy difference between the initial and the final level involved in the transition. For the same spectral line (e.g. $K\alpha$) the wavelength decreases uniformly with the atomic number (figure 3) as defined by the Moseley equation:

FIGURE 2

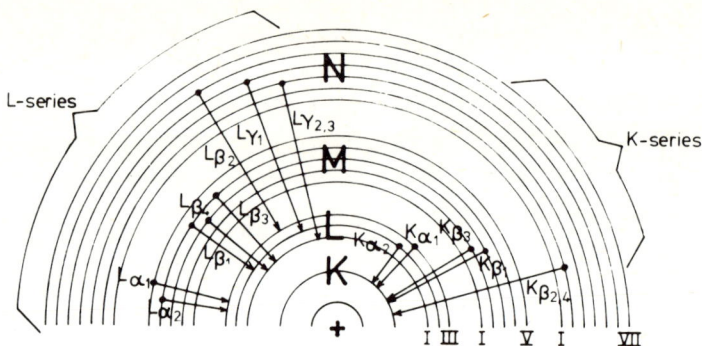

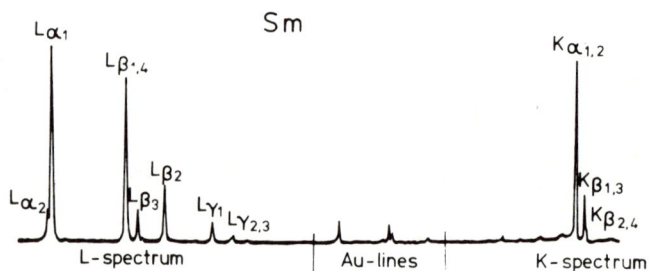

K and L spectrum of samarium
with schematic plot of the electron transitions

$$\sqrt{1/\lambda} \propto \text{prop } Z^2$$

Not all transitions give rise to emission of X-rays. The transition energy
can also be consumed by internal conversion with release of an Auger electron (figure 4). This effect limits the fluorescence yield (ω) and consequently the sensitivity. The variation of the fluorescence yield with atomic number is shown in figure 4. For the L-series of the lanthanides the
fluorescence yield is of the order of 0.2, only.

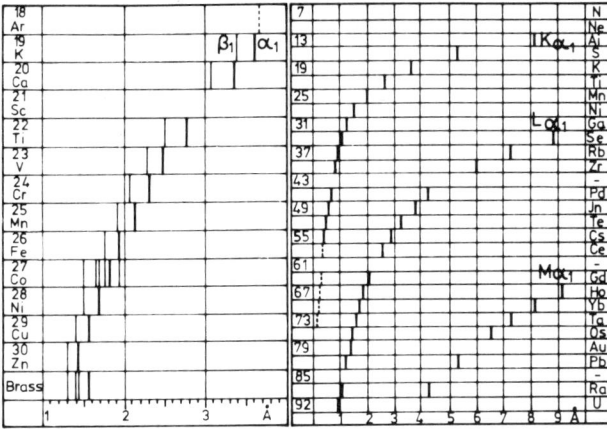

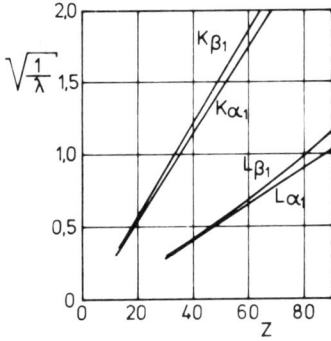

FIGURE 3

Moseley-law diagrams

for selected X-ray spectral lines

The required excitation energy increases with atomic number. The lowest energy (or longest wavelength) that can expel an electron from a given level in an atom of a given element is known as the absorption edge of that level and element. In contrast to optical spectra where absorption and emission are complementary the wavelength of the absorption edge is

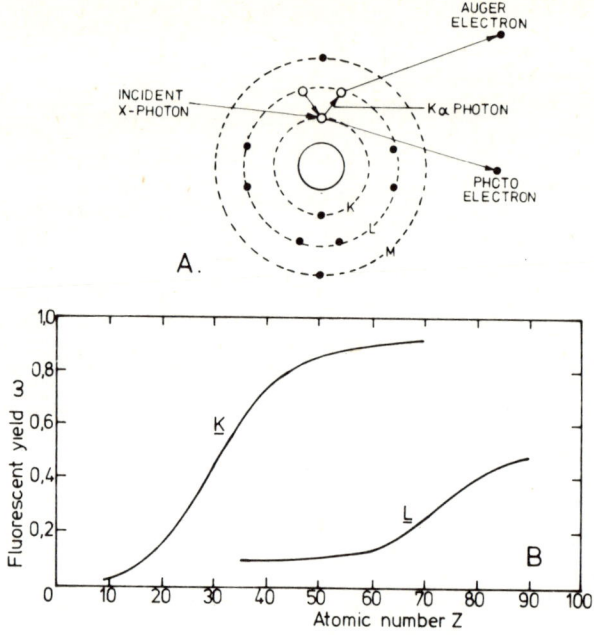

FIGURE 4

A: The Auger effect. B: K and L fluorescent yields

FIGURE 5

Comparison of (A) X-ray and
(B) optical absorption

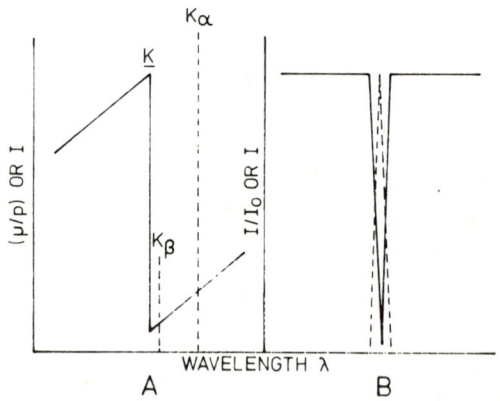

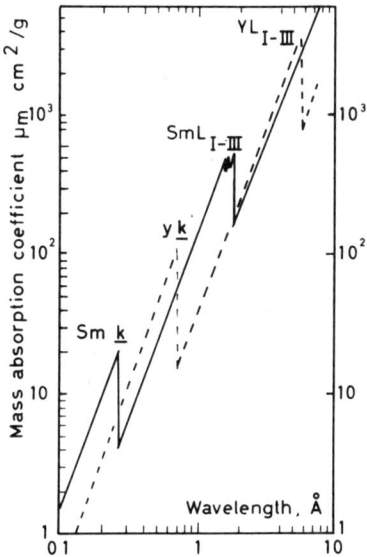

FIGURE 6
X-ray absorption curves of samarium and yttrium

shorter than the shortest line wavelength in the same series (figure 5).

The intensity of the primary and secondary X-rays is reduced by absorption in the sample. The absorption coefficient (μ) increases with wavelength and atomic number. Typical absorption curves are shown in figure 6.

3. MEASUREMENTS AND INSTRUMENTATION

Figure 7 illustrates the essential features of an X-ray spectrometer.
A number of useful accessories are available to increase the versatility of
the instrument and it can also be operated in a variety of modes. Possibilities therefore exist to attack even such a complicated problem as the analysis of all RE elements in a RE mixture.

3.1 Sample Preparation

Solid samples may be analysed in their original form or powdered to

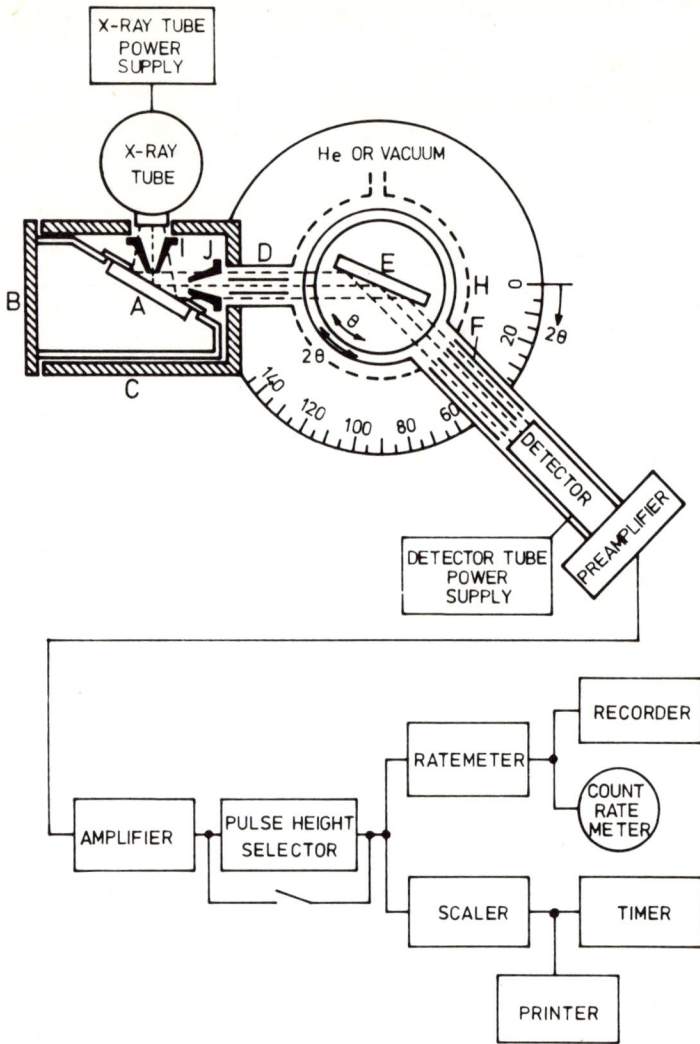

FIGURE 7

X-ray secondary emission spectrometer. A: Sample. B,C: Sample holder
and compartment. D,F: Collimators. E: Analyser crystal.

< 50 u and measured as such. They may also be briquetted or fused in a
flux, or dissolved and measured in solution. Solutions, slurries or powders
can also be measured absorbed, or deposited on filters or membranes. Cross
controls based on different preparation techniques are useful for the elimination of systematic errors. Ores, minerals and oxides are mostly measured
as packed or briquetted powders. For process solutions we have found a filter technique to be useful.

3.2 Excitation

A tungsten or gold tube operated at 50 kV and 30 - 50 mA is mostly used
to excite the L-spectra of the lanthanides and the K-spectra of yttrium.
Both the continuum and the characteristic lines of the target material contribute to the excitation (figure 8). The Y spectrum is therefore more
efficiently excited with molybdenum radiation, which has characteristic
lines near the Y K absorption edge.

3.3 Dispersion

In dispersive analysis the lines of the emitted secondary beam are
"sorted" according to wavelength by diffraction using an analyser crystal
with suitable interplanar spacings (d) and diffractive properties. The
principle is the same as for the diffractometer and is based on the Bragg
equation:

$$n\lambda = 2d \sin \theta$$

LiF_{200}, $topaz_{303}$ or LiF_{220} crystals with 2d values of 4.067, 2.7120 and
2.8480 Å, respectively, are used for RE analysis. LiF_{200} gives the highest
diffracted intensity and is preferred for low RE concentrations. The resolution is better for the closer spaced crystals which are mostly used for
complex RE mixtures. Some gain in intensity is achieved by use of LiF_{220}
instead of $topaz_{303}$.

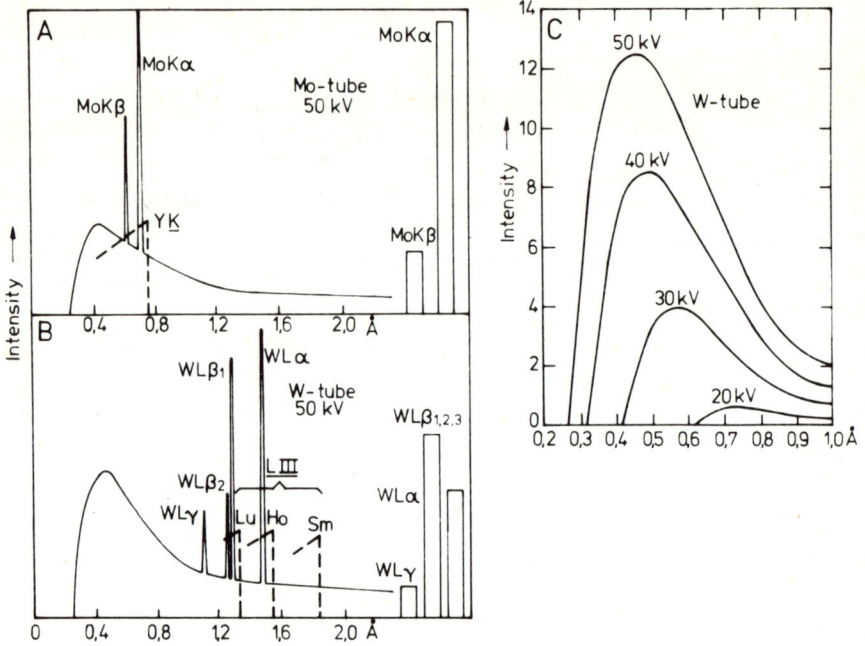

FIGURE 8

Emission spectra of X-ray tubes used for RE analysis

4. CONVERSION OF MEASURED DATA TO ANALYTICAL RESULTS

A qualitative analysis of the RE composition can be obtained from X-ray scannings. Excellent conversion tables and line diagrams listing all possible interferences are available for the commonly used crystals. With a few standard measurements and some estimates of absorptive properties even semiquantitative results are easily obtained. At IFA such semiquantitative lanthanide analyses have been used for preliminary evaluations of 98 - 99.5% pure yttrium oxides. For quantitative analyses rather elaborate calibrations and precautions may be required to avoid spectral interferences and to correct for absorption-enhancement effects.

4.1 Spectral Line Interferences

In a series of consecutive elements like the RE there are many possibilities for overlapping of spectral lines, as seen in figure 9. The analytical lines must be chosen with great care taking into account relative concentrations and intensities. The measurements are carried out with as high
resolution as compatible with the required sensitivity. The chosen lines
will vary with sample composition and experimental conditions. Therefore
no general list of preferred lines can be given. In agreement with other
laboratories we have the experience that suitable first or second order
lines for all RE elements except Eu can be found for practically all types
of RE mixtures. Further improvements may be obtained by selective excitation, use of wavelength filters or by computer programmed spectrum analysis.
Beside the mutual interferences of the RE elements even interferences from
other elements in the matrix, sample holder, crystal and target material
must be taken into account.

4.2 Absorption - Enhancement Effects

.   Both the primary and secondary beams are absorbed in the sample. The
intensity of a spectral line is therefore not only a function of the concentration of the emitting element (C_A) but also of the sum of the mass absorption coefficient (μ_m) of all components in the sample for the wavelengths
involved.

$$I \simeq \text{prop } C_A / \Sigma \, \mu_m$$

Besides this general absorption effect selective absorption-enhancement, i.e.,
interelement excitation of the characteristic lines of an element by secondary X-rays with a wavelength slightly shorter than its absorption edge, may
be a serious problem in RE analysis because of a whole series of overlaps
between absorption edges and spectral lines, as shown in figure 10 [5].

4.3 Other Sources of Error

When solid samples are measured, inhomogeneities, surface structure,
particle and grain size of the samples may affect the results. Counting

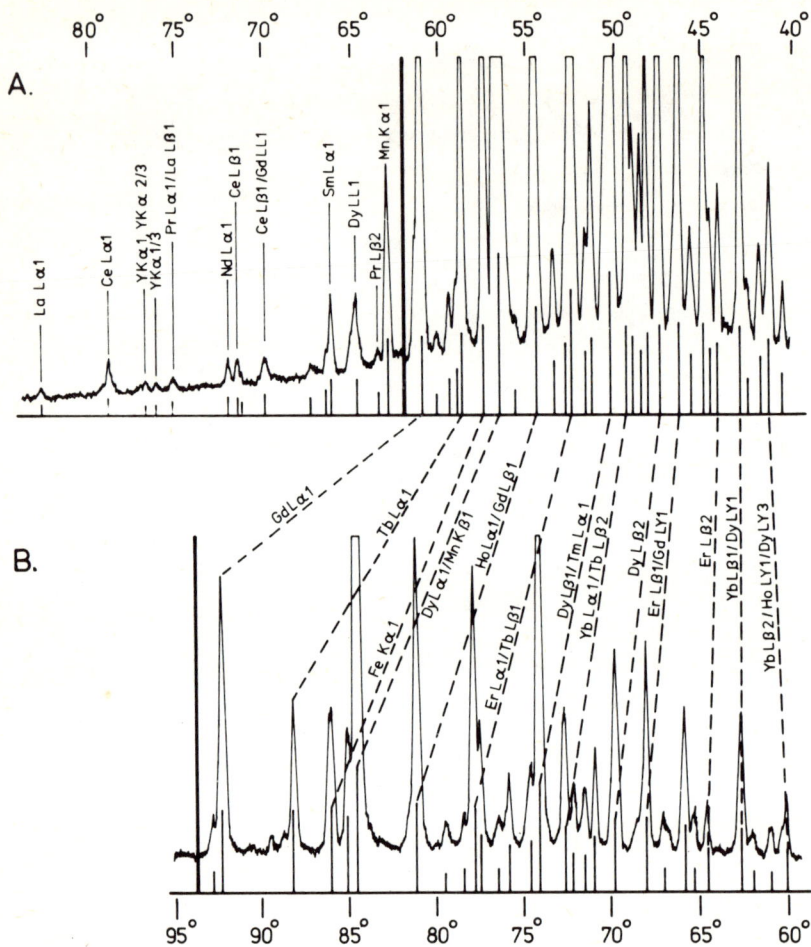

FIGURE 9

X-ray scans of an yttrium concentrate containing 63% Y_2O_3 and other
RE oxides in concentrations ranging from 0.05 to 20%. X-ray tube
with gold target operated at 50 kV and 40 mA. Analyser crystal:
A - LiF_{200} (2d = 4.0267). B - LiF_{220} (2d = 2.8480).

statistics and instrumental stabilities must also be considered.

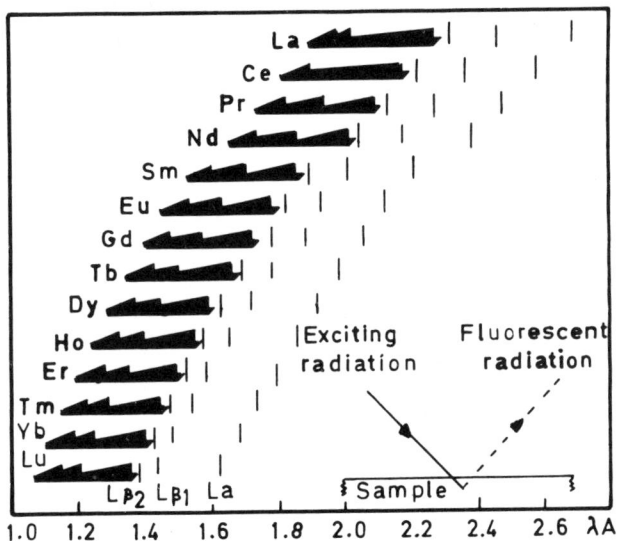

FIGURE 10

L absorption edges and L_{α_1} - L_{β_2} spectra of the rare earth elements

4.4 Calibration and Correction for Errors

There is a variety of methods to deal with these sources of error. Beside the above mentioned precautions against spectral interferences the main remedies are:

1. Internal standardization by addition of known amounts of an element with characteristic lines close to those of the element to be determined, or by comparison with matrix, target or background lines.

2. Use of calibration curves based on standards simulating sample compositions.

3. Spike methods, i.e. addition of known amounts of the sought element
   - preferably in 2 - 3 different concentrations - and subsequent extrapolation, if necessary.

4. Dilution methods which correct for absorption-enhancement effects
   by levelling the absorption coefficients of samples and standards
   to a common value determined by the added diluent.

5. Cross control with other methods

6. Mathematical approaches.

The method to be preferred depends upon the required accuracy and sensitivity, the concentration, composition and number of samples to be analysed. A
combination of several of the above mentioned methods is often useful.

1. Internal standard methods provide an efficient correction for most types
of error. Fassel [6] describes the use of strontium as internal standard
for the determination of yttrium in RE mixtures. Similar methods are in
general not suited for the analysis of all RE elements, because no common
internal standard for all RE is available, and the addition of several internal standards further increases the problems connected with spectral
interferences. However, A.G. Herrman and K.H. Wedepohl [7] report a combination of internal standardization with a technique similar to isotopic
dilution for the analysis of ppb-ppm concentrations of RE in geological
materials. Based on the chemical similarity of the RE elements, Dy and Nd
are used to determine the chemical yield of the RE in a rather elaborate
separation procedure. Simultaneously they serve as internal standards in
the final X-ray fluorescence and optical emission spectrographic analysis.

2. The use of calibration curves requires the rather elaborate preparation
of a large number of standards. A very good accuracy can be obtained with
sufficient similarity between standards and samples. The systematic errors
introduced by deviations in standard and sample composition depend mainly

on the absorptive properties. The method seems to be applied by most laboratories where large series of samples are analysed for their RE compositions. Reference can be made to the detailed publications by F.W. Lytle,
H.H. Heady and other authors from the U.S. Bureau of Mines [2-4] on the
analysis of RE in minerals, concentrates and pure RE oxides (figure 11).
Similar techniques are also applied to the analysis of geological samples
at the Geological Museum in Oslo and at Institutt for Atomenergi.

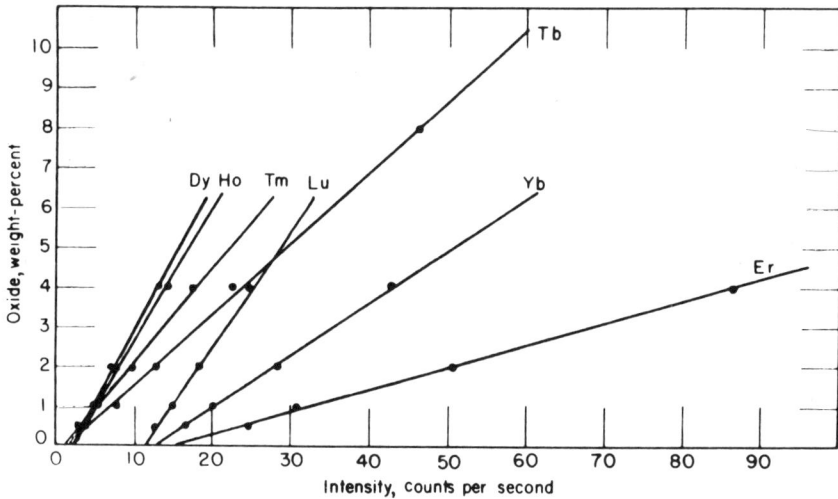

FIGURE 11

Samarium standard series

for analysis of bastnaesite rare earth elements

3. Spike methods (figure 12) are mainly applied to low RE concentrations,
where an increase of the concentration by a factor of 2 - 4 does not affect
the linearity of the concentration curves. It is desirable to control the
linearity by measurement of at least two spike concentrations. The method
corrects for most systematic errors and gives accurate results provided the

background corrections are accurate and homogeneous mixing is ascertained.

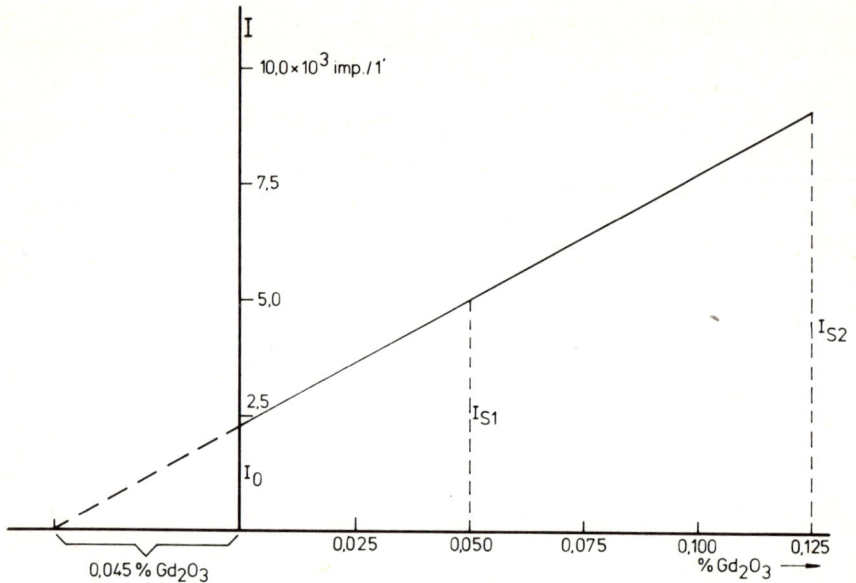

FIGURE 12

Spike method. Analysis of Gd in a rock sample - Rødberg (Red rock). Au target, LiF$_{200}$ crystal. The Gd L$_{\alpha_1}$ line measured at 61.05°, the background at 61.80°. 0.05 (S1) and 0.125 (S2) % Gd$_2$O$_3$ added as spikes.

4. Dilution methods include both dry mixing and dissolution. They are only applicable to relatively high RE concentrations. If sufficiently high dilution factors can be used so that the absorption coefficients of samples and standards are levelled to a common value, efficient corrections for absorption enhancement effects and accuracies of the order of 1% can be obtained.

5. Cross control both with various X-ray fluorescence techniques and principally different methods as neutron activation, atomic absorption or mass

spectrometry have been extensively applied at IFA, and have made it possible
to correct for systematic errors in all methods involved.

6. Mathematical approaches are much applied in X-ray fluorescence analysis,
and a variety of programmes for data processing are available.

5. EXAMPLES OF RE ANALYSIS

5.1 The Determination of Minor Concentrations of RE in Oxides and Geological Samples

Lytle and Heady [4] have described the analysis of 0.01 - 1% of RE in
so-called high purity RE oxides. They state that at these low concentrations
interelement effects between the RE are negligible. This agrees with observations at IFA where Gd and Y oxides have been analysed. The determinations
are based on spike addition, or on calibration curves using standards of individual RE in sufficiently pure oxides. For the lowest concentrations the
accuracy is essentially influenced by the background corrections. The best
results are obtained if the background can be determined in a pure oxide
matrix.

Series of samples from geological surveys have been analysed by similar
techniques. Representative matrix (non RE constituents) compositions determined by optical emission spectroscopy serve as a basis for the standard
matrix. A statistical control of results based on calibration curves is
obtained by parallel spike analysis of every 5 or 10 samples. Some samples
have also been dissolved and measured on filters as described for process
solutions. Furthermore, cross controls have been carried out with neutron
activation analysis. Some typical examples are presented in tables 1 - 3.

5.2 Analysis of RE Minerals and Concentrates

For analysis of higher RE concentrations several calibration-dilution
methods can be applied to level out the absorption coefficients. Lytle et
al. have used lithium carbonate-quartz sand [2]. At IFA a heavier diluent
is preferred, and dilutions of 1 : 9 and 1 : 19 with $CaSO_4$ are used. A
similar approach using iron oxide as a diluent has been recommended by

TABLE 1

Analysis of RE in various geological samples by X-ray fluorescence

Sample	% Y		% La		% Ce		% Sm			% Gd	
	Calibr. curve	Spike	Calibr. curve	Spike	Calibr. curve	Spike	Calibr. curve	Spike	N.A.*	Calibr. curve	Spike
Red Rock	0.015	0.020	0.25	0.23	0.70	0.69	0.023	0.028	0.020	0.035	0.037
Carbonatite with some hematite, magnetite and pyrite	0.026	0.028	0.29	0.28	0.71	0.72	0.030	0.032	0.032	0.039	0.037
Carbonatite	0.022	0.024	0.30	0.33	0.71	0.80	0.021	0.027	0.024	0.032	0.033
Carbonatite with pyrite and magnetite	0.026	0.026	0.24	0.22	0.58	0.56	0.021	0.022	0.027	0.032	0.026
Carbonatite	0.025	0.026	0.22	-	0.53	-	0.022	0.022	0.026	0.028	-
Carbonatite	0.019	0.020	0.19	0.23	0.56	0.65	0.019	0.022	0.023	0.028	0.029
Red Rock	0.031	0.039	0.38	0.38	1.22	1.20	0.050	0.053	-	0.063	0.061
Red Rock, dark	0.024	-	0.43	0.48	1.34	1.50	0.068	0.065	-	0.069	0.074
Red Rock, light	0.062	0.068	0.21	0.20	0.69	0.63	0.045	0.041	-	0.049	0.048

* Neutron activation analysis

Rødberg/Red Rock
Standard major constituents: 60% $CaCO_3$, 20% Fe_2O_3, 10% MgO, 6.9% SiO_2, 2.0% MnO_2, 1.0% Al_2O_3, 0.1% TiO_2.

Carbonatite = carbonate/hematite
Standard major constituents: 60% $CaCO_3$, 40% MgO.

TABLE 2

Analysis of rare earth elements in apatite

RE element	Analytical line	X ray fluorescence			Neutron activation analysis %
		Calibration curve %	Spike %	Solution %	
Y	Kα1	0.30	0.35	0.35	-
La	Lα1	0.074	0.074	0.080	0.072
Ce	Lα1	0.35	0.29	0.25	-
Pr	LB1	0.040	0.039	0.050	-
Nd	Lα1	0.19	0.21	0.22	-
Sm	LB1	0.060	0.066	0.060	0.053
Gd	Lα1	0.082	0.084	0.075	-
Tb	Lα1	0.010	0.010	~0.010	0.0098
Dy	LB1	0.062	0.073	0.070	-
Ho	LB1	0.012	0.013	-	0.0096
Er	LB1	0.037	0.046	0.035	-
Yb	Lα1/2	0.022	0.022	-	0.022

Apatite: $Ca_5(F,Cl)(PO_4)_3$
Standard major constituents: 90% $Ca_3(PO_4)_2$
10% NaCl

Wong Yew Choong [8] for the accurate analysis of Y in xenotime and monazite concentrates. With these techniques accuracies of the order of 1% can be obtained.

5.3 Process Solution

In connection with the process development work at IFA a large number of process solutions had to be analysed for 5 - 12 RE elements. For this purpose a filter technique has been developed [9]. Aliquots of sample and standard solutions are transferred to a defined measuring area on a filter, the area being limited by a water- and acid-resistant China ink circle, as

TABLE 3

Analysis of RE elements in geological samples from
Bolladalen by X-ray fluorescence and neutron activation analysis

Sample no.	% Y X-ray	% La X-ray	NA[*]	% Sm X-ray	NA[*]	% Eu NA[*]	% Yb X-ray
1	0.0098	0.34	0.29	0.036	0.032	0.0066	< 0.0010
2	0.018	0.24	0.24	0.029	0.034	0.0062	0.0010
3	0.015	0.29	0.26	0.028	0.029	0.0059	0.0025
4	0.026	0.45	0.47	0.055	0.057	0.011	0.0040
5	0.012	0.28	0.21	0.031	0.034	0.0068	< 0.0010
16	0.026	0.29	0.33	0.027	0.037	0.0062	0.0030
17	0.012	0.23	0.25	0.036	0.034	0.0071	0.0035
27	0.020	0.30	0.31	0.033	0.045	0.0090	0.0025
29	0.016	0.11	0.12	0.021	0.014	0.0041	0.0030
30	0.032	0.20	0.22	0.038	0.045	0.0087	< 0.0010
31	0.021	0.49	0.43	0.027	0.028	0.014	< 0.0010
35	0.0084	0.046	<0.040	0.012	0.0085	~0.0010	0.0050
36	0.019	0.094	0.10	0.017	0.025	0.0052	0.0025
37	0.015	0.35	0.32	0.024	0.027	0.0065	0.0025
38	0.037	0.42	0.37	0.049	0.050	0.012	0.0070
39	0.013	0.20	0.16	0.052	0.050	0.0075	0.0025
42	0.016	0.24	0.24	0.025	0.021	0.0045	0.0040
43	0.020	0.26	0.21	0.021	0.022	0.0064	0.0045
44	0.021	0.14	0.13	0.028	0.029	0.0061	0.0020
45	0.038	0.54	0.53	0.055	0.069	0.015	0.0025

[*] Neutron activation analysis

Samples from Bolladalen, "Red rock", are carbonatite/hematite (15% Fe)

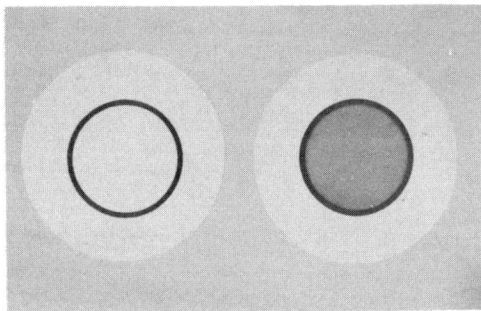

FIGURE 13

A Whatman 40 filter paper provided with a 25 mm inner diameter circle of water resistant printing ink, and the same paper impregnated with 50 μl sample solution

shown in figure 13. With this technique about 100 RE determinations per day have been carried out.

References

[1] H.W. Dunn, ORNL-1917 (1955).

[2] F.W. Lytle, J.I. Botsford, H.A. Heller, Bureau of Mines, Report of Investigations 5373 (1957).

[3] F.W. Lytle, K.R. Stever, H.H. Heady, A/Conf. 15/P/1425 (1958).

[4] F.W. Lytle, H.H. Heady, Anal. Chem. 31, 809 (1959).

[5] V.A. Fassel, Anal. Chem. 32, 19 A (1960).

[6] R.H. Heidel, V.A. Fassel, ibid. 30, 177 (1958).

[7] A.G. Herrmann, K.H. Wedepohl, Z. Anal. Chem. 225, 1 (1966).

[8] Wong Yew Choong, Extract from Transactions/Section B of the Institution of Mining and Metallurgy, 77, B 105 (1968).

[9] A. Follo, IFA Work Report CH-68 (1969).

TRACE LEVEL RARE EARTH DETERMINATIONS BY X-RAY EXCITED
OPTICAL FLUORESCENCE (XEOF) SPECTROSCOPY

by

Velmer A. Fassel, Edward L. DeKalb and Arthur P. D'Silva

Ames Laboratory USAEC and Department of Chemistry,
Iowa State University, Ames, Iowa 50010, U.S.A.

ABSTRACT

X-ray excited optical fluorescence (XEOF) has emerged during the past
decade as a remarkably sensitive technique for the detection and quantitative determination of rare earth impurities in a variety of matrices. Although not all aspects of the excitation phenomena are understood, the basic
mechanism apparently involves the generation of secondary excitants by the
incident X-ray quanta followed by the highly selective transfer of the energy residing in the excitants to rare earth ion impurities. Although some
metallic oxides serve as analytically useful hosts, it is often necessary
to convert the sample into a chemically different but effective host system.
Ternary and quaternary oxide systems and some phosphates provide detectability of rare earth impurities in the 1 in 10^9 concentration range. The
classical internal reference (standard) principle effectively compensates
for the enhancement and depression effects on the fluorescent intensities
induced by changing concentrations of other residual impurities in the sample.
To date, activated fluorescence of trace rare earth impurities has been verified in hosts prepared from 55 elements in the periodic system. These observations portend an increasing use of XEOF as a general analytical technique
for the ultratrace determination of the rare earth elements. A selected
number of applications of XEOF to the determination of residual rare earth
impurities in fertile nuclear materials and in highly purified rare earth
compounds will be discussed.

1. INTRODUCTION

Radiation induced line fluorescence of rare earth ions in solids was first investigated by Makovsky, Low and Yatsiv [1] using X-rays and by Derr and Gallagher [2] using both X-rays and nuclear radiation (α, β, γ and n). The potential of these observations for the determination of ultratrace amounts of rare earth impurities was first indicated by Low, Makovsky and Yatsiv [3] and applied by Linares, Schroeder and Hurlburt [4] to the determination of traces of Eu, Gd, Tb, and Dy in Y_2O_3, and later by other investigators to the direct determination of trace rare earth impurities in Y_2O_3 [5-10], La_2O_3 [10,11], Gd_2O_3 [6,7,9,10], Al_2O_3 [12], ThO_2 [13], and CaF_2 [14].

In view of the excellent detectabilities reported in the early studies, DeKalb, D'Silva and Fassel [15] explored the scope of application of this technique to the practical determination of trace rare earth impurities in other matrices. Their investigations led to the development of a variety of analytically useful phosphor hosts. To date, activated fluorescence of trace rare earth impurities has been verified in compounds prepared from 55 elements in the periodic system. These observations portend increasing use of X-ray excited optical fluorescence (XEOF) as a general analytical technique for the determination of trace rare earth constituents.

2. ORIGIN OF X-RAY EXCITED OPTICAL EMISSION

Many inorganic and organic compounds luminesce when they are irradiated with X-rays [16,17]. Several authors have suggested mechanisms for the excitation and emission processes [1,3,18-28]. Although all aspects of the phenomenon do not appear to be clearly understood, it is possible to draw useful generalizations from the experimental data and suggested mechanisms.

The direct absorption of the primary incident X-rays by the impurity rare earth ions can in principle lead to ejection of inner shell photoelectrons and subsequent emission of characteristic rare earth fluorescent X-rays or Auger electrons. None of these processes are detectable at the low concentration levels at which optical fluorescence is readily detectable. The basic mechanism apparently involves the concept of secondary excitants

generated by absorption of X-rays by the host atoms, followed by the highly
selective transfer of the energy residing in the secondary excitants to the
rare earth impurities.

The photoelectrons or Auger electrons ejected from the host atoms by
the incident primary X-rays may interact with extranuclear electrons in the
host atoms. When they do, the photoelectrons may lose their excess energy
in random sized bits ranging upward from several electron volts to form the
secondary excitants. The formation process of the latter may be illustrated,
in a highly schematic way, with the aid of the energy level diagram shown in
figure 1. This diagram is a somewhat fictional one-dimensional presentation
of energy states in a crystalline insulator or semiconductor. The valence
band represents the range of energy states of the valence electrons that bind
the atoms in the solid state, all energy levels being occupied. The conduction band of energies is initially empty and is separated from the valence
band by an energy gap. The absorption of bits of energy from the photoelectrons (or Auger electrons) ejected from the host atoms by the incident X-rays
may result in the excitation of an electron from the valence band into the

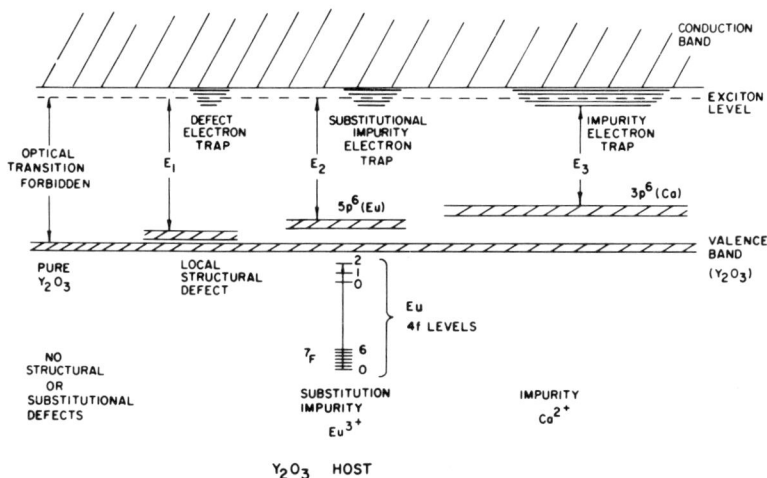

FIGURE 1

Schematic one-dimensional presentation of energy states in a crystalline insulator or semiconductor

conduction band. The "free" electron in the conduction band and its associated "hole" in the valence band may wander in the crystal as a unit called
an "exciton". The "exciton" may be trapped at crystal sites containing
structural or compositional defects. At these sites, recombination of an
electron in the conduction band with a hole in the valence band may be an
allowed process, resulting in the release of energy. If the impurity at a
compositional defect is a rare earth ion with its own unoccupied excited
states (in the 4f subshell), the energy released may excite electrons from
the ground state of the 4f system into excited states from which sharp line
spectra are emitted as the electrons return to the ground state. The combination of mobility of the "free" electron and "hole" secondary "excitants",
the relatively short lifetime of the 4f excited states, and the repeated re-
excitation of the same ion impurity may therefore lead to a highly selective
energy transfer process.

A schematic diagram of a typical experimental facility for observing
XEOF spectra is shown in figure 2.

FIGURE 2

A schematic diagram of the
experimental facility for
observing XEOF spectra

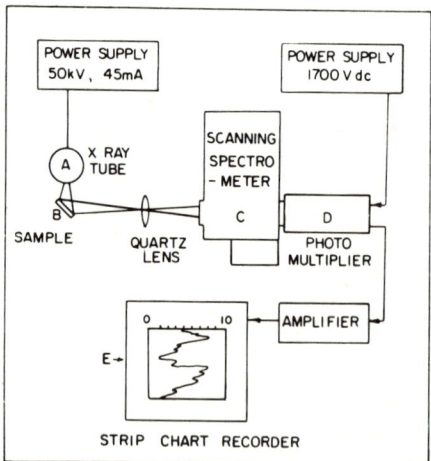

3. ANALYTICAL APPLICATIONS

3.1 The Internal Reference (Standard) Principle

The intensities of the optical fluorescent spectra emitted under X-ray irradiation may be markedly influenced by the chemical nature and concentration of other compositional impurities in the host material and by structural imperfections. Thus, under many circumstances, a prior knowledge of the total impurity content and structural purity of a sample would be required for precise quantitative analyses. Fortunately, this is not so. A characteristic commonly exhibited by both the enhancement and suppression effects is the nearly parallel behaviour of the intensities among the rare earths [29]. Thus, though the individual relative intensities undergo wide excursions, they do so in consort. A typical example of this behaviour is shown in figure 3. These observations immediately suggested the application of the internal standard principle so widely used in optical emission spectroscopy to these radiations as well. The merit of employing intensity ratio measurements to compensate internally for impurity enhancement or depression effects is excellently illustrated in figure 4. Day to day variations in

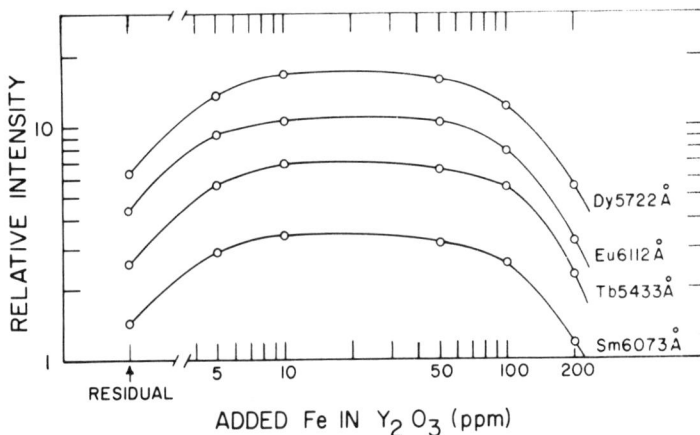

FIGURE 3

Effect of Fe addition on the fluorescent emission of 200 ppm of Sm, Eu, Tb and Dy in Y_2O_3

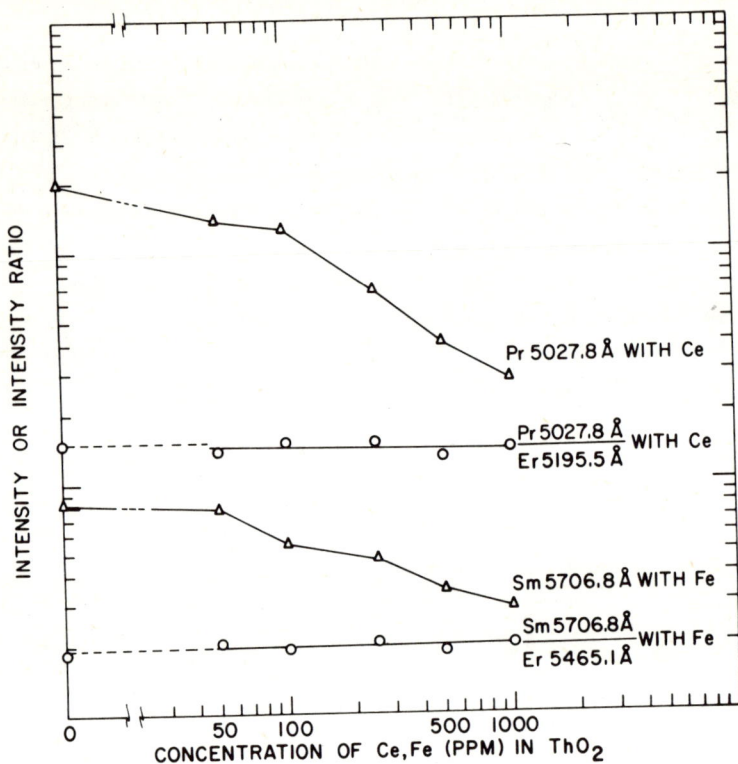

FIGURE 4

Effect of impurities on fluorescent intensities and on
intensity ratios of rare earths in ThO_2

the preparation of the phosphor host may also affect the emission intensities to a significant degree. These variations, too, may be effectively compensated by application of the internal standard principle. Its effectiveness is illustrated in figure 5. For these experiments eight identical
samples were prepared and analysed on different days, with all operating
conditions maintained as nearly constant as possible. Although the standard

deviations in intensity of both the terbium and samarium fluorescent lines are greater than 16%, the intensity ratio is much less variable, with only a 4% standard deviation.

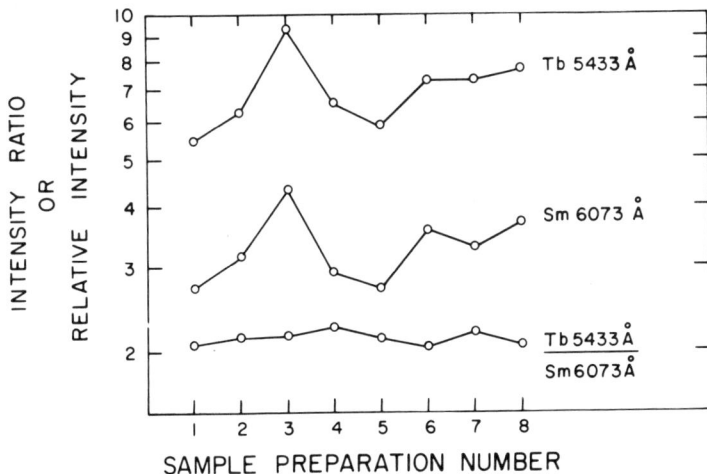

FIGURE 5

Effect of internal standardization in compensating for day to day fluctuations in relative intensities of separately prepared Y_2O_3 samples

3.2 Genesis of Ultrasensitive Phosphor Hosts

Most of the early analytical studies of XEOF employed oxide phosphor hosts. Unfortunately the oxides of many elements do not support useful fluorescence of trace rare earth impurities. However, the extensive studies by DeKalb, D'Silva and Fassel [15] have shown that host systems able to support fluorescence of rare earth impurity "activators" at low concentration levels may be prepared from most elements in the periodic system. The periodic table shown in figure 6 indicates that analytically useful fluorescence has been observed in compounds prepared from most of the elements in

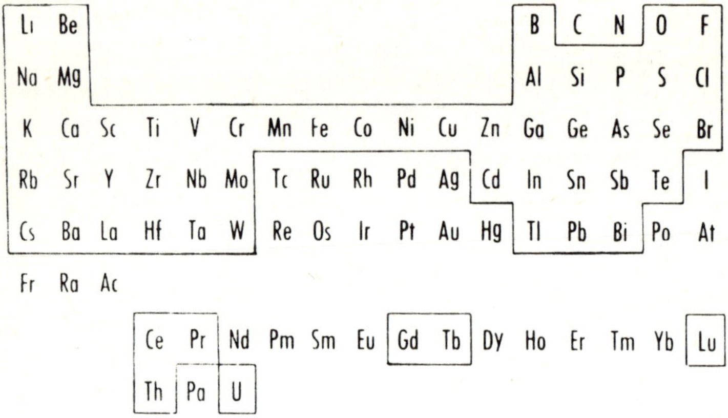

Li	Be											B	C	N	O	F
Na	Mg											Al	Si	P	S	Cl
K	Ca	Sc	Ti	V	Cr	Mn	Fe	Co	Ni	Cu	Zn	Ga	Ge	As	Se	Br
Rb	Sr	Y	Zr	Nb	Mo	Tc	Ru	Rh	Pd	Ag	Cd	In	Sn	Sb	Te	I
Cs	Ba	La	Hf	Ta	W	Re	Os	Ir	Pt	Au	Hg	Tl	Pb	Bi	Po	At
Fr	Ra	Ac														

| Ce | Pr | Nd | Pm | Sm | Eu | Gd | Tb | Dy | Ho | Er | Tm | Yb | Lu |
| Th | Pa | U | | | | | | | | | | | |

FIGURE 6

Elements with boxes have been found to form compounds which will support analytically useful X-ray excited optical fluorescence of trace rare earth impurities

PHOSPHOR GENESIS

$$b\ R_2^{3+}O_3 \cdot cWO_3 - \text{RARE EARTH TUNGSTATES}$$

$\downarrow$

$(R^{2+} = Ca, Sr, Ba)$ $\quad b\ R^{2+}O \cdot b R^{4+}O_2 \cdot cWO_3$ $\quad (R^{4+} = Ti, Zr, Hf, Pu, U, Th)$

$Sr^{2+} \downarrow U^{4+}$

$SrO \cdot UO_2 \cdot 2WO_3$

$aR_2^{1+} \downarrow$ $\quad (R^{1+} = Li, Na, K, Rb, Cs)$

$2Li_2O \cdot SrO \cdot UO_2 \cdot 2WO_3$

FIGURE 7

Quaternary oxide phosphor genesis

the periodic system. Quaternary oxide host systems have been found by D'Silva, DeKalb and Fassel [30] to support fluorescence of the rare earth impurity activators at strikingly low concentrations. The genesis scheme for systems of the type $a R_2^{1+} O \cdot b R^{2+} O \cdot b R^{4+} O_2 \cdot c WO_3$ is shown in figure 7. These phosphors are easily prepared by grinding together stoichiometric

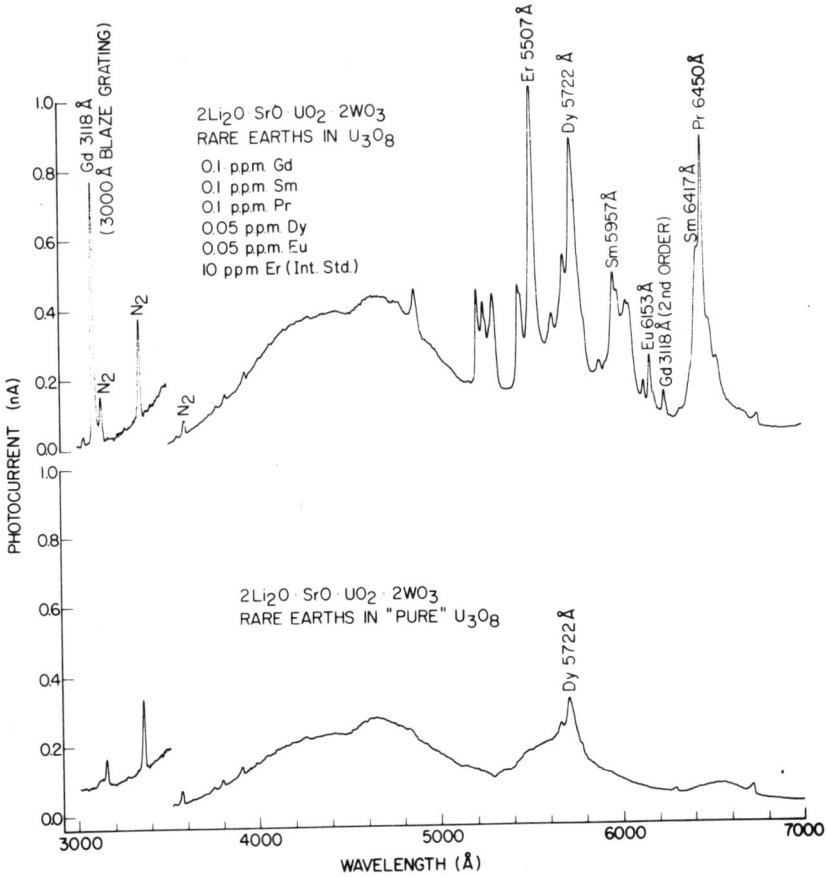

FIGURE 8

X-ray excited optical fluorescence spectra of rare earths in UO_2
phosphor: $2 Li_2O \cdot SrO \cdot UO_2 \cdot 2WO_3$

amounts of the component compounds and heating the mixture on an optimum time-temperature schedule. The spectrum obtained from trace rare earths in UO_2 incorporated into this host is shown in figure 8; the analytical curves shown in figure 9 indicate that the lowest detectable concentrations are in the part per billion (1 in 10^9) range. Similar phosphors containing Ti, Hf and Th have been prepared. It is expected that a similar phosphor containing PuO_2 could provide a host for the direct determination of rare earths and actinides in PuO_2.

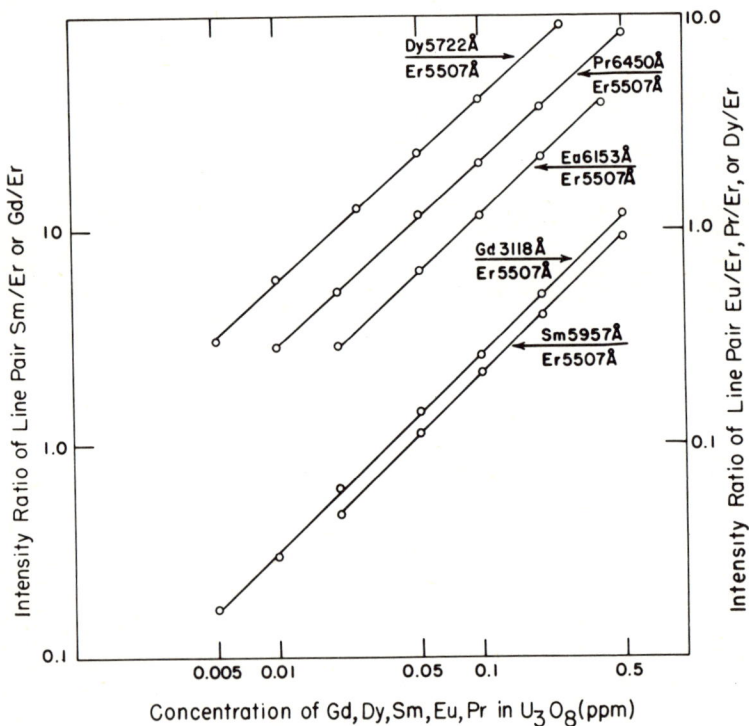

FIGURE 9

Analytical curves for the direct determination of rare earths in uranium

3.3 Determination of Residual Rare Earth Impurities in High Purity Rare Earths

The unique optical fluorescent spectra of rare earth activated materials have found widespread application in diverse industrial products such as solid state lasers, colour television tubes and image conversion devices. Since the luminescent efficiency of these materials may be adversely affected by the presence of ultratrace-level, rare earth impurities, there has existed

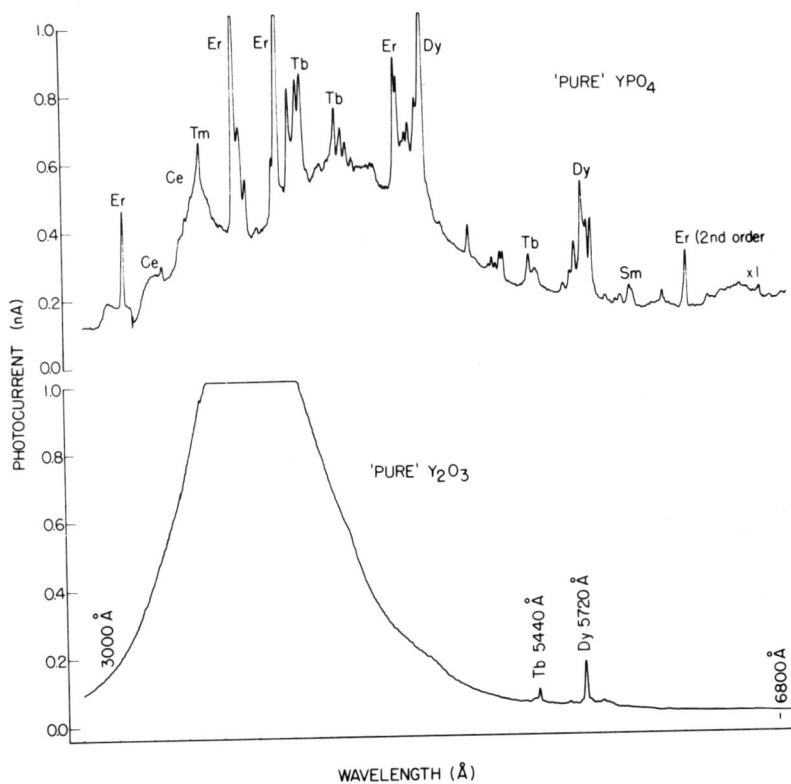

FIGURE 10

A comparison of X-ray excited optical fluorescence spectra of residual rare earth impurities in Y_2O_3 and YPO_4 hosts.

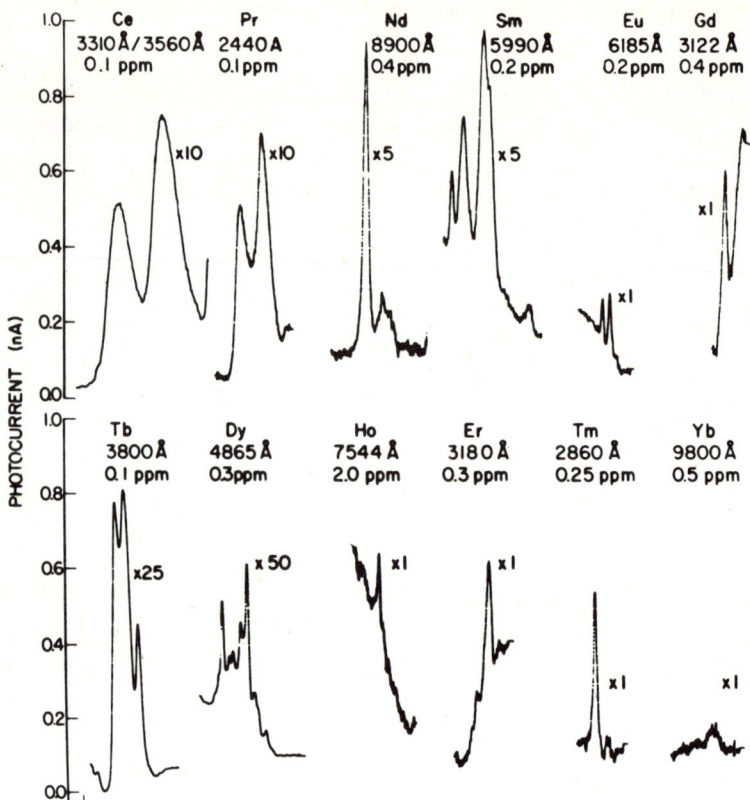

FIGURE 11

X-ray excited optical fluorescent spectra of rare earths in YPO$_4$ and YVO$_4$.
The signals are generally for concentrations nearly equal to the detection
limits obtained by SSMS using the photographic technique.
Ce, Pr, Sm, Gd, Tb, Dy, Er, Tm in YPO$_4$ host
Nd, Eu, Ho, Yb in YVO$_4$ host

a pressing need for an analytical technique capable of assessing this purity.
A comparative evaluation of various rare earth compounds has revealed that
the rare earth phosphate and the isostructural vanadate may provide superior
hosts for the ultratrace level determination of rare earth impurities in

high purity rare earth oxides. A specific application is shown in figure
10; it is seen that the Y_2O_3 host emits only very weak lines of Tb and Dy
residual impurities, whereas the YPO_4 host prepared from the same sample
emits a complex spectrum of lines and bands originating from Ce, Eu, Dy, Tb,
and Sm impurities. These observations have been utilized in the development of a quantitative procedure for the determination of rare earth impurities at the part per giga $(1 \text{ in } 10^9)$ level in Y_2O_3. The powers of detection
obtained using this YPO_4 -YVO_4 host system are clearly illustrated in figure
11, which is a composite of the strip chart recordings for the rare earth
impurities in either YPO_4 or YVO_4 obtained at or near the limits of detection by spark source mass spectroscopy using the photographic technique.
The attenuation factors needed to keep these signals on scale are also indicated.

References

[1] J. Makovsky, W. Low and S. Yatsiv, Phys. Letters 2, 186 (1962).

[2] V.E. Derr and J.J. Gallagher, "Quantum Electronics-Paris 1963 Conference",
     Columbia University Press, New York, N.Y., p. 817, 1964.

[3] W. Low, J. Makovsky and S. Yatsiv, ibid, p. 655.

[4] R.C. Linares, J.B. Schroeder and L.A. Hurlburt, Spectrochim. Acta,
     21, 1915 (1965).

[5] J.F. Cosgrove, D.W. Oblas, R.M. Walters and D.J. Bracco, Electrochem.
     Technol. 6, 137 (1968).

[6] W.E. Burke and D.L. Wood in "Advances in X-Ray Analysis", Vol. 11,
     J.B. Newkirk, G.R. Mallett and H.G. Pfeiffer, eds., Plenum Press,
     New York, 1968, pp. 204 - 213.

[7] R.J. Jaworowski, J.F. Cosgrove, D.J. Bracco and R.M. Walters, Spectro-
     chim. Acta, 23B, 751 (1968).

[8] N. Sasaki, Bunseki Kagaku, 17, 1387 (1968).

[9] M. Sato, H. Matsui and M. Tadao, Jap. Anal. 20, 70 (1971).

[10] H. Ratinen in Acta Polytechnica Scandinavia, "Chemistry including
     Metallurgy Series No. 107" (The Finnish Academy of Technical Sciences,
     Helsinki, 1971), ch. 107.

[11] T. Nakajima, H. Kawaguchi, K. Takashima and Y. Ouchi, Bunko Kenkyu, 18, 210 (1969).

[12] K. Takashima, T. Nakajima, H. Kawaguchi and Y. Ouchi, Bunko Kenkyu, 18, 262 (1969).

[13] T.R.Saranathan, V.A. Fassel and E.L. DeKalb, Anal. Chem. 42, 325 (1970).

[14] W.A. Shand, J. Mater. Sci. 3, 344 (1968).

[15] E.L. DeKalb, A.P. D'Silva and V.A. Fassel, Anal. Chem. 42, 1246 (1970).

[16] H.W. Leverenz, "An Introduction to Luminescence of Solids", Wiley, New York, 1950, pp. 421 - 427.

[17] D. Curie, "Luminescence in Crystals", Wiley, New York, 1960, pp. 298 - 301.

[18] H.W. Leverenz, "An Introduction to Luminescence in Solids", Wiley, New York, 1950, pp. 34 - 58, 316 - 318.

[19] P.R. Thornton, "Scanning Electron Microscopy", Chapman and Hall, London, 1968, ch. 10.

[20] L.G. Van Uitert in "Luminescence of Inorganic Solids", P. Goldberg, ed., Academic Press, 1966, ch. 9.

[21] L.G. Van Uitert, J. Electrochem. Soc. 114, 1048 (1967).

[22] F. Moatossi and S. Nudelman in "Methods of Experimental Physics", Vol. 6, Part B, K. Lark-Harovitz and V.A. Johnson, eds., Academic Press, New York, 1959, pp. 293-4.

[23] G. Blasse and A. Bril, J. Electrochem. Soc. 115, 1067 (1968).

[24] G. Blasse, J. Chem. Phys. 45, 2356 (1966).

[25] L.G. Christophorou and J.G. Carter, Nature, 209, 678 (1966).

[26] H.N. Hersh and H. Forest, J. Luminescence, 1,2, 862 (1970).

[27] E.R. Ilmas and T.I. Savikhina, J. Luminescence, 1,2, 702 (1970).

[28] I.A. Parfianovich, E.I. Shuraleva and P.S. Ivakhnenko, Bull. Acad. Sci. USSR, Phys. Series, 31, 838 (1967).

[29] E.L. DeKalb, V.A. Fassel, T. Taniguchi and T.R. Saranathan, Anal. Chem. 40, 2082 (1968).

[30] A.P. D'Silva, E.L. DeKalb and V.A. Fassel, Anal. Chem. 42, 1846 (1970).

CATHODE-RAY-EXCITED EMISSION (CREE) SPECTROSCOPIC ANALYSIS
OF TRACE RARE EARTHS

by

Simon Larach

RCA Laboratories, Princeton, New Jersey 08540, U.S.A.

ABSTRACT

Line emissions from f-f transitions of trace amounts of rare earths
contained in a matrix are obtained by cathode-ray excitation. The distinctive energy of these emissions is utilized to detect the presence of such
trace rare earths, and the intensity of the emission is proportional to the
amount of rare earth present.

This non-destructive method is applicable to the analysis of rare earths
in any matrix which supports cathodoluminescence. Results are discussed for
rare earths in yttrium oxide, gadolinium oxide, and zinc sulphide. In the
case of thulium, comparisons are made with analytical results obtained with
solid state mass spectrometry.

Equipment required for CREE analysis is described, and specific examples
are given.

1. INTRODUCTION

With the increasing utilization of rare earths in materials, especially
electronically-active solids, it becomes important to be able to perform
analyses for small traces of rare earths quickly and efficiently, and preferably in a non-destructive way. While neutron activation analysis and
solid state mass spectrometry can be used, these are complex techniques
which are destructive and quite time-consuming.

In analyses for rare earths, it is extremely useful to take advantage
of the "shielded" 4f shell of the rare earth ion, particularly in exciting
and detecting luminescence in this shell, where we define luminescence as

that photon emission, after excitation, of non-thermal energy. In the rare earths, such luminescence emission, being of the non-interacting f-f type, is essentially a narrow band or line, or series of such lines, whose energy is indicative of the particular rare earth being excited, and whose emission intensity is a function of the number of such rare earth ions. Conventional optical excitation requires a detailed knowledge of the excitation spectrum of each rare earth, whereas excitation with energy large enough to excite across the band gap of the matrix, followed by energy transfer to the rare earth ions, would provide a general means of exciting f-f luminescence. For this reason, cathode-ray-excited emission (CREE) is used in the presently discussed analysis. Thus, CREE analysis can be used for any rare earth in any matrix which will support cathodoluminescence, i.e., luminescence excited by cathode rays.

2. CATHODE-RAY-EXCITED LUMINESCENCE (CATHODOLUMINESCENCE) [1]

The process of cathodoluminescence is highly complex, consisting of several sub-processes which take place between the incidence of the initial high-energy electron, and the resultant emission of luminescence photons. For example, after incidence of the primary electron, many secondary electrons can be produced, whose energy is transferred to the rare-earth ions, which emit specific energies due to their characteristic f-f transitions. The cathodoluminescence emission intensity is a function of the potential through which the primary electrons are accelerated. The specific energy loss has been derived by Garlick [2] from Bethe's non-relativistic equation for the interaction of charged particles with matter, as being:

$$- \frac{dE}{dX} = \frac{2\pi N Z e^4}{E} \ln \frac{E}{E_1} \tag{1}$$

where E is the electron energy, X is the depth of penetration measured along the electron path, N is number of bound electrons cm^{-3}, Z is the mean atomic number, e is electron charge, and E_1 is the mean ionization energy for all the electrons.

At potentials higher than about 3000 eV, the Thomson-Whiddington Law may be held to apply:

$$X = k(E_o^2 - E^2) \tag{2}$$

where E_o is the initial electron energy, E is the energy after penetration to the distance X, and k is a constant roughly inversely proportional to the density of the material.

If cathodoluminescence emission intensity is plotted as a function of accelerating potential, a typical curve is obtained, as shown in figure 1.

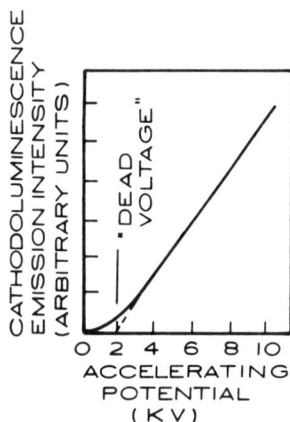

FIGURE 1
Cathodoluminescence Intensity as a Function of Accelerating Potential.

Practically all luminescent materials show a threshold when the linear portion of the curve is extrapolated to the abscissa; this threshold is often referred to as the dead-voltage. Gergely [3] has proposed a mechanism to account for the dead-voltage which involves carrier diffusion to the surface where radiationless recombination can take place.

In the present case, excitation with high-energy electrons creates carriers which can transfer their recombination energy to a rare-earth ion. Thus, by utilizing cathode-ray-excited emission (CREE) spectroscopy, luminescence, at room temperature, was obtained for all of the pertinent rare earths.

As was pointed out in the introductory comments, CREE analyses can be

performed so as to detect the energy (wavelength) of the emitted luminescence, i.e., qualitative detection, and/or detecting the intensity of the
emitted luminescence, i.e., quantitative determination of rare earth impurities. Both of these types of CREE analyses will be described in this
chapter.

2.1 Qualitative Detection [4]

CREE analyses, as in other spectroscopic methods, require comparison
standards, consisting of rare earths in the matrix of interest. For the
detection of trace amounts of rare earths, matrix materials are required
which contain little or no rare earth ions. Such materials were in use in
these laboratories, as for example, ultrapure yttrium oxide (99.9999% purity with respect to rare earths), and luminescent grade zinc sulphide,which
contained no detectable rare earths, and these materials were used as primary matrices. Thus, series of 1, 10, and 50 ppm of separate rare earths
(99.9%-99.99% purity, with respect to rare earths) were added as the sul-
phates to Y_2O_3, and these samples were fired at $1100^{o}C$ in silica crucibles,
in which blank charges were fired first. Additional series of these rare
earths together, and samples containing all of the pertinent rare earths together, were prepared as standards. Rare earths in zinc sulphides were
charge-compensated with Li^+, as has been discussed [5].

2.1.1 Measurements

CREE analysis requires basically a source of electrons and a detector.
We have found it most useful to use a demountable cathode-ray equipment, as
shown schematically in figure 2. In order to avoid the dead voltage effect,
an acceleration potential of the order of 8 KV is recommended. The samples
are placed, as a series, on a rotatable metal disk, so that successive samples can be excited by the electron beam. No effect due to sample geometry
was observed, and a disk containing 24 samples can easily be prepared in
less than one hour.

"Demountable" refers to the tube's capability of being repeatedly opened to the atmosphere for the purpose of changing samples, electron-gun
structures and type of window-material (quartz, sapphire, etc.). Especially

important is the capability of being able to accommodate, at one pumping, all samples of a series, in order to minimize the error due to carrying over a standard from one pump to another. The unit used in most of our CREE work can accommodate 29 samples.

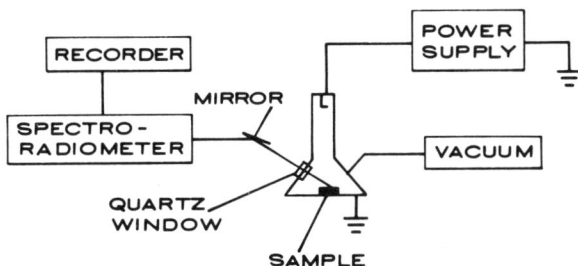

FIGURE 2

Schematic Representation of Experimental Equipment

The luminescence emission excited by the electron beam can be analyzed with a grating spectrometer and appropriate multiplier phototube as for example, RCA 1P21 for wavelengths from 300 nm to 700 nm, and a cooled RCA 7102 for wavelengths from 700 nm to 1200 nm. Particularly for qualitative analysis, far simpler detection methods can be used; for example, interference filters as spectral analyzers in conjunction with a multiplier phototube, are quite feasible, particularly when only the presence or absence of a particular line is required.

Rare-earths impurities were easily detected in Y_2O_3 by this type of spectroscopy[*]. At 1-50 ppm concentrations of rare earths, identifying line spectra were obtained for all except cerium, which emitted in a broad band.

---

[*] It is interesting that this technique, utilizing Crooke's tubes, had been tried in the early 1900's by Urbain and others. H.V.F. Little has reviewed the early use of this technique as "cathodic phosphorescence analysis", to detect rare-earth impurities of the order of 1% in, mainly, pure lime. See A Textbook of Inorganic Chemistry, edited by J.N. Friend, Vol. 4, Griffin and Co., London, 1921, p. 305.

TABLE 1

Wavelengths and Intensities of Trace Rare Earths Obtained by

Cathode-Ray-Excited Emission Spectroscopy

(Y_2O_3 matrix)

Rare earth	λ(nm)	Sensitivity (rel.no. photons)	Rare earth	λ(nm)	Sensitivity (rel.no. photons)
Ce	375	(broad band)	Eu	611	54
Pr	630	27		587	18
	619	24		593	17
	635	19		582	16
	645	5		599	6
	597	5		631	3
	592	3		533	22
				538	14
Nd	878	8		553	6
	893	17		551	5
	913	13		467	19
	938	8		473	12
	946	8		464	10
	1054	9		499	9
	1076	17		488	5
	1103	10		416	6
				425	5
Sm	565	30		414	3
	607	28		427	3
	569	17			
	576	14	Ho	550	0.9
	585	10		536	0.3
	617	9			
	597	4	Er	564	1.1
	623	3		553	0.7
	556	3		539	0.5
	655	2		522	0.5
	668	1			
			Tm	453	1.6
Gd	631	3		460	1
	629	3		463	1
	315	3		361	0.3
Tb	543	27	Yb	950	1
	551	20		976	19
	483	11		1030	8
	492	9		1078	4
	488	9			
	501	4			
	583	2			
	569	1			
	579	1			
	588	1			

(cont.)

Table 1 (cont.)

(Y_2O_3 matrix)

Rare earth	λ (nm)	Sensitivity (rel.no. photons)
Dy	572	292
	582	50
	567	24
	559	10
	486	65
	483	50
	491	49
	480	47
	476	34

Table 1 lists the line[*] obtained for each rare earth, and the relative intensities. With respect to mutual interferences, between the rare earths, erbium emission is masked by terbium and holmium by europium. In the latter case, however, it is interesting to note that band shape can also be utilized in overcoming interference effects, as shown in figure 3, where (a) shows the 553 nm emission from europium in Y_2O_3; (b) that of holmium, and (c) shows the emission from the pair.

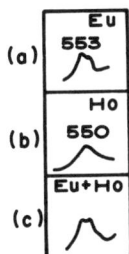

FIGURE 3

Emission Line Shapes of (a) Eu, (b) Ho, and (c) Eu+Ho in Y_2O_3 Matrix.

[*] The number of minor lines appearing is, of course, a function of the rare-earth concentration.

Low [6] has discussed the use of x-ray excitation, and this method was used by Linares [7], and by Walters [8]. However, with the x-ray method, sample geometry must be controlled, some rare earths do not fluoresce at room temperature, and some colour centre formation in the matrix was found to occur after long periods of X-irradiation. Thus, cerium and praseodymium were reported not to emit in Y_2O_3 at room temperature with x-ray excitation [8] and only a weak broad band was obtained for praseodymium in Y_2O_3 at $77^{\circ}K$, also with x-ray excitation [7]. However, with the cathode-ray excitation, praseodymium emitted its characteristic spectrum at $300^{\circ}K$, and a broad-band emission, possible f-d in nature, was obtained for cerium.

The stick spectrum of rare-earth emission is given in figure 4, for Y_2O_3 matrix, which shows relative peak height and wavelength of the cathode-ray-excited rare earth lines. Rare earths shown in parentheses indicate transitions of about equivalent energy which can be masked by the rare earths listed without parentheses. If the spectrum for a specific matrix is used

FIGURE 4

Hauptlinien Spectrum of Trace Rare Earths in Y_2O_3 Matrix

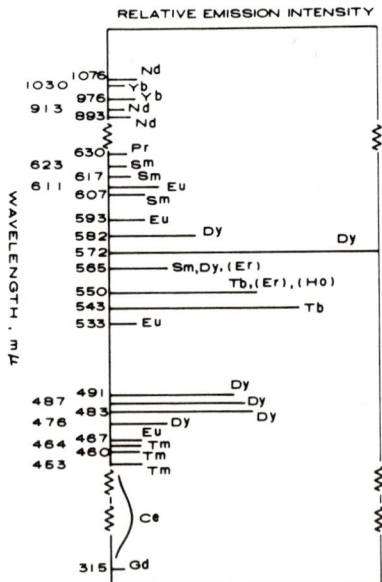

as a standard, it is a simple matter to detect qualitatively the presence of trace amounts of rare earths.

2.1.2 Matrix effects

Although our discussion has been mainly on rare earths in an yttrium oxide matrix, the method is by no means limited to this particular matrix material. Thus, cathode-ray-excited emission spectroscopy may be utilized for rare earth analysis in any matrix which will incorporate the rare earth, and which will support cathodoluminescence. In some instances, changing the matrix can improve the sensitivity of the rare earth emission, or even result in emission lines appearing which may not be evident in other matrices. As an example of matrix effects, figure 5 shows a spectrum for rare earths in a zinc sulphide matrix. Whereas neodymium in Y_2O_3 required detection at 900 nm, neodymium in zinc sulphide was apparent at 600 nm and 607 nm. It is of interest that for rare earths in the zinc sulphide matrix, Ho is masked by Tm, Eu emits in a broad band, probably as divalent Eu, and Ce and Gd also emit in broad bands.

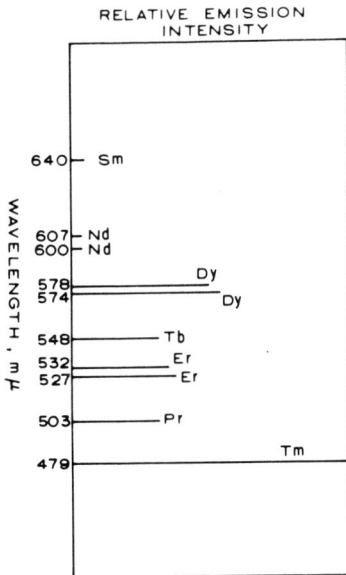

FIGURE 5

Hauptlinien Spectrum of Trace Rare Earths in ZnS Matrix.

2.2   Quantitative Determinations

CREE analysis can be adapted easily to the quantitative determination
of trace rare earths in solids. By preparing a series of standards, in
which a particular rare earth is synthesized as in the unknown, a standard
curve can be obtained, against which the concentration of the trace rare
earth impurity can be determined. Where the trace is found in a matrix of
Y_2O_3 or ZnS, the lines listed in table 1 can be used as a guide; for rare
earths in other matrices which support luminescence, the most sensitive
emission lines can be easily determined by incorporating particular rare
earths in a matrix, exciting with cathode rays, and determining the emission
characteristics. Several examples are given below of CREE analysis used for
quantitative determination of trace rare earths.

2.2.1   Europium [9]

The problem was to detect trace amounts of europium in an yttrium oxide
matrix. From table 1 it is seen that the 611 nm line is most sensitive, and
this emission $(^5D_0 \rightarrow {}^7F_2)$ was chosen for one of the detection lines. The
533 nm emission $(^5D_1 \rightarrow {}^7F_1)$ was selected as a second line.

A series of standards in quadruplicate was prepared, where europium,
from 5×10^{-4} % to 0.1%, was incorporated in yttrium oxide, by the described
firing procedure. "Blanks" of yttrium oxide alone were prepared in identical fashion. Similar procedures were used in preparing trace amounts of
europium in a gadolinium oxide matrix.

Figure 6 shows the CREE emission intensity at 611 nm as a function of
concentration of trivalent europium in the yttrium oxide matrix. These data
were obtained at an acceleration potential of 8000 volts, and a beam current
of 1 μA. The "blank" reading of a sample included in the same series showed
a residual emission at 611 nm equivalent to 0.25 ppm of europium, and this
"blank" was subtracted from the series. It is seen in figure 6, that linearity occurs roughly between 0.5 ppm and 100 ppm; at 500 ppm (and beyond),
the intensity-concentration curve becomes non-linear. This is in approximate agreement with Brown et al. [10], who reported rare earth ion-ion
interactions appearing at about 1000 ppm.

The results for gadolinium oxide are shown in figure 7. Again non-linearity of the log intensity vs. the concentration curve occurred in the region of 1000 ppm. That a change in the matrix can have beneficial results is demonstrated by the fact that the CREE intensity from europium in gadolinium oxide was about seven times greater than that in yttrium oxide.

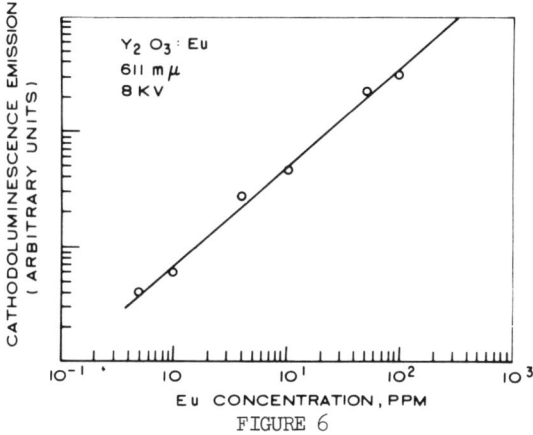

FIGURE 6

Cathodoluminescence emission intensity at 611 nm as a function of Eu concentration in Y_2O_3 matrix, 8 kV

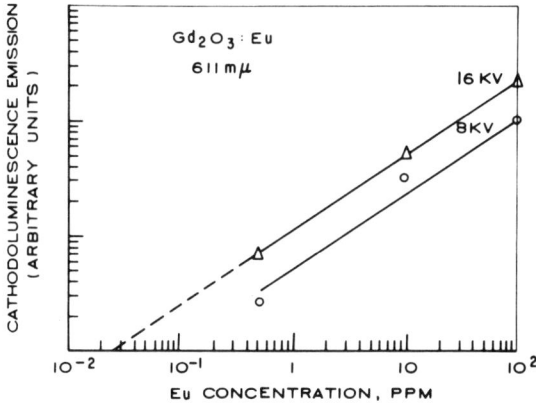

FIGURE 7

Cathodoluminescence emission intensity at 611 nm for two accelerating potentials as a function of Eu concentration in Gd_2O_3 matrix

As we have seen, the cathodoluminescence emission intensity varies directly with the potential through which the primary electrons are accelerated. This fact can be utilized to increase the emission intensity, particularly where small traces of europium are involved. Cautions to be observed include (1) measuring the "blank" under the higher voltage beam, and (2) measuring the complete series under the higher-voltage condition.[*] Figure 8 shows intensity of the secondary line at 533 nm as a function of europium concentration, in a gadolinium oxide matrix, for two acceleration potentials.

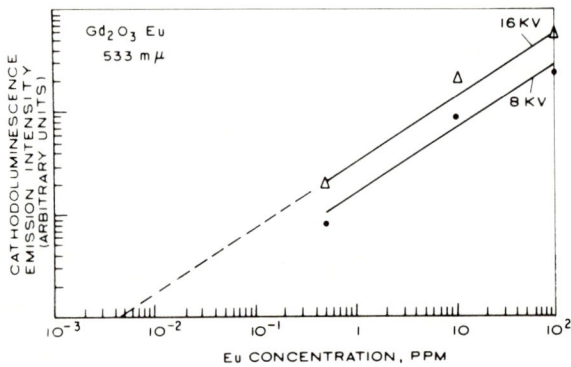

FIGURE 8

Cathodoluminescence emission intensity at 533 nm for two accelerating potentials as a function of Eu concentration in Gd_2O_3 matrix

2.2.1.1 Detection limits and analyses

Europium, in a pure yttrium oxide matrix, can be detected to a level of about 0.1 ppm, when the 611 nm emission is used as the detection line. However, by taking advantage of matrix effects, and by increasing the electron beam accelerating potential, the detection limit can be lowered. Thus, for a matrix of gadolinium oxide, using 16 KeV electrons, and the 533 nm emission, the detection limit for europium, by extrapolation, is about 0.005 ppm. It

Samples in a particular series should be analyzed at the same beam voltage and current. Results for one matrix do not necessarily duplicate these of another matrix, particularly at high voltages, where differences in secondary emission may take place.

is thus apparent that the CREE method should be applicable to trace rare earths in the ppb range, and this will be discussed later in the chapter.

Typical "added-found" analyses for europium in yttrium oxide, calculated from the standard curve of figure 6, are shown in table 2.

TABLE 2

CREE Analyses of Eu in Y_2O_3 at 611 nm

Eu Added (ppm)	Eu Found (ppm)
0.70	0.67
2.3	2.2
14.	13.
60.	57.

2.2.1.2 Interferences

It is well known that different species of rare earth ions in the same matrix may interact, and in doing so, affect the luminescence emission intensity. Such interactions have been shown to be dependent on (a) ion type, (b) concentration, and (c) matrix type [11]. However, in general, such interactions are not taken as applying to any large extent where the concentration of the trace ion is much less than about 0.1%. In the general cathode-ray-excited emission method being described, interferences can manifest themselves in either or both of two ways. Type I interfering ions may emit spectrally in a region which overlaps the emission from the desired ion, thereby reinforcing the emission by increasing the number of photons available to the detector and giving rise to a positive error. The Type II interfering ion can interact with the ion under investigation in such a way that the number of radiative transitions of the latter ions is effectively decreased thereby yielding a negative error.

In the absence of large amounts of interfering ions, the possibilities of spectral reinforcement can be avoided by making use of a secondary line for the rare earth.

Examples of the two types of interference were demonstrated in the following experiments. As can be seen from the spectrum of rare earths in yttria, the 611 nm emission from europium is flanked by the 607 nm and 617 nm emissions from samarium, while the closest line to the europium 533 nm line is at 565 nm. A series of samples was prepared, with yttrium oxide as the matrix, containing 5 ppm of europium, together with various amounts of samarium. The results are shown in table 3.

TABLE 3

Relative Cathodoluminescence Emission Intensities at 611 nm and at 533 nm

Material	611 nm	533 nm
(a) Y_2O_3 :Eu (5 ppm)	100	100
Y_2O_3 :Eu (5 ppm) :Sm (10 ppm)	187	100
Y_2O_3 :Eu (5 ppm) :Sm (10^2 ppm)	800	67
:Sm (10^3 ppm)	1400	45
(b) Y_2O_3 :Eu (100 ppm)	100	100
Y_2O_3 :Eu (100 ppm) :Sm (10^2 ppm)	107	96
:Sm (10^3 ppm)	60	32
(c) Y_2O_3 :Eu (1000 ppm)	100	100
Y_2O_3 :Eu (1000 ppm) :Sm (10^3 ppm)	47	46
(d) Y_2O_3 :Eu (5 ppm)	100	100
Y_2O_3 :Eu (5 ppm) :Dy (10 ppm)	100	85
:Dy (10^2 ppm)	-	62
:Dy (10^3 ppm)	69	32
(e) Y_2O_3 :Eu (100 ppm)	100	100
Y_2O_3 :Eu (100 ppm) :Dy (10^2 ppm)	78	75
:Dy (10^3 ppm)	9	13

As can be seen in table 3 (a), Type I spectral reinforcement by samarium of the europium 611 nm emission can lead to very large errors in the true reading of this line, particularly as the interfering samarium content is increased to 10 ppm. However, the 533 nm line of europium is unaffected by small (ca. 10 ppm) amounts of samarium, and only slightly affected by larger amounts; even 10^3 ppm of samarium yields an error of only 55% as against the enormous error at the 611 nm line. It can also be seen from table 3 (b) that increasing the amount of europium to 100 ppm, although this gives greater tolerance to the amount of samarium interference, still shows spectral reinforcement of the 611 nm line, compared to the 533 nm emission. In table 3 (c), the large concentration of rare earths would indicate that Type II interference (ion-ion interaction) would be expected to predominate over spectral reinforcement; that this is the case is shown by the identical intensity readings of the 611 nm and 533 nm lines of europium.

As another Type II interfering ion, dysprosium was chosen, since this showed the highest cathode-ray excited intensity. The results for dysprosium given in table 3 (d-e), bear out the conclusions arrived at with the Type I interfering ion, samarium, namely that when the interfering trace is low (i.e., of the same magnitude as the europium trace, both being less than 100 ppm), the intensity of the lines will give a valid reading.

It is also of interest to consider interferences of the first transition group, and for this purpose, samples of Y_2O_3:Eu (100 ppm) were prepared with 100 ppm each of iron, manganese and nickel. The results are shown in table 4.

TABLE 4

Cathodoluminescence Emission Intensities at 611 nm and at 533 nm
for Y_2O_3:Eu, Doped with Iron, Nickel and Manganese

Material	611 nm	533 nm
Y_2O_3:Eu (100 ppm)	100	100
Y_2O_3:Eu (100 ppm):Fe (100 ppm)	88	92
Y_2O_3:Eu (100 ppm):Mn (100 ppm)	77	80
Y_2O_3:Eu (100 ppm):Ni (100 ppm)	97	92

It is evident that ions of the first transition group can interfere in this determination of europium, which indicates the possibility of coupling

between 4f and 5d electrons. Manganese, the ion showing the largest interference effect on europium, can itself be "poisoned", luminescence-wise, by
iron and by nickel in group II - group VI compounds [12].

2.2.2 Terbium [13]

For the analysis of trace amounts of terbium in yttrium oxide, the 544
nm emission $(^5D_4 \rightarrow {}^7F_5)$ was selected as the primary detection line. Preliminary analysis of the "pure" yttrium oxide showed no lines attributable
to terbium and a faint trace of dysprosium. However, high-resolution CREE
analysis, as will be discussed, showed the presence of about 16 ppb of terbium in the "blank" yttrium oxide.

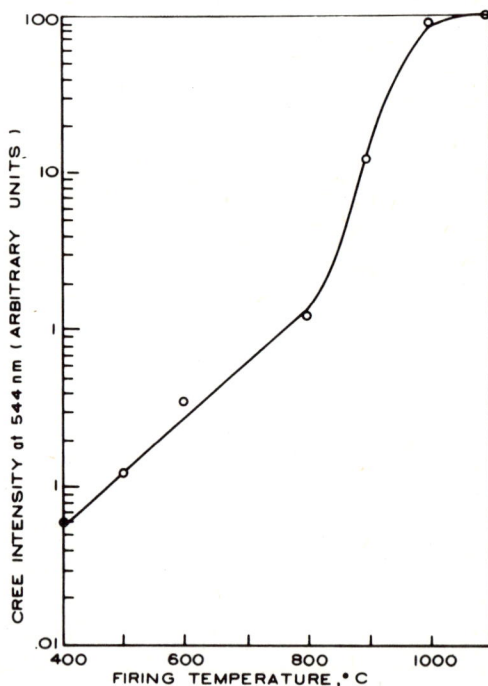

FIGURE 9
Intensity at 544 nm
as a function of
firing temperature

It is rather apparent that the firing conditions in which samples are
prepared for analysis must be constant for all samples. In particular, it

is essential that the firing <u>temperature</u> is carefully controlled. The effect of firing temperature on the intensity of the 544 nm emission is shown in figure 9; it is seen that diffusion and incorporation of the terbium is complete at about 1000°C, and 1100°C was selected as the sample firing temperature.

CREE intensity as a function of terbium content is shown in figure 10, for an accelerating potential of 16 KV. With this type of measurement, the detection limit is about 20 ppb. Up to the present, CREE analyses were measured with a spectrometer having a band-pass of about 5 nm. By going to a band-pass of 1 nm or less, i.e., higher resolution, thereby rejecting the background of broad-band luminescence, the signal-to-noise aspect can be improved, and high resolution CREE analyses can be carried out.

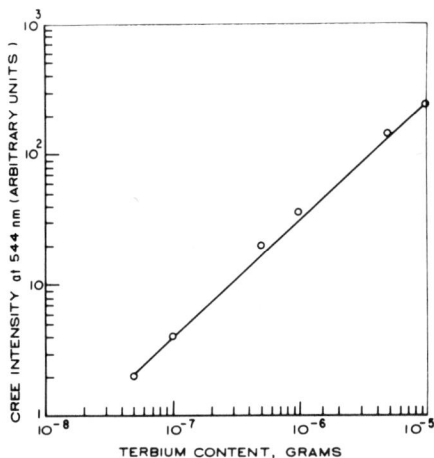

FIGURE 10
Intensity at 544 nm as a function of Tb concentration, Y_2O_3 matrix, 16 kV

Figure 11 shows the actual traces from the high-resolution spectrometer (0.5 nm band-pass), for these samples of terbium, 0, 10 and 50 ppb in an yttrium oxide matrix. An analysis of these curves yields an estimated 16 ppb of terbium in the "ultrapure" starting yttrium oxide. As operated for these measurements, the total noise level is equivalent to about 1 ppb of terbium; thus, for a signal-to-noise detection limit of two, 2 ppb of terbium would be detectable. With signal-averaging equipment, the ultimate detection-limit could probably be much less than 1 ppb, but no attempt was made to determine this, because of the unavailability of yttrium oxide completely free of terbium.*

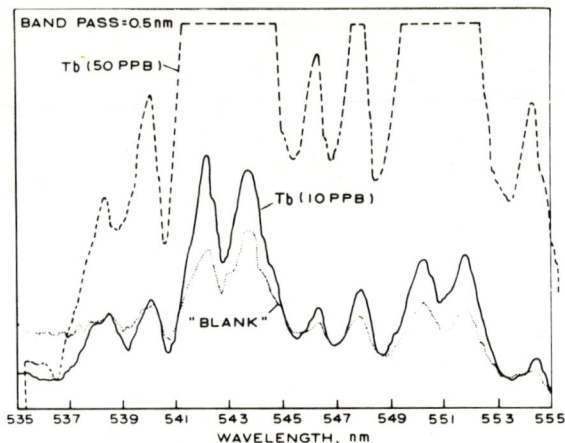

FIGURE 11

High resolution spectrometer traces for Y_2O_3 with 50 ppb and 10 ppb of Tb, Trace for "blank" Y_2O_3 shows as estimated 16 ppb present in the raw material

"Added-Found" data, analyzed by means of the standard curve of figure 10, are given in table 5.

2.2.2.1 Interferences

As has been discussed, other rare earth ions in the same matrix may interact with the ion under investigation, thereby affecting the CREE intensity. These interactions depend on concentration, ion-type, and matrix-type. Where an interfering ion emits in a spectral region which overlaps that from the desired ion, thereby resulting in a positive error, use can be made of a secondary line. If the interaction results in a decreased CREE emission,

---

* Very recent work with a new source of pure yttrium oxide has indicated a detection limit of the order of 0.05 ppb terbium in yttrium oxide.

the materials for the standard curve should be prepared so as to contain
the interfering impurities, and the CREE analyses can still be carried out.

TABLE 5

Analysis of Terbium in Yttrium Oxide Matrix

Tb added (ppm)	Tb found (ppm)
0.07	0.06
0.40	0.43
6.0	5.2

In addition to rare-earth-interferences some ions of the first transition group can also act as "poisons" of luminescence. As shown in table 6,
the extent of such an effect varies with the ion, iron being the worst, for
this case.

TABLE 6

Effect of Fe, Mn and Ni on Tb-CREE Emission from Y_2O_3:Tb

Material	CREE Intensity
Y_2O_3:Tb (0.5 ppm)	100
Y_2O_3:Tb (0.5 ppm):Fe (100 ppm)	42
Y_2O_3:Tb (0.5 ppm):Mn (100 ppm)	76
Y_2O_3:Tb (0.5 ppm):Ni (100 ppm)	100

Iron interference was investigated in greater detail for Y_2O_3:Tb (5 ppm),
the result being shown in figure 12, for iron concentrations of 1-100 ppm.
We observe a monotonic decrease in CREE intensity __versus__ iron concentration,
over the range examined.

2.2.3 Dysprosium [14]

The CREE intensity from dysprosium in zinc sulphide either containing,
or without, lithium as a charge-compensating ion, is relatively weak and un-

- 142 -

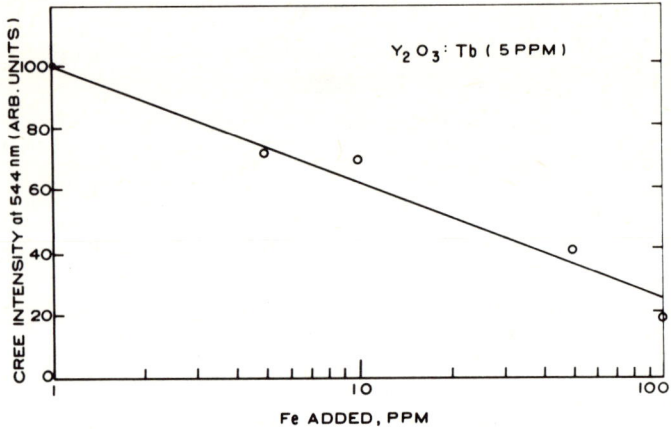

FIGURE 12

Intensity at 544 nm as a function of iron addition

structured, resulting in a high detection threshold. Previous studies [15] of the luminescence properties of dysprosium in zinc sulphide had shown that both the intensity and emission spectrum of dysprosium were enhanced by silver ions. Figure 13 shows the spectral distribution of the CREE from (a) 100 ppb dysprosium in zinc sulphide containing 100 ppm lithium, and (b) under identical operating conditions, 100 ppb dysprosium in zinc sulphide containing 100 ppm lithium and 100 ppm silver ion. The relatively unstructured curve of (a) is believed to be related to a high site-multiplicity factor; the development of a strong band centered at 5750 Å, as shown in (b), suggests that the presence of silver ions may affect the occupancy of certain sites.

The effect of sample preparation temperature is shown in table 7, and 1100°C was selected as the crystallization temperature for the materials constituting the working curve and unknowns. Figure 14 is a plot of CREE intensity as a function of dysprosium concentration, in materials containing both silver and lithium, and table 8 compares the "added-found" results for a series of samples.

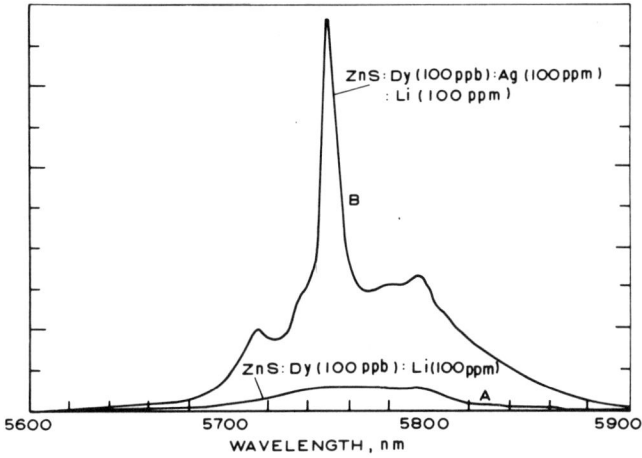

FIGURE 13

Spectral distribution curves of the cathodoluminescence emission from
ZnS:Dy:Li, without Ag (curve A), and with Ag (curve B)

TABLE 7

Dy CREE Intensity vs. Firing Temperature

Firing Temperature	Relative Intensity
800°C	0.5
900	1.5
1000	5.4
1100	96.0
1200	32.0

TABLE 8

Analysis of Dysprosium in Zinc Sulphide Matrix

Dy added	Dy found
20 ppb	17 ppb
50	46
50	40
200	170
500	420
500	520

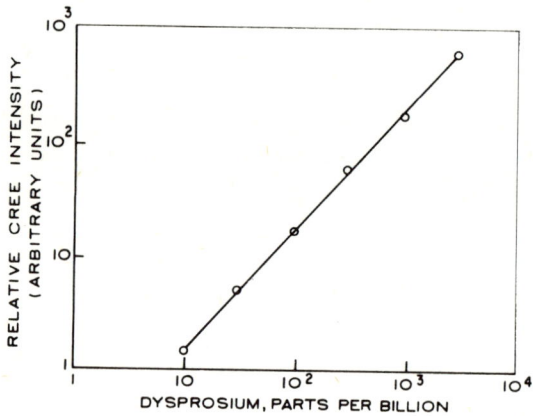

FIGURE 14

Intensity as a function of Dy concentration in ZnS matrix, 8 kV

Since the overall CREE intensity is increased many-fold by the addition
of silver, at a given emission peak, the combination of these effects results
in a detection limit for dysprosium, in a zinc sulphide matrix, of 1-10 ppb.

2.2.4 Thulium [16]

Trace amounts of thulium were determined by CREE analysis, in a zinc sulphide matrix. Alkali metal ion charge compensation [17] was used. In the case of thulium, it had been reported [17] that lithium enhanced the visible emission, and for this reason, lithium ion was used as the charge compensating species.

The matrix material was "luminescent grade" zinc sulphide. Firings were carried out in Vitreosil fused quartz vessels, which had been prefired with pure zinc sulphide. The firing atmosphere was hydrogen sulphide. To improve sample uniformity, firings were carried out in the apparatus shown in figure 15. The firing tube was preflushed by passing the hydrogen sulphide through the inlet tube. During the firing, the gas flow was shunted through the cap of the firing tube, thus maintaining a positive pressure of hydrogen sulphide in the tube, while flow around the sample was avoided. The samples were fired at 800° for 15 min and at 1100° for 15 min.

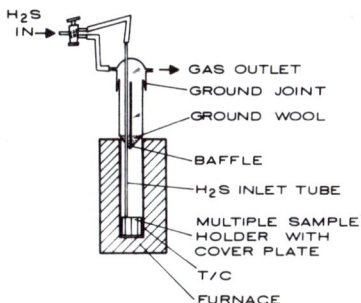

FIGURE 15
Apparatus for firing samples
for CREE analysis

With the low rare-earth content investigated in this work, it was found that sequential firings introduced expected variations from run to run. For this reason, a multiple sample-holder was designed, shown in figure 15, which allowed six samples to be fired simultaneously in the same ambience.

Sample mixes in each firing were repeated in subsequent firings, thereby providing a method of normalizing the various firings.

A Spex grating monochromator (Model 1700), fitted with a 7265 multiplier phototube was used for the thulium analyses. It was found desirable to scan the monochromator over a narrow range which included the rare earth line, 470-485, while recording the photocurrent (or detector output) to determine the amount of background luminescence. Obviously the resolution (or band-pass) should be adequate to distinguish the line emission from the background. Since the spectral character of the rare earth emission is quite insensitive to concentration below 100-500 ppm, one needs to read only a peak-height above background for each sample. Availability of zero-offset for the background before amplification is helpful in providing accuracy when a weak line is superimposed on a strong background.

2.2.4.1 Results

Thulium in zinc sulphide showed detectable luminescence at or above $800^{\circ}C$, and table 9 shows CREE intensity as a function of firing temperature, for one composition.

TABLE 9

Emission Intensity of Thulium vs. Firing Temperature

Firing temperature ($^{\circ}C$)	Relative intensity
800	3.8
900	9.0
1000	14
1100	51

The visible emission from thulium in zinc sulphide has been reported [17] to consist of two spectral complexes, the major being at about 470-485 nm, and the minor at about 630-665 nm. The spectral distribution curves are shown in figure 16.

The CREE emission intensity of the 470-485 nm emission from thulium, as a function of concentration (between one and 300 ppb), in zinc sulphide, is shown in figure 17. A detection limit of about 0.01 ppb is indicated for this type of analysis. "Added-Found" data are given in table 10. For

- 147 -

FIGURE 16

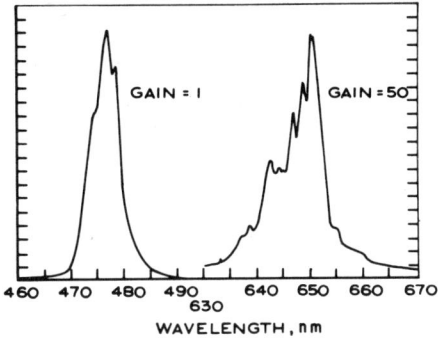

Cathodoluminescence spectral
distribution curves of blue
and red emission bands from
Tm in ZnS (note gain change)

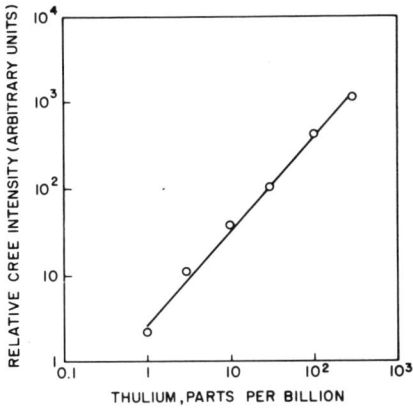

FIGURE 17

Intensity as a function of Tm concentration in ZnS matrix, 8 kV.

Tm Found, two columns of data are shown, one for CREE analysis, and the
other for solid state mass spectrometry (SSMS), details of which are presented in reference [16].

TABLE 10

Determination of Thulium in Zinc Sulphide Matrix with Charge Compensation

Tm added (ppb)	Tm found (ppb)	
	CREE	SSMS
6	6.2	6.8
20	24	23
20	19.5	
60	54	64
60	54	
200	220	222
200	205	

2.2.4.2 Interferences

Silver ion in zinc sulphide promotes a blue, broad-band, cathodoluminescence emission, and the interference effect of silver was therefore investigated. Various concentrations of silver were incorporated into zinc
sulphide containing 5 ppm of thulium. As seen in figure 18, silver ions
begin to decrease the emission intensity from thulium at a concentration of
about 1 ppm. With 10 ppm of silver, the thulium emission is decreased by a
factor of about five, and at 50 ppm of silver, the thulium emission is only
about 2% of the original level.

The interference effects of twelve rare earths were also investigated
by using 100 ppm of rare earth in the analysis of 5 ppm thulium in zinc
sulphide (lithium). The results are shown in figure 19. The slight general
trend, with the exception of europium appeared to be decreasing emission
intensity from thulium with increasing atomic number of the interfering rare
earth. With europium, no cathode-ray-excited emission was detected from
thulium, which may be related to the fact that europium is divalent in zinc
sulphide [18], and whose excitation spectrum may overlap the emission from
thulium.

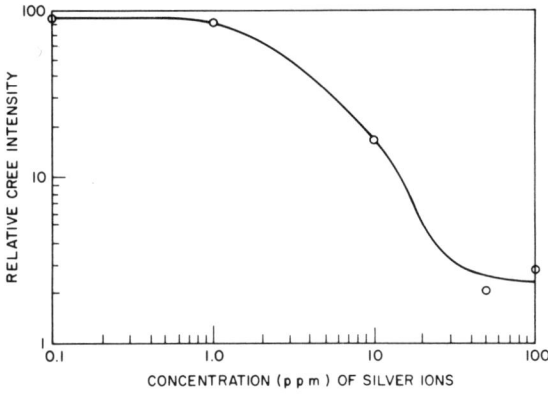

FIGURE 18

Intensity as a function of Ag concentration
for ZnS containing 5 ppm Tm

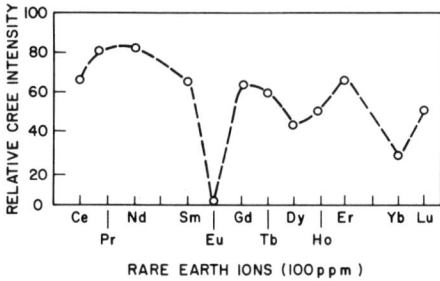

FIGURE 19

Intensity from ZnS:Tm (5 ppm)
as a function of rare earth
interferences

Acknowledgements

The author gratefully acknowledges assistance in materials preparation from R.A. Kauffunger and J. McGowan; in spectral measurements from Drs. W.H. Fonger and R.A. Shrader; in solid state mass spectrometry from Dr. W.L. Harrington, as well as valuable discussions with Drs. S.J. Adler, M.R. Royce, A.L. Smith and P.N. Yocom.

References

[1] Cathodoluminescence is reviewed in: S. Larach, ed., Photoelectronic Materials and Devices, D. Van Nostrand, New York (1965); H.W. Leverenz, Introduction to Luminescence of Solids, Wiley, New York (1950); G.F.J. Garlick, chapter in Luminescence of Inorganic Solids, P. Goldberg, ed., Academic, New York (1966).

[2] G.F.J. Garlick, Brit. J. Appl. Phys. 13, 541 (1962).

[3] G. Gergely, J. Phys. Chem. Solids 17, 112 (1960).

[4] S. Larach, Anal. Chim. Acta. 41, 189 (1968).

[5] S. Larach, Proc. Intl. Conf. on Luminescence, Budapest (1966).

[6] W. Low, J. Makovsky and S. Yatsiv, Quantum Electronics III, P. Grivet and N. Bloembergen, eds., Columbia, New York (1964).

[7] R.C. Linares, J.B. Schroeder and L.A. Hurlbut, Spectrochim. Acta 21, 1915 (1965).

[8] R.M. Walters, J.F. Cosgrove and D.J. Bracco, Sympos. on Trace Charac., pp. 174-180, October 1966.

[9] S. Larach, Anal. Chim. Acta, 42, 407 (1968).

[10] M.R. Brown, J.S.S. Whiting and W.A. Shand, J. Chem. Phys. 43, 1 (1965).

[11] L.G. VanUitert and R.R. Soden, J. Chem. Phys. 36, 1289 (1962).

[12] R.H. Bube, S. Larach and R.E. Shrader, Phys. Rev. 92, 1135 (1953).

[13] S. Larach and R.E. Shrader, Anal. Chim. Acta 45, 227 (1969).

[14] R.E. Shrader, S. Larach and R.A. Kauffunger, Anal. Chim. Acta 54, 162 (1971).

[15] R.E. Shrader, S.M. Thomsen and P.N. Yocom, Abstr. 35, Electrochem. Soc., San Francisco 1965.

[16] S. Larach, R.E. Shrader and R.A. Kauffunger, Anal. Chim. Acta 51, 393 (1970).

[17] S. Larach, Proc. Int. Conf. Luminescence 1966, ed. by G. Szigetti, Budapest 1968, p. 1549.

[18] S. Larach, unpublished EPR data.

TRACE ELEMENTS CHARACTERIZATION IN Y_2O_3 BY MASS SPECTROSCOPY AND ISOTOPIC DILUTION

by

John Haaland and Derk E. Stijfhoorn

Institutt for Atomenergi, Kjeller, Norway.

ABSTRACT

Rare earth elements at concentrations down to 0.1 ppm have been deter-mined in pure yttrium oxides using isotopic dilution and the ATLAS CH4 single focussing instrument with a double rhenium filament ion source. The eight poly-isotopic elements were determined in one run after addition of a master-spike. Oxide peaks from the lighter elements were removed effective-ly by passing benzene through the source.

The AEI MS 702 double focussing spark source mass spectrometer with photographic registration was used to give a general survey of the distribu-tion of both RE and non-RE elements. The Y_2O_3 samples were mixed with 60% pure graphite to make them conducting and pressed into a pair of tiny elec-trodes. A series of spike isotopes were again mixed in for calibration. A simplified plate calibration procedure was used. The peaks of the spike isotopes served for quantitative determinations, either by isotopic dilution or by precalculated sensitivity factors for other elements. A background spectrum from carbon, yttrium carbide, etc. interferes in some cases with analysis lines. The general limit of determination is also in this case 0.1 ppm.

1. INTRODUCTION

The two types of ion sources most frequently used for mass spectrometry of solids are the surface ionization filament and the high frequency spark. Due to a rapid decrease in surface ionization efficiency with increasing ionization potentials, the first mentioned source is not generally applicable

for sensitivity reasons. In the present work, the application was limited
to the rare earth group. These elements constitute a uniform group with low
ionization potentials and with reasonably similar evaporation and dissociation properties. For this reason, the eight poly-isotopic RE elements could
be determined in one run by addition of a "masterspike"containing enriched
isotopes of the eight elements according to table 1. The oxides LaO^+, CeO^+,
NdO^+, PrO^+, and partly SmO^+ and GdO^+ remain practically undissociated under
normal vacuum conditions. These oxide peaks will interfere with the heavier
elements sixteen mass units up. It proved possible to remove this interference completely by passing a flow of benzene through the source to maintain a pressure of 0.5 torr during registration (see figure 1).

TABLE 1

Masterspike for RE determinations

Spike isotope	μg spike (y) per 500 mg sample	Spike isotope	μg spike (y) per 500 mg sample
Ce 142	4.2	Gd 160	19.5
Nd 145	7.3	Dy 161	4.2
Sm 149	10.5	Er 170	14.7
Eu 151	2.0	Tb 173	5.5

Contrary to surface ionization, the high frequency spark ionized all
elements with a high and uniform efficiency. The background spectrum from
carbon (C_1 to C_{20}), the yttrium carbides and oxides may cause trouble.
Likewise, the hydrocarbon spectra interfere to the extent they are not
resolved. The background from hydrocarbons is effectively reduced by careful drying of the electrodes in vacuum. Spike isotopes are added for important non-RE elements like Fe and Ca, and for elements which represent groups
with similar properties. The non-spiked elements, which first of all include all the monoisotopic ones, are estimated by means of precalculated
sensitivity factors.

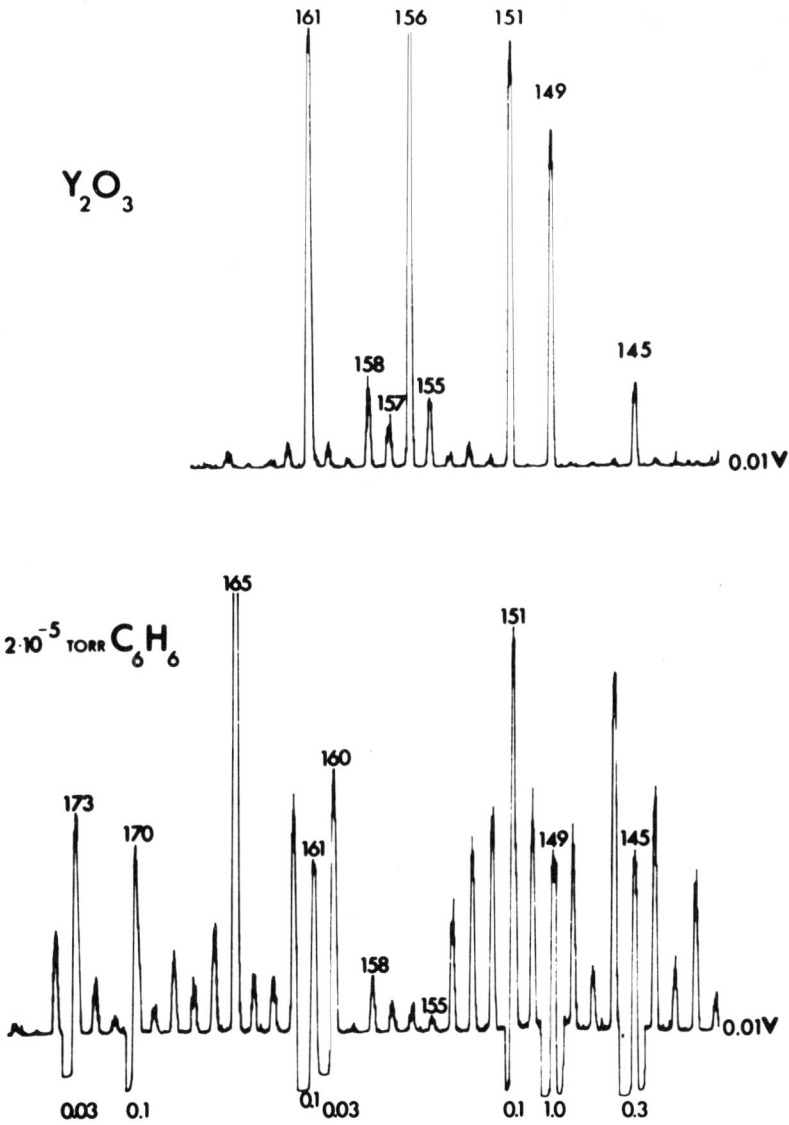

FIGURE 1

Reduction of oxide peaks by means of C_6H_6

2. SAMPLE PREPARATION

To obtain a sample solution suitable for the rhenium filament source, 500 mg of a homogeneous sample were dissolved in 3 - 4 ml suprapure nitric acid together with 100 λ "masterspike" (table 1). The 10 ml beaker was placed on a hot plate, the sample dissolved, and the excess nitric acid evaporated to almost dryness. A few ml of distilled water were added and again evaporated. At the end, the sample residue was taken up in a few ml quartz distilled water. From this solution a few microlitres were transferred to the centre evaporation filament by means of a small "throw-away" pipette. The samples were run at filament currents between 3 and 5 amp, and both filaments therefore preheated at a higher current for cleaning and removal of possible background interferences. The sample was dried at 1 - 2 amp in air before degassing in the ion source. Rather small samples were applied to obtain stable ion currents.

A mixture of Y_2O_3 sample powder and Ringsdorff RBS graphite powder (2:3 by weight) was transferred to a 10 ml beaker with a small steel ball in it and mixed well in a "Turbula" for one hour. The electrodes were then pressed in an AEI Moulding Die at 10 - 12 tons and dried overnight in vacuum. The Y_2O_3 sample with appropriate spikes was prepared as above except for a final ignition at 800°C.

3. INSTRUMENTATION

An ATLAS CH4/58 single focussing mass spectrometer was provided with a double filament ion source and electron multiplier registration with pen recorder. A vacuum lock allowed the cages with sample filaments to be transferred into and out of the ion source without breaking the vacuum. To attain a satisfactory reproducibility a rather strict filament heating programme was established. The current through the ionization filament (i.f.) was slowly brought to 4A in 15 minutes. The current through the evaporation filament (e.f.) was increased in steps of 0.1 to 0.2 A every 5 minutes. At 2.0 - 3.5/4.0 - 4.5 A (e.f./i.f.) Ce is recorded as CeO^+ at masses 156 and 158. At 3.5 - 3.6/5.0 A, Nd, Sm, Eu, and Dy are recorded. After opening the valve to the benzene inlet the rest of the RE can be determined at 4.5/5.0 A. A

considerable increase in sensitivity was obtained when benzene was admitted
to the source.

The AEI MS702 spark source instrument was equipped with an "AUTOSPARK"
control unit and photoplate (Ilford Q2) registration. The spectra were evaluated in an ARL microphotometer with an 18 micron fixed slit and pen recorder.

4. THE ISOTOPIC DILUTION METHOD

When a known amount y of a spike isotope is added to a sample, the unknown amount x can be calculated from the formula

$$x_{\mu g} = y_{\mu g} \cdot \frac{B_{1k} - C_{1k}}{C_{1k} - A_{1k}} \cdot \frac{\sum_{1=1}^{n} A_{1k} M_{1}}{\sum_{1=1}^{n} B_{1k} M_{1}}$$

where A_{1k}, B_{1k}, and C_{1k} are the ratios between an arbitrary isotope i and
the spike isotope k in the original sample, the spike material, and the mixture respectively. M_{1} is the mass number of isotope i. The last term is
constant for a certain spike material. The term $B_{1k} - C_{1k}/C_{1k} - A_{1k}$ varies
with the measured isotopic ratio C_{1k} in the mixture or analysis spectrum.
C_{1k} depends on the added amount y, and it is important to adjust this added
spike so that C_{1k} does not come too close, either to A_{1k} or B_{1k}. In such
cases, the reproducibility drops drastically as indicated in figure 2, This
is especially important when intensities are measured from photoplates.

5. STANDARDIZATION IN THE SPARK SOURCE MASS SPECTROMETRIC (SSMS) METHOD

The adopted method of standardization in SSMS is to calculate certain
sensitivity factors for each element in relation to an added internal standard element. These factors are matrix dependent. In the present investigation, a more extensive use of isotopic compositions effectively removes

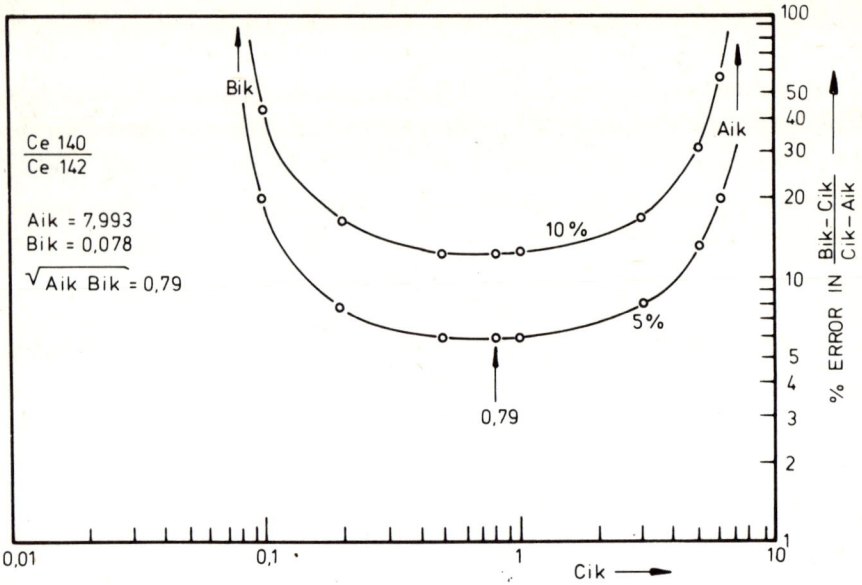

FIGURE 2

Effects of C_{ik} values and errors on final results

matrix effects and affords a close control on residual impurities in materials for standard preparations.

6. PHOTOPLATE CALIBRATION

To obtain a relationship between the relative intensity of the incident ion beam and the corresponding photoplate response, a series of calibration marks is set on the photoplate emulsion, and the per cent transmission of light through the line is measured in a micro densitometer. The calibration marks are obtained by exposing the photographic emulsion until a preset electrical charge ranging from one to a few hundred nano-coulombs(nC) is read on the Monitor Collector which intercepts half of the ion current through the

spectrometer. Another method used alone or in combination with the first
one is to select certain known isotopic ratios which are quite constant for
natural elements. In the latter case one can make a plate calibration from
a single mass spectrum. The calibration curve shown in figure 3 is obtained
by the first method, i.e., the charge collection. The flat portion of the
curve corresponding to transmissions in the range 50 - 100% is replaced by
a 100% transmission curve branch. The resulting curve is the heavy drawn
one which is almost symmetrical around a vertical line. It is our experience that the Ilford Q2 photoplate calibration curves vary only little from
plate to plate, and that the curve in figure 3 can be considered an average
one.

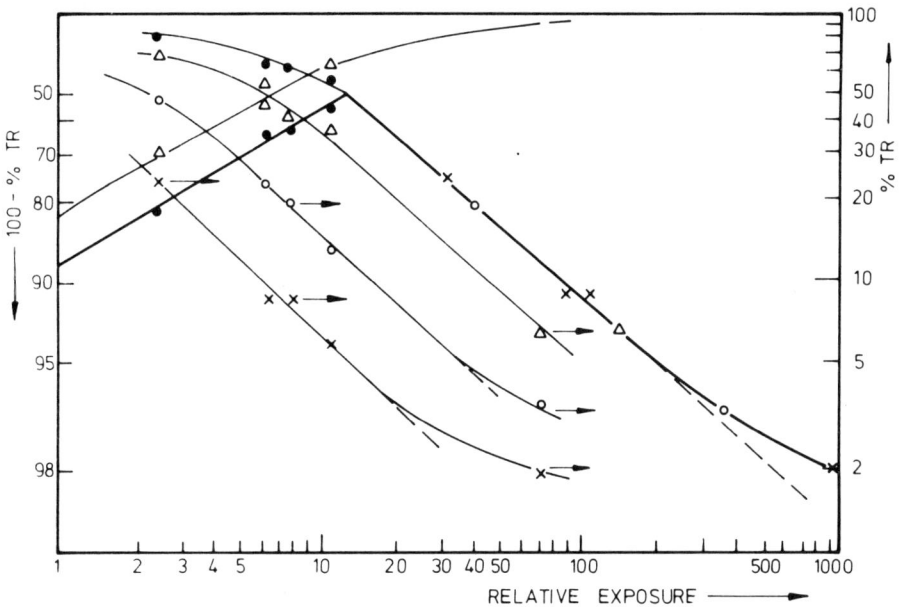

FIGURE 3

Calibration curve based on barium isotopes

7. INTENSITY MEASUREMENTS AND SPECTRUM EVALUATIONS

The diagram in figure 4 is based on the average calibration curve and can be used to transfer peak heights, i.e., per cent transmission on a strip chart, directly into relative intensities as read from the vertical scale.

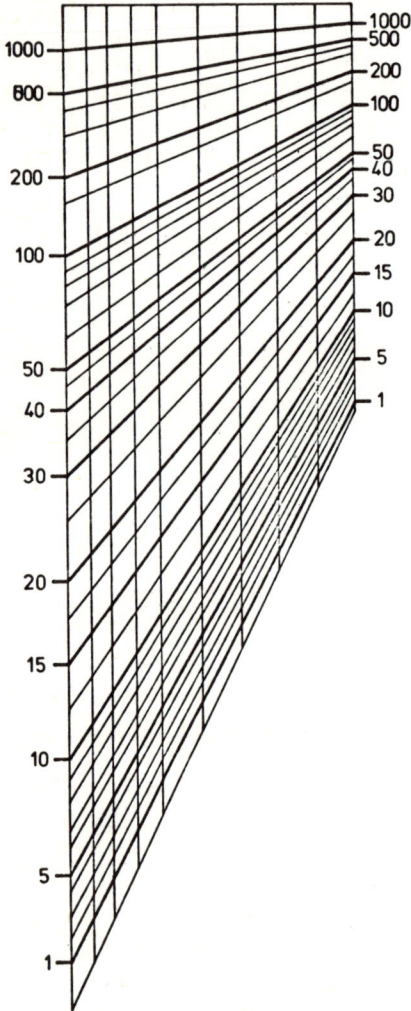

FIGURE 4

Intensity evaluation nomogram

The length of the vertical lines corresponds to the separations between zero transmission and the "100% transmission" corresponding to the background density of the photoplate. The relative intensities measured in this way are quite consistent in the range 1 - 200.

Let S_o be the selected spike or internal standard isotope with known concentration $C_{s_o} = C_s \cdot A_{s_o}$. Further, X_o is the calibration isotope with known concentration $C_{x_o} = C_x \cdot A_{x_o}$ of the element X to be determined. A_{s_o} and A_{x_o} are the isotopic abundances of S_o and X_o respectively. The isotopic ratios:

$$R_o = \frac{(X_o)}{(S_o)} = \frac{C_{x_o}}{C_{s_o}} \cdot f_x , \quad \text{and}$$

$$R_1 = \frac{(X_1)}{(S_o)} = \frac{C_{x_1}}{C_{s_o}} \cdot f_x = \frac{C_x}{C_s} \cdot \frac{A_{s_o}}{A_{x_1}} \cdot f_x$$

where X_1, which means any isotope of element X, are calculated from intensity evaluations by means of the graph in figure 4.

The sensitivity factor f_x, characteristic for element X, is then given by

$$f_x = R_o \cdot \frac{C_s}{C_x} \cdot \frac{A_{s_o}}{A_{x_o}}$$

and the unknown concentration

$$C_x = R_1 \cdot C_s \cdot \frac{1}{f_x} \cdot \frac{A_{x_1}}{A_{s_o}} = R_1 \cdot C_s \cdot F_{x_1}$$

A Megon Y_2O_3 product labelled C_6B_4 was doped with RE spikes according to table 1. A series of SSMS spectra were recorded at 100, 30, and 10 nC. The resulting sensitivity factors with reference to Nd-145 and Er-170 as internal standard spike isotopes are given in tables 2 and 3.

TABLE 2

Sensitivity factors for the light RE fraction
using Nd-145 as spike isotope

$\dfrac{X_o}{S_o}$	R_o			$\dfrac{R_o}{M}$	$\dfrac{C_s}{C_x}$	$\dfrac{A_{s_o}}{A_{x_o}}$	f_x	$\dfrac{A_{x_1}}{A_{s_o}}$	F_{x_1}
	100 nC	30 nC	10 nC						
$\dfrac{\text{Ce-142}}{\text{Nd-145}}$	0.95	1.13	1.36	1.14	0.70	0.96	0.77	0.115	1.50
$\dfrac{\text{Pr-141}}{\text{Nd-145}}$	1.37	1.26	1.24	1.31	0.73	0.87	0.82	1.04	1.27
$\dfrac{\text{Sm-149}}{\text{Nd-145}}$	1.75	1.75	1.53	1.68	0.69	1.00	1.16	0.144	0.125
$\dfrac{\text{Eu-151}}{\text{Nd-145}}$	0.18	0.27	0.16	0.21	3.65	0.99	0.78	0.50	0.64

TABLE 3

Sensitivity factors for the heavy RE fraction
using Er-170 as spike isotope

$\dfrac{X_o}{S_o}$	R_o			$\dfrac{R_o}{M}$	$\dfrac{C_s}{C_x}$	$\dfrac{A_{s_o}}{A_{x_o}}$	f_x	$\dfrac{A_{x_1}}{A_{s_o}}$	F_{x_1}
	100 nC	30 nC	10 nC						
$\dfrac{\text{Gd-160}}{\text{Er-170}}$	1.30	1.17	1.13	1.20	0.755	1.06	0.96	0.228	0.24
$\dfrac{\text{Tb-159}}{\text{Er-170}}$	2.92	3.30	2.87	3.03	0.37	0.96	1.08	1.05	0.97
$\dfrac{\text{Dy-161}}{\text{Er-170}}$	0.26	0.29	0.29	0.28	3.50	1.07	1.04	0.20	0.19
$\dfrac{\text{Ho-165}}{\text{Er-170}}$	0.73	0.73	0.73	0.73	1.47	0.96	1.03	1.05	1.01
$\dfrac{\text{Tm-169}}{\text{Er-170}}$	0.75	0.72	0.86	0.77	1.47	0.96	1.07	1.05	0.98
$\dfrac{\text{Yb-173}}{\text{Er-170}}$	0.21	0.28	0.30	0.26	2.67	1.13	0.78	0.167	0.22
$\dfrac{\text{Lu-175}}{\text{Er-170}}$	0.50	0.47	0.53	0.50	1.47	0.96	0.71	1.05	1.47

8. CONCLUSION

The sensitivity factors f_x for RE elements as given in tables 2 and 3 are close to unity. This is also expected from a group of elements with very similar properties. Further, it indicates that the "short-cut" intensity evaluation procedure adopted in this work is consistent. Figures 5 and 6 show intensity variations for a number of elements in relation to nano-coulomb exposures. Where curves for two or more isotopes of an element are drawn, a certain parallelism indicates good reproducibility in isotopic ratio measurements.

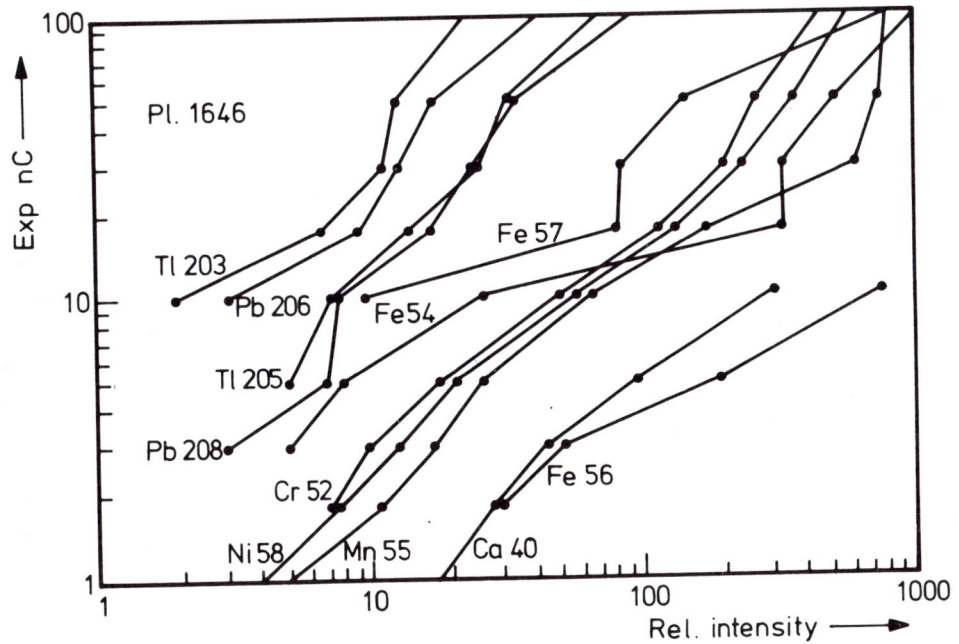

FIGURE 5

Intensity variations

in relation to nano-coulomb exposures

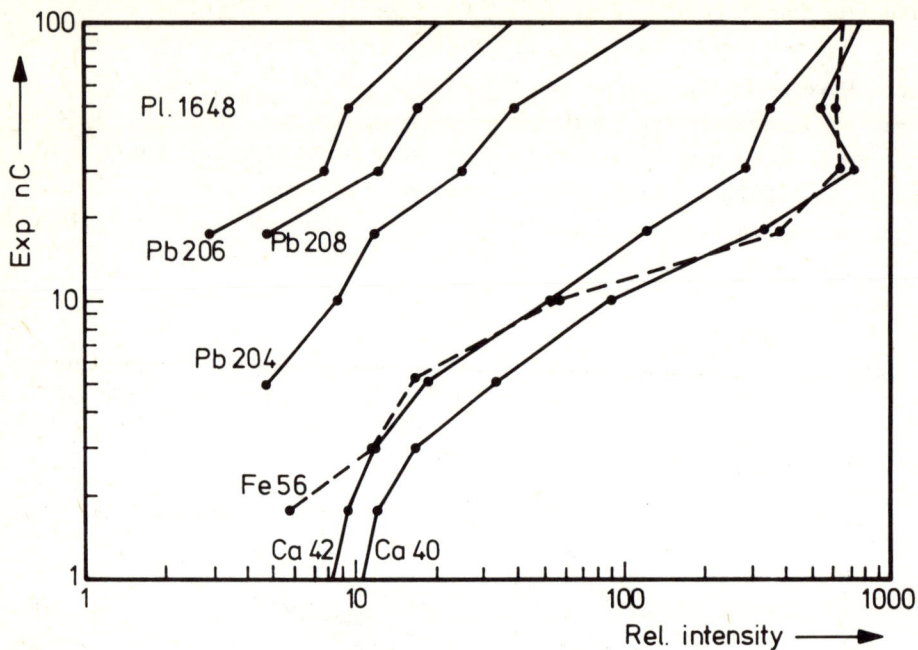

FIGURE 6

Intensity variations

in relation to nano-coulomb exposures

We will again underline the importance of keeping down the hydrocarbon background spectrum and to avoid contamination from the ion source (memory), from beakers, moulding die, graphite admixtures etc.

RARE EARTH DETERMINATION BY NEUTRON ACTIVATION ANALYSIS

by

Eiliv Steinnes

Institutt for Atomenergi, Isotope Laboratories, Kjeller, Norway.

ABSTRACT

Advantages and disadvantages of neutron activation analysis are discussed, and the techniques used at present for the determination of rare
earth elements are described. Methods based on a chemical separation of
the rare earth group following the activation process, and γ-spectrometry
with solid state detectors as the final step will in most cases yield high
sensitivity and results of good accuracy. Purely instrumental activation
analysis permits rapid analyses, and is feasible for the determination of
certain rare earths in a number of matrices.

Applications of thermal and fast neutron activation analysis are briefly reviewed. The analysis of rare earth matrices for traces of other rare
earths is associated with particular difficulties, and requires in most
cases the use of methods based on chromatographic separations.

1. INTRODUCTION

Neutron activation analysis is a sensitive tool for the determination
of a great number of elements, including most of the rare earth ones (REE).
Besides that, the specificity is good and interfering effects are relatively few, which means that accurate results may be obtained even at very low
concentration levels, provided that a proper selection of methods is made.
Neutron activation analysis has therefore been extensively used in recent
years for the determination of REE in a variety of matrices. A review ar-
ticle on the use of activation analysis for rare earth determination has
recently been published in Journal of Radioanalytical Chemistry [1]. Geological samples is obviously an important area of application, and is dealt

with more specifically in another paper at this conference [2]. Other
application areas include reactor materials, high purity chemicals, metals,
biological tissue, and water.

The present paper attempts to give a brief survey of the procedures
used at present for the determination of REE (here defined as the fourteen
naturally occurring lanthanides plus yttrium) by neutron activation. Activation analysis using projectiles other than neutrons has found limited
application for RE determination so far, and will not be included here. In
view of the character of this meeting, the determination of REE in rare
earth materials will be discussed specifically.

2. METHOD

Activation analysis can be described as consisting of two discrete
steps:

a) Production of a radioactive nuclide from the element to be determined by some kind of nuclear reaction.

b) Measurement of the amount of induced radionuclide.

The disintegration rate (D) of an induced radionuclide at the end of
the irradiation period is given by

$$D = N \cdot \Phi \cdot \sigma \cdot \theta (1 - e^{-0.693 \ t/t_{\frac{1}{2}}})$$

where

N = number of target atoms available for activation

Φ = flux of activating particles

σ = activation cross-section of the target nuclide for
the nuclear reaction in question

θ = fractional isotopic abundance of target nuclide

t = duration of irradiation period

$t_{\frac{1}{2}}$ = half-life of induced radionuclide

The main factors determining the sensitivity of an activation analysis are connected with the activation step; the particle flux available, the activation cross-section and the irradiation time chosen. Some factors associated with step (b) above will also be of significance in sensitivity considerations, such as the half-life of the induced radionuclide, the modes of decay, and the energy of the associated radiation. The sensitivity and energy resolution of the counting equipment also has to be considered.

2.1 Irradiation

Most activation analyses are carried out according to the relative method, which means that a standard containing the element to be determined, in a known amount is irradiated together with the samples to be analysed. In the case of short-lived nuclides, samples and standards may be irradiated alternately provided constant irradiation period and flux. It is also possible to calculate the amount of element present from the measured activity using the equation given above. This would, however, require knowledge of the activation cross-section, the particle flux in the irradiation position, the decay scheme of the radionuclide concerned, and the efficiency of the detector system, and would at best give results accurate to 10 - 20%. The relative method is therefore to be preferred in most cases.

The physical and chemical form and geometrical shape of the standards to be used in neutron activation analysis need not be the same as for the samples. Dilute solutions are frequently used as standards. Sometimes such solutions are used after evaporation to dryness. For materials that do not contain elements with high absorption cross-section, the accuracy of the results is not critically dependent on the sample size, shape, and composition. The factor limiting the sample size is in many cases the flux gradients in the irradiation position. Typically samples of 0.1 - 1 g are used. The materials used for sample containers must become only moderately radioactive during the irradiation. For thermal neutron activation, foils or tubes of polyethylene, aluminium or quartz are used, depending on the samples involved and the irradiation time employed.

Once the radioactive atoms have been formed, chemical contamination is not likely to influence the results. The pre-irradiation treatment of samples is therefore normally kept to a minimum. In certain cases, however, a chemical separation prior to the activation may be convenient, or even necessary, in order to avoid interfering effects or increase the sensitivity.

2.2 Post-Irradiation Treatment

Considering the post-activation procedure, activation analysis can be conveniently divided into two different categories:

A. Instrumental activation analysis, in which case the irradiated samples are subjected to radioactivity measurements without any intermediate treatment. In this case, the sensitivity is very much dependent on the concentration of other elements in the samples under investigation.

B. Radiochemical activation analysis, which involves dissolution of the irradiated sample and removal of interfering activities by some sort of chemical separation. If a known amount of the element to be determined (carrier) is added, and a chemical exchange between radioactive and carrier atoms is brought about, the separation need not be quantitative, since the recovery of the carrier can be established after the activity measurements. The introduction of radiochemical separations makes the analysis more work-consuming, but will in many cases be necessary in order to attain very low concentration levels.

The separation techniques most frequently used in activation analysis for rare earth elements are listed in section 3.3.

2.3 Radioactivity Measurements and Data Handling

In earlier days of activation analysis, simple counting devices based on proportional or Geiger-Mueller counters for β-particles, or NaI(Tl) scintillation detectors for γ-rays, were employed. In recent years, γ-spectrometry systems based on Ge(Li) solid state detectors have been extensively used. The superior energy resolution of such systems compared to equipment previously available has greatly increased the potentialities of purely instrumental activation analysis. The introduction of Ge(Li) detectors has also to a considerable degree made it possible to simplify the chemistry that is necessary to optimize the analytical procedure; this is convincingly demonstrated in the case of the RE elements.

The interaction of a γ-ray with a Ge(Li) detector causes an electronic pulse, the amplitude of which is proportional to the energy absorbed. After proper amplification and shaping, the pulse is fed into an analogue-to-digital converter (ADC). The ADC generates a number which defines the channel in which the pulse is to be counted in the memory of the analyser. Normally a 4096-channel system will be sufficient to make full use of the excellent resolution of modern Ge(Li) systems, at least for most activation analysis applications. The output of data from the memory may be effected by means of a printer, a punched-tape or a magnetic-tape recorder. A system for Ge(Li) γ-spectrometry is shown in figure 1. In this system, a small digital computer is used both for the data accumulation and for calculations on the basis of the recorded γ-ray spectra.

A typical energy spectrum of a complex mixture of γ-emitters is presented in figure 2. The peaks shown in the spectrum correspond to full-energy absorption of characteristic γ-rays in the detector, which is mainly due to the photoelectric effect. Unfortunately a considerable fraction of the γ-rays will only be partially absorbed due to the Compton effect. This is the reason for the large continuum shown in the spectrum underlying the peaks. Another effect causing incomplete energy absorption is the pair production process. These effects can be much reduced by means of an anti-coincidence coupling of the Ge(Li) detector and a NaI(Tl) crystal surrounding the detector. A review of the present state of Ge(Li) γ-spectrometry systems has been given by Heath [3].

Data processing by means of computers has proved to be of great use in activation analysis when working with complex γ-spectra, for problems such as location and identification of peaks, resolving of complex peaks, peak area calculation according to various procedures, spectrum smoothing, decay curve analysis, etc. The various methods and applications of computer data reduction in activation analysis have been reviewed by Yule [4,5].

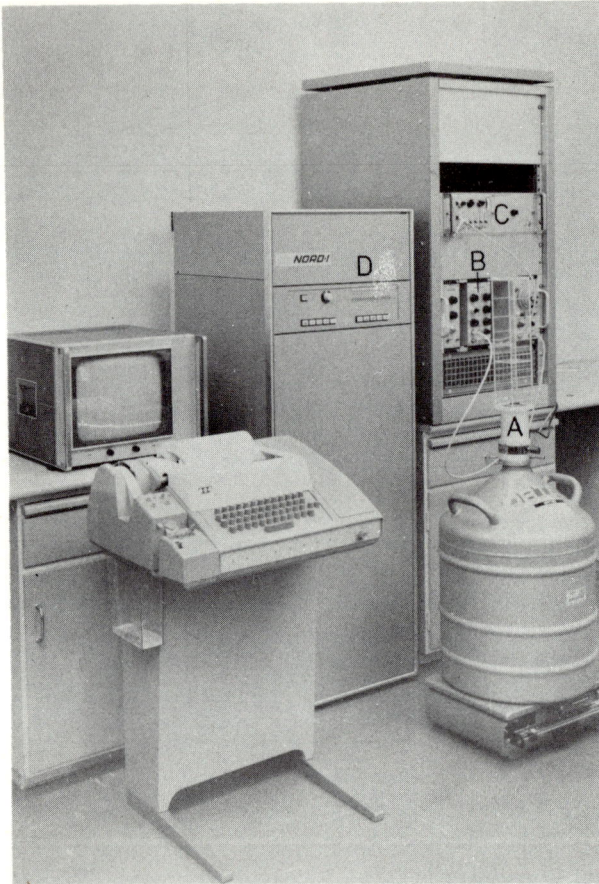

FIGURE 1

Equipment for gamma-ray spectrometry. A. Ge(Li) detector with cryo-
stat and preamplifier. B. Main amplifier. C. Analogue-to-digital
converter. D. Small digital computer for accumulation and processing
of data.

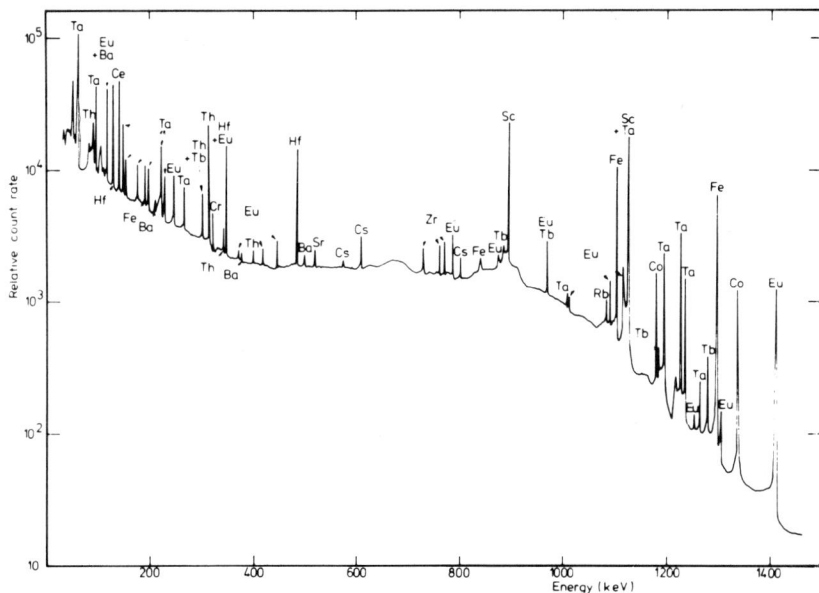

FIGURE 2

Gamma-ray spectrum of a neutron-activated silicate rock, recorded 5 weeks after the end of the irradiation with a Ge(Li) detector system. Peaks due to 15 different elements are clearly identified.

3. DETERMINATION OF REE BY THERMAL NEUTRON ACTIVATION ANALYSIS

Thermal neutrons are unique in activation analysis due to the high cross-sections for (n,γ) reactions induced in a great number of elements, which means that very high sensitivities can be obtained provided that a high neutron flux is available. With a medium-size research reactor, at least 50 elements can be determined in amounts of 1 nanogram or less, including most of the REE. Sources of thermal neutrons other than nuclear reactors are unimportant as far as the sensitivity aspect is concerned.

TABLE 1

Some Data for Thermal Activation Analysis of Rare Earth Elements

Element	Thermal neutron absorption cross-section[a] (barn)	Stable isotope	Abundance (%)	Thermal neutron activation cross-section[a] (barn)	Resonance activation integral[b] (barn)	Half-life of induced radioisotope	Detection limit[c] (μg)	Approximate activity induced in 100 mg of element, mCi (10^{13} n cm^{-2}s^{-1}, 24h)	Most useful γ-energies of radioisotope (keV)
La	8.9	^{139}La	99.9	9.6	11.8	40.2 h	0.0001	$3 \cdot 10^2$	487 , 1597
Ce	0.73	^{140}Ce ^{142}Ce	88.5 11.1	0.54 0.95	0.48 -	32.5 d 33.4 h	0.02	0.7 7	145 293
Pr	11.6	^{141}Pr	100	10.9	14.1	19.2 h	0.005	$7 \cdot 10^2$	1576 (3.7%)
Nd	48	^{146}Nd ^{148}Nd	17.2 5.7	1.3 2.5	3.2 $\sim$ 12	11.1 d 1.73 h	0.02	0.2 24	91 , 531 211
Sm	5820	^{152}Sm	26.7	210	2530	46.8 h	0.00003	$1.3 \cdot 10^3$	103
Eu	4400	^{151}Eu ^{151}Eu	47.8 47.8	3100 5700	$\sim$ 1700 -	9.3 h 12.4 y	0.000002	$6 \cdot 10^4$ 60	122 , 841 122 , 1408
Gd	49000	^{152}Gd ^{158}Gd	0.20 24.9	1100^b 3.5	3000 84	242 d 18.6 h	0.001	0.8 60	98 , 103 363
Tb	22	^{159}Tb	100	22	365	72.1 d	0.001	27	299 , 879
Dy	930	^{164}Dy ^{164}Dy	28.2 28.2	2200 2700	- $\sim$ 370	1.3 m 2.35 h	0.000002	$6 \cdot 10^4$	108 , 515 95 , 361
Ho	67	^{165}Ho	100	63	710	26.7 h	0.00002	$3 \cdot 10^3$	81
Er	160	^{166}Er ^{168}Er ^{170}Er	33.4 27.1 14.9	10 2.0 9	- - -	2.3 s 9.5 d 7.5 h	0.001	4 - $1.2 \cdot 10^2$	208 - 308

cont.

Table 1 (cont.)

Element	Thermal neutron absorption cross-section[a] (barn)	Stable isotope	Abundance (%)	Thermal neutron activation cross-section[a] (barn)	Resonance activation integral[b] (barn)	Half-life of induced radioisotope	Detection limit[c] (µg)	Approximate activity induced in 100 mg of element, mCi (10^{13} n cm^{-2} s^{-1}, 24h)	Most useful γ-energies of radioisotope (keV)
Tm	106	169Tm	100	106	1550	130 d	0.0005	70	84
Yb	37.5	168Yb	0.135	5500	~ 20000	32 d	0.0001	30	64 , 198
		174Yb	31.8	55	30	4.2 d		$2.7 \cdot 10^2$	396
Lu	108	175Lu	97.4	18	~ 450	3.68 h	0.00003	$3 \cdot 10^3$	88
		176Lu	2.6	2100	< 1000	6.7 d		$5 \cdot 10^2$	208
Y	1.3	89Y	100	1.3	~ 0.8	64.1 h	0.0005	50	

a. Data from Chart of the Nuclides, 3. Edition 1968, Der Bundesminister für Wissenschaftliche Forschung, Bonn.

b. Data from work performed in the author's laboratory.

c. Induced activity 40 disintegrations s^{-1}; neutron flux $1 \cdot 10^{13}$ n cm^{-2} s^{-1}; irradiation for 7 days or to saturation, whichever is shorter.

In table 1 some data of interest in activation analysis for REE are given. The values given for limits of detection are based on a minimum detectable activity of 40 dps. This seems to be a moderate limit; considerably lower activities can be measured with high-sensitivity/low-background counting systems. Furthermore, the figures are based on a flux of $1 \cdot 10^{13}$ n cm^{-2} s^{-1} and maximum seven days' irradiation. In many cases, the integrated neutron flux can be considerably increased without appreciable difficulties being introduced. Even with Ge(Li) detectors, which have comparatively low efficiencies, an activity of 40 dps should be within reach, provided that limited amounts of interfering activities are present.

In the following, some typical procedures used for REE determinations by thermal neutron activation are discussed.

3.1 Instrumental Activation Analysis

The feasibility of purely instrumental activation analysis is very much dependent on the composition of the sample. This is illustrated in figure 2 which shows a γ-spectrum of a silicate rock recorded with a Ge(Li) detector five weeks after the end of the irradiation. In this case, where the major elements do not contribute appreciably, the spectrum consists of about 65 clearly distinguishable peaks representing 15 different elements, most of which are present in trace amounts. If measurements of shorter-lived nuclides had been made on the same rock, a total number of about 25 elements, including the REE La, Ce, Sm, Eu, Tb, Dy, Yb, and Lu might have been determined [6,7]. In certain materials, such as graphite, high-purity aluminium, fresh water, etc., the determination of some RE is possible in the sub-ppm range by this technique.

Instrumental activation analysis may compete favourably with other instrumental analytical techniques also in certain cases where higher concentrations of REE are to be determined. In the author's laboratory, a large number of europium determinations (concentration range 10 - 100 ppm Eu) have been carried out in connection with a survey of Norwegian mineral resources. The time necessary for a simultaneous determination of Eu and Sm in one sample is at present about ten minutes.

3.2 Determination after Group Separation

The procedures used for the separation of the REE as a group from fering elements have in most cases been based on hydroxide-fluoride or hydroxide-oxalate precipitation cycles using carriers. Several methods for RE in silicate rocks based on group separation and subsequent Ge(Li) γ-spectrometry have been published [8-11]. By this means, all 14 RE may be determined [9], with a precision and accuracy of $\pm 5\%$ or better for 10 elements (La, Ce, Nd, Sm, Eu, Tb, Dy, Ho, Yb, Lu) [11]. Activation with epithermal neutrons followed by group separation seems to improve the conditions for the determination of Nd, Gd, Ho and Er [12]. Recent experiments indicate that yttrium can be determined in group-separated fractions by means of β-counting, using appropriate absorbers and making corrections for certain interfering RE nuclides, determined simultaneously by γ-spectrometry (E. Steinnes, unpublished work).

Application of methods based on group separation of REE and Ge(Li) γ-spectrometry has so far found limited use for non-geological samples. It seems quite obvious, however, that the methods developed for rocks may be used for a variety of other materials without major modification. We have employed a similar method for the determination of RE in human enamel and dentine at the ppb level (E. Steinnes and S. Dahm, to be published).

3.3 Separation of Individual REE

Before the advent of Ge(Li) detectors, methods based on group separation and subsequent chromatographic separations had to be used in order to obtain good data for many REE simultaneously. Separations based on elution from a cation exchange column with a suitable complexing agent have been most popular [13-16], but also other types of separations such as anion exchange in the presence of EDTA [17], anion exchange in mixed solvents [18, 19] and reversed phase partition chromatography [20] have been used. In most cases these quite tedious and time-consuming procedures can now be replaced by methods based on group separation, except for cases concerning the determination of traces of REE in RE matrices, and perhaps cases where extremely low concentrations are involved, such as in sea water samples [16].

3.4 Problems in Thermal Neutron Activation Analysis

Although neutron activation analysis is relatively free from matrix effects, there are certain difficulties which should not be overlooked:

a) If the matrix elements have high absorption cross-sections, errors may arise due to shielding effects. This type of errors can be avoided by irradiating a sufficiently dilute aqueous solution of the sample, or separating the elements having high absorption cross-section prior to the irradiation.

b) In some cases the samples become highly radioactive upon irradiation. If so is the case, the post-irradiation treatment has to be performed behind proper shielding, or sensitivity has to be sacrificed using shorter irradiation time or lower flux.

c) The nuclide of interest may be formed by competing nuclear reactions. In the light RE group, most nuclides that are usable for neutron activation determination, may also be formed as uranium fission products. This interference is especially serious for Ce and Nd, in which case the uranium interference will be about 10% at the same concentration of elements. In a work concerning RE impurities in metallic uranium [21] this problem was circumvented by removing U by cation exchange before the irradiation.

Another type of interfering nuclear reaction is second order interference, for example of the type

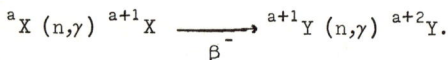

$$^{a}X \ (n,\gamma) \ ^{a+1}X \ \xrightarrow[\beta^-]{} \ ^{a+1}Y \ (n,\gamma) \ ^{a+2}Y.$$

Interferences of this kind have been calculated by Op de Beeck [22]. In cases where the nuclide ^{a}X has a large activation cross-section, appreciable difficulties are encountered if trace amounts of the element Y are to be determined _via_ the radioisotope ^{a+2}Y in a sample of element X. If, for example, a dysprosium sample is irradiated for 22 hours at a flux of $5 \cdot 10^{12}$

n cm^{-2}s^{-1} for the determination of holmium, the ^{166}Ho activity produced by
second order interference corresponds to 120 ppm Ho. Consequently this
type of interference can be serious in cases where an element is to be determined in a matrix of an adjacent element.

4. DETERMINATION OF REE BY 14 MeV NEUTRON ACTIVATION ANALYSIS

14 MeV neutrons are obtained from neutron generators based on the nuclear reaction

$$\underset{1}{\overset{2}{}} H + \underset{1}{\overset{3}{}} H \longrightarrow \underset{2}{\overset{4}{}} He + \underset{0}{\overset{1}{}} n$$

The nuclear reactions produced by 14 MeV neutrons are of the types (n,p),
(n,γ) and (n,2n) and have fairly low cross-sections compared to most (n,γ)
reactions. Furthermore, the available neutron fluxes are a factor of 10^3 -
10^6 lower than the thermal neutron fluxes available in most nuclear reactors.
This technique is therefore not very sensitive. Its justification is due to
the fact that rapid analyses based on short-lived nuclides can be made. The
elements that can be favourably determined include Y, Ce, Pr, Nd, Sm, Tb,
and Er [23,24]

5. DETERMINATION OF REE IN REE MATRICES

The foundation of activation analysis dates back to 1936 when Hevesy
and Levy [25] determined the dysprosium content of yttrium oxide using a
Ra-Be neutron source. After nuclear reactors became available, a number of
applications of activation analysis on REE matrices have been reported [1].
Trace determinations of REE in REE matrices by neutron activation is not
always a simple task, however, for the following reasons:

a) As is evident from table 1, a number of the REE has high thermal
    neutron absorption cross-sections, and some have also large reson-
    ance absorption integrals. Shielding effects are therefore likely
    to occur in many cases if precautions are not taken to avoid this.

b)   In many cases the neutron capture process results in formation
     of a radioactive nuclide, which means that the matrix may become
     extremely radioactive, as indicated in the table.

c)   Second order interfering reactions are likely to cause trouble
     in certain cases.

d)   The radiochemical separations which are in most cases necessary
     in order to separate the matrix element, are relatively compli-
     cated.

Instrumental activation analysis is really useful only in a limited
number of cases concerning trace determinations in REE matrices. Yttrium
is probably the most favourable matrix. Experience in the author's laboratory has shown that Dy and Eu may be determined in Y_2O_3 down to about 0.1
ppm, and La and Sm down to about 1 ppm by means of a Ge(Li) detector. Elements producing isotopes with long half-lives (Ce, Gd, Tb, Yb) may be determined down to very low levels in Y_2O_3, but this requires several weeks' de-
lay in order to obtain a sufficient reduction of the ^{90}Y activity. Okada
[26], using very short-lived isotopes, was able to determine Yb and Dy in
Tm_2O_3, and Er in Ho_2O_3 non-destructively.

If chromatographic separation techniques are introduced, the potentialities are considerably increased. Several authors have successfully employed a chromatographic method or a combination of different separation
systems for post-irradiation separation of trace impurities in RE oxides
[17,27,28]. The most promising methods, however, seem to be those based on
pre-irradiation separation of the matrix element. If this is done, the
difficulties pointed out in a) - c) above are avoided, and sensitivities
approaching those theoretically calculated can be obtained. Alimarin et
al. [21] used reduction of europium(III) with zinc and subsequent separation of the trivalent REE by extraction chromatography for the pre-irradiation separation of impurities in Eu_2O_3. Molnar and Lebedev [29] discussed
methods based on anion exchange in the nitrate/methanol system for the
separation of REE impurities from cerium sub-group matrix elements, which

should be applicable for pre-irradiation separation. In the case of chemical treatment before the activation, however, a blank problem is introduced which does not exist with post-irradiation procedures.

If neutron activation analysis is to be used for analysis of REE ores or concentrates, a group separation of the REE by solvent extraction [30] or precipitation [31] followed by activation of an appropriate dilute solution of the REE fraction [32] seems to be a reasonable procedure in order to avoid shielding effects due to excessive neutron absorption.

Finally, it should be mentioned that activation with high-energy photons seems to be a promising technique for yttrium, which is difficult to determine by thermal neutron activation. Determination of Y in various RE oxides in the ppm range has been shown to be feasible [33].

6. CONCLUDING REMARKS

In order to facilitate the comparison with other analytical techniques discussed at this meeting, an attempt is made here to summarize those advantages and drawbacks associated with neutron activation analysis which are the most significant ones in the author's opinion. The advantages include:

High sensitivity for a great number of elements.

Good specificity - both half-life and γ-energy may be used for identification.

Freedom from reagent contamination (provided that no pre-irradiation chemical separation is carried out).

Relative freedom from matrix effects.

Possibility of multi-element determination.

The most obvious drawbacks seem to be the following:

Nuclear reactors are not easily accessible to all laboratories.

The necessary equipment is relatively complex and expensive.

In some cases one has to work with considerable amounts of radioactivity.

It is sometimes necessary to wait for quite a long time before results are available.

For a more detailed and comprehensive treatment of activation analysis the reader is referred to one of the textbooks on this topic [34-36].

References

[1]   T. Bereznai, J. Radioanal. Chem. 9, 81 (1971).

[2]   L.A. Haskin and R.L. Korotev, this volume, p. 183.

[3]   R.L. Heath, Modern Trends in Activation Analysis, Nat. Bur. Stand. (U.S.) Spec. Publ. 312, Vol. II, 959 (1969).

[4]   H.P. Yule, ibid, vol. II, 1155 (1969).

[5]   H.P. Yule, in Activation Analysis in Geochemistry and Cosmochemistry, A.O. Brunfelt and E. Steinnes, eds., Universitetsforlaget, Oslo (1971), p. 145.

[6]   G.E. Gordon, K. Randle, G.G. Goles, J.B. Corliss, M.H. Beeson and S.S. Oxley, Geochim. Cosmochim. Acta 32, 369 (1968).

[7]   A.O. Brunfelt and E. Steinnes, Talanta 18, 1197 (1971).

[8]   K. Tomura, H. Higuchi, N. Miyaji, N. Onuma and H. Hamaguchi, Anal. Chim. Acta 41, 217 (1968).

[9]   F.M. Graber, H.R. Lukens and J.K. MacKenzie, J. Radioanal. Chem. 4, 229 (1970).

[10] S. Melsom, ibid., 4, 355 (1970).

[11] E.B. Denechaud, P.A. Helmke and L.A. Haskin, ibid., 6, 97 (1970).

[12] A.O. Brunfelt and E. Steinnes, ibid., in press.

[13] F.W. Cornish, AERE C/R 1224 (1956).

[14] A.W. Mosen, R.A. Schmitt and J. Vasilevskis, Anal. Chim. Acta 25, 10 (1961).

[15] L.A. Haskin, T.R. Wildeman and M.A. Haskin, J. Radioanal. Chem. 1, 337 (1968).

[16] O.T. Høgdahl, S. Melsom and V.T. Bowen, in Trace Inorganics in Water, R.F. Gould, ed. Advances in Chemistry Series No. 73 (American Chemical Society, Washington D.C., 1968).

[17] J. Minczewski and R. Dybczynski, Chem. Anal. (Warsaw) 10, 1113 (1965).

[18] H.B. Desai, R. Krishnamoorthy Iyer and M. Sankar Das, Talanta 11, 1249 (1964).

[19] A.O. Brunfelt and E. Steinnes, Analyst 94, 979 (1969).

[20] H. Grosse-Ruyken and J. Bosholm, Kernenergie 8, 224 (1965).

[21] I.P. Alimarin, A.Z. Miklishanskij and Yu.V. Yakolev, J. Radioanal. Chem. 4, 45 (1970).

[22] J.P. Op de Beeck, ibid., 4, 137 (1970).

[23] K.G. Broadhead, D.E. Shanks and H.H. Heady, Phys. Rev. 139, B 1525 (1965).

[24] I.N. Plaksin, L.P. Starchik and V.T. Tustanovskij, Zavod. Lab. 33, 1098 (1967).

[25] G. von Hevesy and H. Levy, Kgl. Danske Vidensk. Selsk. Math.-fys. Medd., 14, No. 5 (1936).

[26] M. Okada, Nature 188, 52 (1960); 191, 1090 (1961).

[27] L.P. Hunt, Thesis, Iowa State University, Report IS-T-120 (1967).

[28] I.P. Alimarin, Yu.V. Yakolev, A.Z. Miklishanskij, N.N. Dogadkin and O.V. Stepanets, J. Radioanal. Chem. 1, 139 (1968).

[29] F. Molar, N.A. Lebedev, ibid., 2, 91 (1969).

[30] F.M. Graber, H.R. Lukens and K. Heydorn, Trans. Am. Nucl. Soc. 9, 87 (1966).

[31] D.L. Massart and J. Hoste, Anal. Chim. Acta 42, 7 (1968).

[32] F. van den Bergh, F. Adams and J. Hoste, J. Radioanal. Chem. 4, 347 (1970).

[33] G.J. Lutz and P.D. La Fleur, Talanta 16, 1457 (1969).

[34] H.J.M. Bowen and D. Gibbons, Radioactivation Analysis, Clarendon Press, Oxford (1963).

[35] P. Kruger, Principles of Activation Analysis, John Wiley & Sons, New York (1971).

[36] J. Hoste, J. Op de Beeck, R. Gijbels, F. Adams, P. van den Winkel, D. de Soete, Instrumental and Radiochemical Activation Analysis, Butterworths, London (1971).

DETERMINATION OF RARE EARTHS
IN GEOLOGICAL SAMPLES AND RAW MATERIALS

by

Larry A. Haskin and Randy L. Korotev

Department of Chemistry, University of Wisconsin, Madison, 53704, U.S.A.

ABSTRACT

The relative abundances of the rare earth elements(REE) in rocks and minerals are related to those of the chondritic meteorites, which are believed to have preserved the REE with the same relative abundances as were present in the primitive matter from which the planetary bodies in the solar system developed. The relationships among relative REE abundances in different materials are most easily seen through the use of comparison diagrams with chondritic meteorites or with more immediately related materials. On such diagrams, differences in relative REE abundances between two materials appear, except for Ce and Eu, as smooth functions of REE atomic number. This is a consequence of the type of chemical separations that accompany rock-forming processes.

Accurate and sensitive geochemical analysis is now done mostly by neutron activation analysis or mass spectrometric isotope dilution, sometimes on nanogram and subnanogram quantities of the individual REE. Precision for isotope dilution is a fraction of a percent, that for neutron activation ranges from about ± 1% to about ± 10%, depending on the element. Useful data are also obtained by spark-source mass spectrometry and X-ray fluorescence, with precision ranging upwards from about ±10%.

Because all rocks and minerals apparently give smooth curves on comparison diagrams with chondritic meteorites, very accurate concentrations for unanalysed REE can be obtained by interpolation on comparison diagrams drawn through ratios for carefully analysed REE. Reasonably good estimates for most of the REE can be obtained from estimated comparison diagrams based on analyses for as few as 4 of the REE.

1. INTRODUCTION

From a geochemical point of view, the rare earth elements (REE) are an especially valuable group. Not only is the geochemical behaviour of the REE intrinsically interesting, but REE concentrations and relative abundances are becoming a useful tool for understanding rock-forming processes. Our present understanding and use of these elements as a geochemical tool have depended on improved techniques for analysis of trace quantities of REE in natural materials. The most accurate data have been obtained by mass spectro metric isotope dilution and by neutron activation analysis. Data of useful quality have been obtained by spark-source mass spectrometry and X-ray fluorescence.

Information stemming from studies of REE in natural materials can be applied to the analysis of REE ores. An outline of REE geochemistry is given below. This is followed by a discussion of the techniques used for analyses of natural materials for REE, with emphasis on those aspects of the techniques that are helpful in analysing complex materials of variable bulk composition. The application of correlations found among the REE in natural materials to analysis of REE ores by conventional methods is then considered.

2. OUTLINE OF REE GEOCHEMISTRY

The chemical similarity of the REE (lanthanides and yttrium) to each other is the characteristic that makes this series of elements geochemically unique. Where one member of the REE group is found in nature, all other members are present too. That in itself is no profound observation; essentially any natural material contains at least a little of every chemical element. The unique quality of the REE is that their relative abundances in natural materials can be rather simply related to those believed to be representative of the primitive solar system. Except for Eu, Ce, and Y, variations of relative REE abundances in natural materials are smooth functions of REE atomic number. Support for this contention will be developed below, or can be found in considerable detail in review articles [1,2,3,4].

The chondritic meteorites are samples of solar matter believed to retain the chemical elements, except for those that are volatile, in approximately their primordial relative abundances. (For a review of meteorites, see ref.[5].) Although this most abundant class of meteorites can be divided into numerous subclasses on the basis of mineralogic and chemical data, compared with other types of meteorites, with terrestrial rocks, or even with any given type of terrestrial rock, it is chemically quite homogeneous. The relative abundances of the non-volatile elements in these meteorites match those in the atmosphere of the sun, within experimental error (e.g., [6]). Variations in REE concentrations in chondritic meteorites range over a factor of 3, but relative REE abundances are much less variable [1,7-9].

The concentrations of the REE as determined by neutron activation analysis on a composite of 9 chondritic meteorites are given in table 1 [10, 11]. These abundances are plotted against REE atomic number in figure 1.

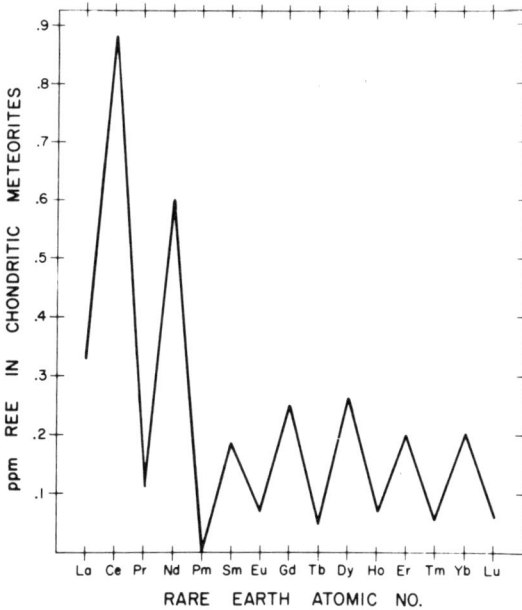

FIGURE 1

The concentrations of the REE in parts-per-million in a composite sample of 9 chondritic meteorites [11] have been plotted against rare earth atomic number.

TABLE 1

Concentrations of REE in parts-per-million in some meteorites, rocks, and minerals*

	A	B	C	D	E	F	G	H	I	J
La	0.33 ± .013	32	2.6	12.8 ± .2	13.0	-	320	17,600	5,990	1.1
Ce	0.88 ± .01	73	5.0	36.4 ± .8	34.5	2,220	990	23,700	11,000	3
Pr	0.112 ± .005	7.9	-	-	-	-	-	-	-	1.7
Nd	0.60 ± .01	33	-	22.0 ± 1	21.9	1,220	870	6,200	4,400	5.4
Sm	0.181 ± .006	5.7	1.06	6.50 ± .06	6.56	235	280	660	980	4.1
Eu	0.069 ± .001	1.24	3.18	1.23 ± .03	1.21	20.8	64	168	87	0.3
Gd	0.249 ± .011	5.2	2.7	8.5 ± .2	8.59	218	320	460	840	6.8
Tb	0.047 ± .001	0.85	0.22	1.62 ± .08	-	-	48	52	170	1.5
Dy	0.317 ± .005	-	1.52	11.1 ± .4	10.5	168	250	220	1,100	14
Ho	0.072 ± .010	1.04	-	2.1 ± .2	-	-	43	36	240	3.4
Er	0.200 ± .005	3.4	-	-	6.51	90.6	101	64	710	16
Tm	0.030 ± .002	0.50	-	-	-	-	-	-	-	2.8
Yb	0.200 ± .007	3.1	0.86	6.1 ± .3	6.00	73.9	52	33	750	33
Lu	0.034 ± .002	0.48	0.113	0.89 ± .01	-	11.8	6.3	4	102	7
Y	1.96 ± .09	27	-	-	-	-	-	-	-	-

* Data of column E determined by mass spectrometric isotope dilution, of column J by X-ray fluorescence spectrometry, of others by neutron activation analysis.

A. Composite of 9 chondritic meteorites, Haskin et al. [11].
B. Composite of North American shales, Haskin et al. [9].
C. 2.9 mg feldspar, lunar breccia 14063, Haskin et al. [38].
D. Lunar basalt 14053, Helmke et al. [40].
E. Lunar basalt 14052, Hubbard et al. [54].
F. Apatite, Torihama, Japan, Nagasawa (1970).
G. Apatite, Skaergaard intrusion, Paster et al. [39].
H. Dysanalyte, Magnet Cover, Arkansas, M. Anderson, analyst.
I. Eudialyte, Kalundborg, Greenland, M. Anderson, analyst.
J. Euxenite, Romteland, Norway, relative values only, Semenov and Barinskii [66].

The outstanding feature of that graph is the odd-even abundance relationship, the classic example of the Oddo-Harkins rule.

Supposedly, the starting material from which the Earth formed had this same set of relative REE abundances, if not also approximately the same REE concentrations as the chondritic meteorites. The concentrations of the REE in materials found at the Earth's surface are much higher than those found in chondritic meteorites, so much so that approximately 75% of the Earth's supply of La would have to reside in the 0.45% of the mass of the Earth that makes up its crust. Such enrichment would require incredibly effici-ent extraction of the REE from the bulk of the Earth's matter.

Of all the REE, La is the most enriched at the Earth's surface, as compared with its abundance in the chondritic meteorites. The relative abundances of the REE as well as the approximate level of their average concentration at the Earth's surface are represented by the values for a composite of North American shales (table 1). On a graph of ppm REE versus atomic number, the most striking feature of the data for the shales would be the odd-even abundance effect. Thus, in order to compare the abundances of the REE in the shales with those in the chondrites, it is convenient to use the following procedure [12-14]. The ratios obtained by dividing the concentration of each of the REE in the shales by that of the corresponding REE in the chondritic meteorites are plotted against REE atomic number. In order that a constant factor between ratios be represented by a fixed linear vertical distance, the ratios are plotted on logarithmic ordinate. If the ratios define a horizontal line, then the relative REE abundances are the same in the two materials being compared. Figure 2 is a comparison dia-gram for the composite of shales and the chondrites.

From figure 2 it is seen that the relative abundances of the elements Ho-Lu are approximately the same as those for the chondrites. There is a smooth increase in relative abundances for the elements Tb to La, except for Eu, whose abundance is only 0.67 times the interpolated value. The concentrations range from about 15 times those of the chondrites for Ho-Lu up to nearly 100 times that of the chondrites for La.

REE abundances in igneous rocks, precursors of sedimentary rocks such as the shales, are highly variable. It is possible to determine and dis-

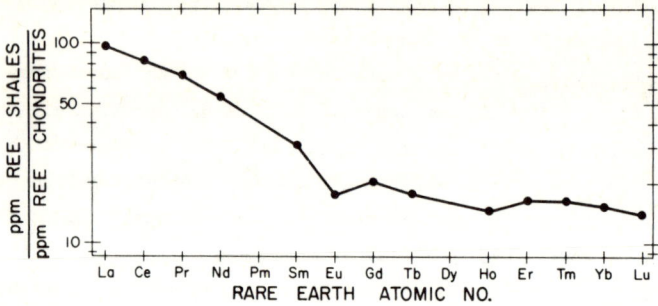

FIGURE 2

Comparison diagram for the composite sample of North American shales [9]
and chondritic meteorites. The ratios of concentration for each of the
REE in the shales to that of the same REE in the chondrites have been plotted
on a logarithmic ordinate against REE atomic number on a linear abscissa.

cuss with some confidence the average relative REE abundances at the
Earth's surface only because the bulk of the REE is distributed among
common types of igneous rocks and sediments and because the processes of
formation of sedimentary rocks tend to average away the more variable dis-
tributions of the igneous rocks. Thus, composites of sediments from North
America, from Europe, and from Russia all have, within experimental uncertainties, the same relative REE abundances [15,16]. Individual samples of
sediments of abundant types have relative REE abundances that are similar
to the averages. Sedimentary processes do cause some differentiation of
the REE, as has been well documented by Balashov et al. [17], but the extent of these separations is relatively small in comparison with the effects
of averaging.

Just why the processes of sedimentation do not more strongly affect
relative REE abundances probably reflects the insolubility of the REE.
Høgdahl et al. [18] and Goldberg et al. [19] have shown that the concentrations of the lanthanides in ocean waters are in the range of 0.1 to 3 nanograms per liter. Thus, the entire amount of REE dissolved in the ocean
barely equals that in the uppermost 0.5 millimeters of sediment lining the

ocean floor [20]. The concentrations of the REE in natural waters are so
low that they are probably controlled by adsorption onto particles such as
colloidal, hydrated iron oxides rather than by the solubilities of their
compounds.

What processes are responsible for the generally much higher concentrations of the REE at the Earth's surface than in the chondrites and for
the differences in the relative REE abundances between these two sources?
The differences quite possibly result from the zoning of the primitive
Earth into its present core, mantle, and crust. Some evidence supporting
this suggestion is considered below. Alternatively, the separations
within the REE group and the concentration of the REE as a group may have
preceded accretion of the Earth. No satisfactory mechanism for this has
been proposed.

Figure 2 shows the average relative REE abundances for the Earth's
surface as determined from sedimentary rocks. Presumably, the same average
would be obtained by thorough sampling of surface igneous rocks. Granite
and basalt are two types of rock that represent a major fraction of the
Earth's crust. The relative abundances for composites of a large number of
granites and of basalts from many sites on the North American continent do
show essentially the same relative REE abundances as the shales, although
composites of granites from a single formation show the unique characteristics of that formation [9]. Suites of basalts from a given area, e.g.
Hawaii, show an orderly increase in the extent of enrichment of the lighter
REE over the heavier REE as basalts that are considered on petrologic
grounds to be relatively primitive grade into more differentiated types
(e.g. [21]). The relative enrichment, to some degree, of lighter REE is
a general property of basalts from continents and oceanic islands (e.g.
[22,23]) (figure 3).

Basalts with REE abundances unlike those found in continental and
oceanic basalts and which suggest that the interior of the Earth may still
retain REE with relative abundances very similar to those of the chondrites
are found to be the principal lavas extruded from the submarine oceanic
ridges and rises [23-25] (figure 3). The abundances of the REE in these
basalts are depleted in La, Ce, and Pr relative to chondrites. The ocean

ridges are those sites from which fresh material from the Earth's mantle presumably comes to the surface; the unusual REE abundances in the rocks have been considered extensively by Kay et al. [25] and Schilling [26].

FIGURE 3

Comparison diagram for an andestitic trachybasalt (BCR-1 [31]), a tholeiitic submarine basalt (KD11 [25]), and a high-temperature, al-pine type peridotite (VH-2D, Helmke and Medaris, unpublished data).

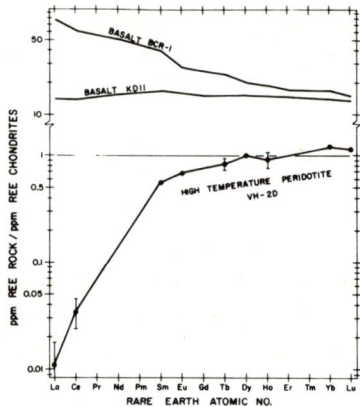

If the Earth as a whole has relative REE abundances like those in chon-drites and if the lighter REE are relatively more abundant at the Earth's surface than the heavier REE,there should exist regions of the Earth's mantle that are relatively depleted in the lighter REE. Considerations of tempera-ture, pressure, and composition suggest that the upper mantle should be com-posed of peridotite rocks [27]. Most peridotites analysed so far have re-lative REE abundances that are enriched, compared with chondrites, in the lighter REE [28]. Only one class of peridotite, the high-temperature alpine type, has REE abundances that are strongly ·depleted in the lightest REE (figure 3). No direct sampling of the Earth's mantle has yet been achieved, however, and it remains an open question whether any of the accessible bodies of peridotite can be considered samples of mantle material essentially unaltered during their transport to the Earth's surface.

The processes of igneous rock formation that can partially separate the REE from each other, or concentrate the REE in one type of rock relative to another, and that have been investigated are partial melting and fractional crystallization (e.g. [29-31].) These processes are described in models on the basis of distribution coefficients for each REE, defined as the ratio of the concentration of that REE in one silicate phase (mineral or melt) to its concentration in another silicate phase that is in equilibrium with the first. The principal minerals that make up common terrestrial rocks are capable of incorporating surprisingly high concentrations of REE but are unable to compete successfully against silicate liquids for them [32,33]. Thus, under most circumstances of formation of silicate rocks, the REE tend to remain concentrated in silicate liquids. Values of distribution coefficients for REE among common minerals have been estimated from a variety of natural and experimental systems (e.g [34-38]). Typical values for distribution coefficients of REE, defined as the ratios of concentration of each REE in the mineral phase to that in the equilibrium silicate liquid, are shown in figure 4. Crystallization of a single mineral, for example, clinopyroxene, from a silicate liquid results in a solid and in a residual liquid both with different relative REE abundances from the parent. Similarly, the concentrations and relative abundances of the REE in silicate liquids produced by partial melting depend on the mineralogy of the residual solid from which the melt was removed. Note the selective concentration of Eu by feldspar.

Because the distribution coefficients for the REE in most rock-forming minerals tend to be low, crystallization of one or more such minerals from a silicate liquid does not appreciably affect the relative REE abundances in the residual liquid until a large fraction of the material has solidified. Nevertheless, careful analysis has made it possible to observe genetic relationships among rocks from a given locality and to identify fractional crystallization as the mechanism that has produced the different, but

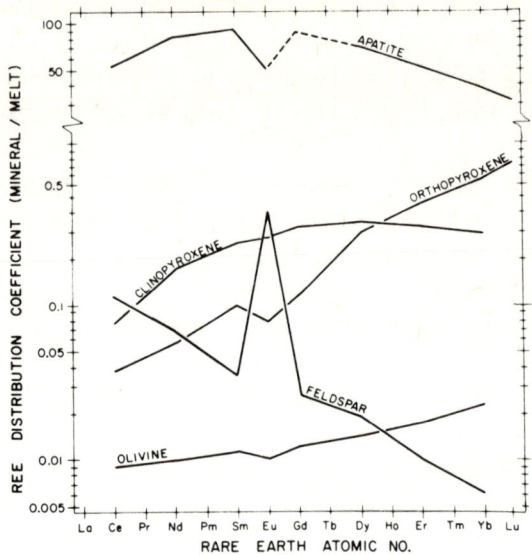

FIGURE 4

Distribution coefficients
for the REE as estimated
from natural phenocryst-host
matrix pairs. Values for
apatite are from the paper
by Nagasawa [35], the rest
are from the paper by
Schnetzler and Philpotts
[34].

related rocks. In the case of the Steens Mountain basalts, a suite of 70
successive lavas from a single volcano, the concentrations of the individual
REE varied by less than a factor of 3 at the extremes [38]. Nevertheless,
8 separate groups of lavas could be recognized and classified according to
whether they were essentially undifferentiated parent liquid , residual
liquid from crystallization of plagioclase, or of clinopyroxene, or both,
or mixtures of accumulated crystals of mineral plus residual liquid. The
REE concentrations for the parent liquid, for a lava rich in accumulated
feldspar crystals, and for a residual liquid after extensive crystallization of feldspar are shown in figure 5.

The models used to describe fractional crystallization can be applied
to large bodies of silicate material (e.g. [39]) or to individual silicate
rocks (e.g. [40]). If a mass of silicate liquid is slowly cooled and
allowed to solidify, the earliest crystals formed take up REE according to
the values of their distribution coefficients and the concentrations of the
REE in the trapped liquid at the time they formed. In the simplest case,

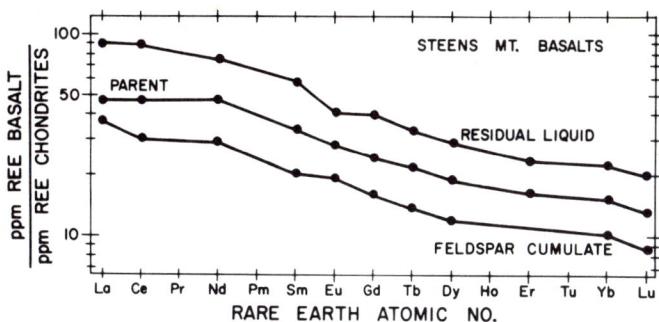

FIGURE 5

Comparison diagram for the REE from genetically related basalts from Steens
Mountain. The middle curve is the average of the "group E" basalts which are
believed to have the same composition as the parent liquids for all the
basalts. The upper curve (basalt no. 50) represents residual liquid after
fractional crystallization of the parent liquid. The lowest curve (basalt no.
13) consists of a mixture of plagioclase crystals and a trapped liquid.

as a crystal grows in size, its interior is effectively shielded from further
interaction with the surrounding liquid phase. The liquid, thus, becomes
increasingly enriched in the REE until, because of changes in bulk composition, it begins crystallizing REE-concentrating minerals such as apatite or
minerals such as monazite in which the REE are essential constituents. At
the same time, the earlier crystals may continue to grow wherever they inter-
face with the last portions of the liquid phase so that, despite the relatively low values of their REE distribution coefficients, they take up very high
concentrations of the REE in their outermost layers. Results for an extreme
case, where no accessory minerals are allowed to form, are shown in figure 6.
Note that the ordinate gives the average concentration for the entire crystal,
not that for the last increment formed, as a function of the fraction of the
original melt that has solidified along the abscissa.

Just as the crystal structure of a major, rock-forming mineral makes
that mineral selective in its uptake of REE, so are the structures of the
REE minerals themselves selective. REE concentrations for several REE

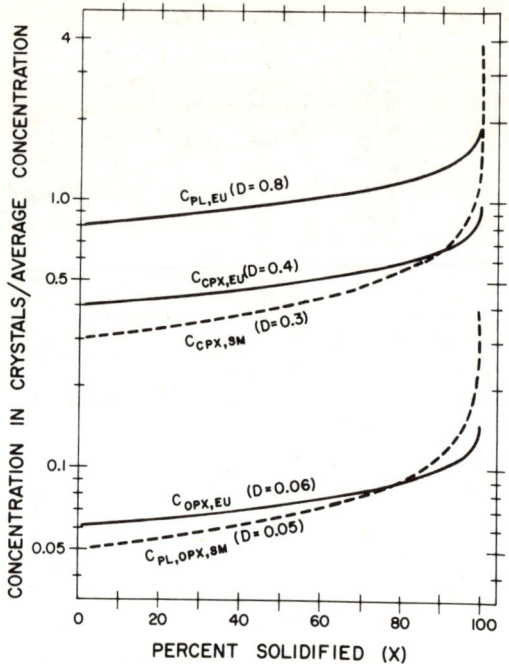

FIGURE 6

The changes in average concentrations of Sm and Eu for growing crystals of feldspar (PL), clinopyroxene (CPX), and orthopyroxene (OPX) are shown as functions of the percent of the parent liquid that has solidified. Taken from Helmke et al. [38].

minerals are compared with those of chondrites in figures 7 and 8. Note the anomalously low concentrations of Eu in several of the minerals.

From a geochemical point of view, Eu is one of the most fascinating members of the REE group. The only mechanism for its selective separation from the rest of the REE for which there is direct evidence is its incorporation into feldspar, presumably as a +2 ion (e.g. [41]). The ratio of the activity of Eu^{3+} to that of Eu^{2+} in a silicate liquid, however, depends on (among other things) the oxygen fugacity of the melt. Thus, separations leaving Eu relatively depleted in various minerals resulting from a low ratio of activity of Eu^{3+} to that of Eu^{2+} have been considered (e.g. [42, 31]).

The moon appears to constitute a system in which the oxygen fugacity

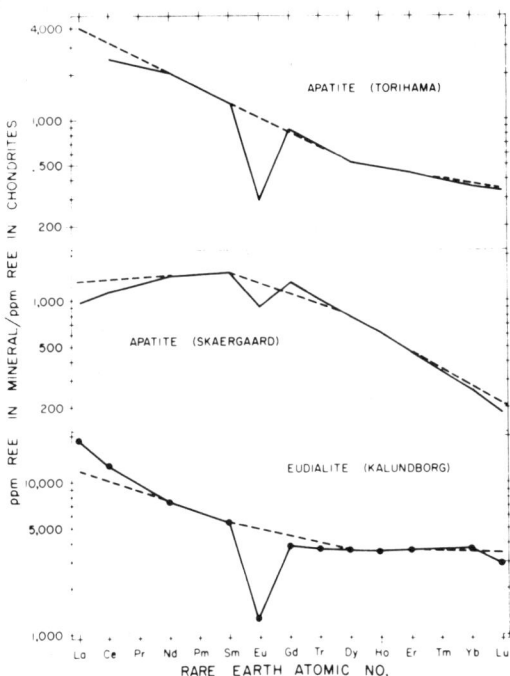

FIGURE 7

Comparison diagrams for
REE in apatites (Nagasawa,
1960; Paster et al. [39])
and eudialite (M.Anderson,
analyst). Dashed lines are
estimates based on hypothetical analysis for Nd, Sm,
Dy, and Er by absorption
spectrophotometry.

is very low. A comparison diagram for lunar basalts is shown as figure 9.
The large relative deficiency in Eu is the most striking feature of the REE
abundances in the basalt. Otherwise, the relative REE abundances are not
strongly differentiated from those in chondrites. The extent to which the
severe depletion of certain lunar basalts in Eu is a consequence of low
oxygen fugacity, as decoupled from selective separation into feldspar, is
uncertain. The lunar highlands are known to be very rich in feldspar, and
a sample of massive anorthosite (almost pure feldspar) shows a very strong
concentration in Eu relative to the other REE [43]. Possibly, the missing
Eu is in the feldspars of the lunar highlands.

The early crust of the Earth had features that may have developed in
ways somewhat analogous to the lunar highlands. Large bodies of Precambrian

FIGURE 8

Comparison diagram for
Dysanalyte (M.Anderson,
analyst) and Euxenite
(Semenov and Barinskii
[66].) Dashed lines are
estimates based on hypo-
thetical analysis for
Nd, Sm, Dy, and Er by
absorption spectrophoto-
metry.

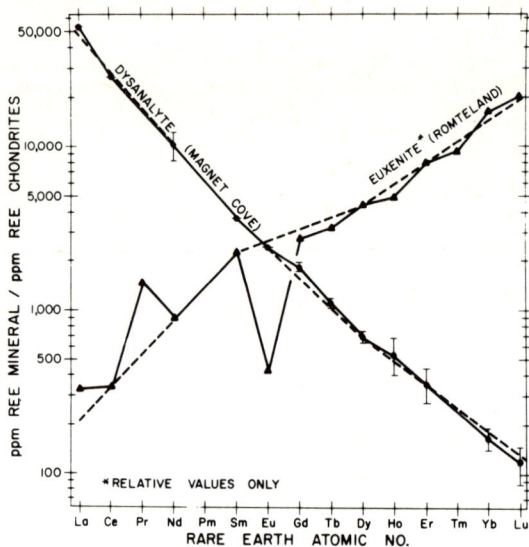

FIGURE 9

Comparison diagram for
two lunar basalts (Helmke
et al.[40]) and a lunar
anorthosite (Hubbard et
al.[43]).

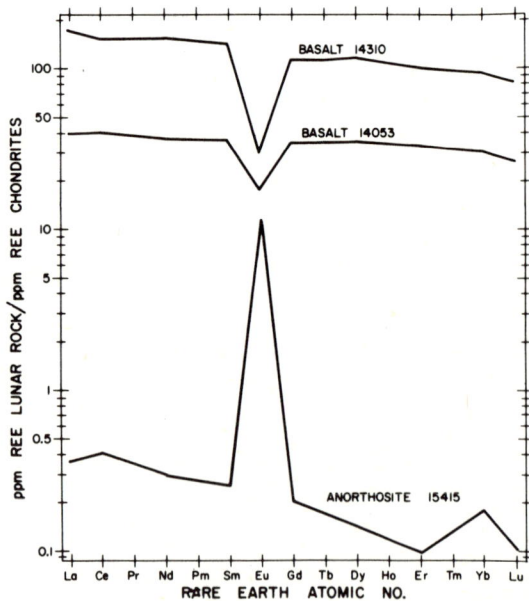

anorthosite are known, and are anomalously enriched in Eu [44,45]. Younger
geologic bodies òf this type are not found. An intriguing and possibly
related phenomenon is a decrease in the relative abundance of Eu in terrestrial sediments during the past 3.5 billion years from an average value at
least as large as the interpolated value on a comparison diagram with chondrites to the value representative of sediments of Paleozoic age and younger
(e.g. those in figure 2) which is equal to 0.67 times the interpolated value
[46].

3. REE ANALYSIS OF GEOLOGIC MATERIALS

Correlations among REE in related series of rocks such as the Steens
Mountain basalts [38] cannot be observed without precise, accurate analysis.
Correlations between Eu and Sr in lunar rocks have been observed because
the analyses for those elements were precise to a fraction of one percent.
(P.W. Gast, private communication). Thus, over the past few years, precision and accuracy of analysis for REE have been emphasized in several geochemical laboratories.

The beginning of precise, accurate analysis for REE at low concentrations in rocks and minerals came with the use of neutron activation analysis
(e.g. [47,10]). These procedures relied on flow-type proportional detectors
for beta radiation and NaI(Tl) detectors for gamma rays. The individual REE
had to be separated from each other as well as from all other elements.
With the advent of Ge(Li) detectors, it became necessary only to separate
the REE as a group away from the other elements in the sample, as enough
lines from each REE could be resolved from the combined spectra for satisfactory analysis of all elements except Pr (e.g. [48,49]). Several of the
REE (La, Ce, Sm, Eu, Dy, Yb, Lu) can be determined with moderate accuracy
in common types of rocks without any chemical separation following activation (e.g. [50,51]).

The particular advantages of neutron activation analysis for the analysis of rocks and minerals centre around the following characteristics of
the method. (A detailed discussion of the method as applied to such materials has been given by Haskin and Ziege [52]).

The method is very sensitive. As an example, consider the analysis (table 1) of 2.9 mg of fragments of feldspar from lunar breccia 14063 [38]. Multiplying the concentrations in table 1 by the sample weight shows that the quantities analysed range from 14 nanograms for Ce to 0.33 nanograms for Lu. Uncertainties range from ±15% for Ce and Yb down to ± 2% for Sm and ± 1% for Eu. The samples were irradiated for 30 hours at a neutron flux of about $1 \cdot 10^{13}$ $n/cm^2/sec$.

If necessary, the present sensitivity could be increased by using higher reactor fluxes and chemical procedures that separate the individual REE, in conjunction with proportional counters which have much higher detection efficiencies than the Ge(Li) detectors used to obtain the data on the feldspar fragments. In fact, however, even the present sensitivity probably exceeds any meaningful control on sampling.

The method is relatively free from matrix effects. The principal requirement is that the matrix of the sample containing the REE does not absorb any significant fraction (less than 0.1%) of the neutrons that attempt to pass through it. Most common rocks and minerals fulfil this criterion. Also, aqueous solution standards, easily prepared to high accuracy, can be used with a wide variety of rock types.

Interferences are few and well-defined. Provided that adequate chemical separations are made, nearly all interferences are caused by nuclear reactions such as (fast n,α), (fast n,p), and (n,fission) that produce the nuclide being assayed from elements of higher atomic number than the (n,γ) parent. Interferences for the REE have been listed by Allen et al. [53].

The activation with neutrons is done before any chemical separations are made. After the activation, accidental addition of some of the element to be analysed does not interfere with the accuracy of the analysis. Thus, the real sensitivity approaches the theoretical, based on the capabilities of the reactor and detectors and the properties of the nuclides. Blank levels are essentially zero. This makes neutron activation the most sensitive technique currently used in geochemical analysis for REE.

Chemical separations, when made, need not be quantitative. Provided that carrier is added for each element analysed, chemical yields for the separation can be determined and corrections made for losses. Thus, the uncertainties associated with quantitative chemical recovery are obviated.

Disadvantages of neutron activation analysis include the specialized and rather expensive equipment required, the amount of time needed for a single analysis, and the minor annoyance of having to work with radioactivity. Samples with high concentrations of REE, including REE minerals, must be dissolved and diluted prior to activation because several of the REE (especially Dy and Gd) have very high neutron cross-sections that make such samples opaque to neutrons.

The accuracy of neutron activation analysis for REE is difficult to demonstrate in any absolute way. It is necessary to analyse materials of known composition. There are no other methods of higher proven accuracy against whose values for a particular sample data obtained by neutron activation analysis can be judged. Thus, the method must be tested on materials made up to known composition, and such materials are of no better quality than the standards made up for use in the activation analysis itself. Thus, estimates of accuracy depend on the precision of the analysis and estimates of accuracy of the standards. In careful work, the precision of the analysis is controlled by counting statistics down to a level of about ± 1%, which is a practical lower useful limit for most geochemical analyses.

Confidence in the method is enhanced by agreement of results with values obtained by other procedures. A sample of lunar rock 14053 was ground to < 200 mesh and split between two laboratories. Results of the analyses for REE from the two laboratories, one using neutron activation [40] and the other mass spectrometric isotope dilution [54] are included in table 1. The agreement is quite good except for Ce.   The present level of accuracy for determination of REE in ordinary rocks and minerals is about ± 10% for Er and Tm, and better than ± 5% for the rest.

Details of procedures for assay of REE in geologic materials by mass spectrometric isotope dilution, the other method that has, so far, provided accurate REE data on rocks and minerals, are given by Schnetzler et al. [55] and by Gast et al. [56]. This method also has the advantage that there are no matrix effects and relatively few, well-defined interferences. Spike compositions can be calibrated against solutions made up to known REE concentrations; as in the case of neutron activation analysis, this limits the accuracy to about 0.5%. Precision is excellent, usually about ± 0.1%.

Chemical separations of the REE from the bulk material are necessary, but need not be quantitative. Some of the REE are better analysed as ions, some as ionized oxides; Ba, when present, interferes with La. Thus, the REE are usually split chemically into two or three groups before they are loaded onto the filaments.

Disadvantages include the following. Only elements with more than one naturally-occurring isotope can be readily analysed. This eliminates only Pr, Tb, Ho, and Tm. The mass spectrometers currently used for isotope dilution measurements on REE are highly specialized, automated, and computer controlled, which makes them expensive to purchase and maintain. The most severe problem is that chemical separations must be done **prior** to analysis, with the consequent hazards of sample contamination. Sample preparation must be done under very clean conditions. Sample blanks for most REE presently range from about 0.1 to 10 nanograms, depending on the element. For a given REE, the blank cannot be relied on to remain constant to better than a factor of 2, so that the quantity of REE from the sample must exceed that of the blank by more than an order of magnitude for accurate results.

The accuracy of mass spectrometric isotope dilution for REE in geologic samples is, at present, about ± 1 to 2%, the same as that of neutron activation analysis for the elements that can be done the best by that technique. Relative REE abundances can be obtained to a precision of within a fraction of a percent by mass spectrometric isotope dilution.

Spark source mass spectrometry for REE has been done with best accuracy so far by Taylor [57], who obtains a precision of about ± 10%. The spark source mass spectrometer simultaneously analyses for very many chemical elements. For highly accurate REE analysis it cannot, at the present stage of its technology, compete with the two above methods. Sample preparation requires no chemical separations, just grinding and mixing of the sample with briquetting graphite and Tm_2O_3 as an internal standard. Thus, blank levels can be kept low. Sensitivity is good. Matrix effects appear to be only a minor problem. Interferences result from multiple ionization, formation of polyatomic ions, amd hydrocarbons. A spark source mass spectrometer is an expensive instrument to buy and to maintain.

More rock and mineral samples for which analyses of REE have been reported in the literature have been analysed by X-ray fluorescence than by
any other method, many by the method of Turanskaya [58]. Data of useful
quality are obtained when concentrations of REE are high enough that clean,
quantitative separations can be made on reasonable amounts of sample. The
accuracy for most samples reported in the literature and obtained by this
method does not match that of the other methods discussed above (e.g. [59,
60].)

4. ESTIMATES OF REE ABUNDANCES IN MINERALS FROM PARTIAL ANALYSES
Concentrations of REE determined carefully in duplicate by accurate,
sensitive methods in natural rocks and minerals have all produced smooth
curves (except for the points for Ce and Eu) on comparison diagrams with
chondritic meteorites. This results from the nature of the partial separations within the REE group that accompany processes of formation of rocks.
Thus, the quality of an analysis for the REE in a rock or mineral can be
judged, in part, by plotting and examining a comparison diagram. Ratios for
elements other than Ce and Eu that do not fall on the smooth curve indicate
incorrect values for those elements. Provided that the position of the
curve is well defined, accurate values for those elements can be obtained by
multiplying the interpolated or extrapolated value taken from the curve by
the concentration of that same element in chondritic meteorites. Similarly,
values can be obtained for elements for which no data were taken. It is
risky to interpolate values for Ce, more so for Eu, because these elements
can be selectively separated from the other REE by natural processes involving changes in oxidation state. Because they can be separated readily from
the other REE through changes in oxidation state in the laboratory as well,
their individual analysis in ores or minerals can be readily done [61].
Extrapolations to obtain values for La are somewhat risky because that element is somewhat variable in its behaviour relative to the rest of the REE
group and because the value for Ce may not really lie on the curve. If
values for Y are to be estimated, account must be taken of the somewhat
variable behaviour of that element in comparison with any particularly heavy

REE. Ratios for Y usually match those of Dy or Er, but may in extreme cases match those of any element between Gd and Lu, and possibly be different from all of these.

Ratios are not needed for all the REE in order to establish the general shape of the curve on a comparison diagram. Thus, if the concentrations of the REE are to be determined in a REE ore or mineral, it may be possible to analyse accurately for a few of the REE, then plot a comparison diagram and obtain data of satisfactory accuracy for the rest, or at least for a number of them. A relatively inexpensive and standard laboratory instrument might suffice to provide such a REE analysis. (Observation of sympathetic variations among REE in minerals is not new and is no doubt already used in mineral analyses, e.g. [62].)

One method of analysis commonly available is spectrophotometry. Absorption lines for individual REE tend to be sharply defined (except for La and Lu, for which there are none). Absorption spectra for the REE are given, for example, by Carnall and Fields [63] and Stewart and Kato [64]. From the graphs provided in those references, relative molar absorptivities and extents of mutual interference among the REE were estimated. Those REE that can be most easily analysed in the presence of the other REE in the natural mixtures of those elements in the minerals shown in figures 7 and 8 are Nd, Sm, Dy, and Er. By using those relative molar absorptivities, it would appear to be possible to obtain values for those four elements to an accuracy of a few percent by spectrophotometry of a solution of pure REE from each of those minerals. An approximate comparison diagram is then plotted, using the measured values for those four elements, plus linear interpolations between adjacent ratios, and linear extrapolations from each pair, Nd-Sm and Dy-Er. The results of such interpolations and extrapolations are shown as the dashed lines in figures 7 and 8.

The procedure for obtaining the values for Nd, Sm, Dy, and Er from the spectrophotometric data depends on corrections for mutual interferences among the REE. In describing this procedure, it is presumed that the instrument has already been appropriately calibrated for absorptivity versus concentration for each of those four elements.

The peak for Dy at 911 nanometers appears to be free from any significant interferences. Thus the peak height for Dy can be used without correction to give the concentrations for that element in the solution and, therefore, in the mineral. The peak for Nd at 580 nanometers has as its only significant interference a small contribution from Pr. Since these two elements are of adjacent atomic number and, consequently, their ratio in natural materials does not vary greatly, it is a satisfactory approximation (at worst, an error of about 3%) to multiply the observed peak height for Nd by 0.96, and obtain the value for the concentration of Nd from the result.

The peak for Sm at 402 nanometers has interferences from Dy and Er. The potential interference from Eu is expected to be unimportant because in all minerals with high concentrations of REE for which analyses have been reported so far the ratio of concentration of Sm to that of Eu exceeds 2.5. The relative molar absorptivities of Sm and Dy at 402 nanometers are approximately in the ratio 27:1 and of Sm and Er approximately 67:1. Assuming a roughly constant value for the ratio of concentrations of Dy to Er, we have constructed a curve (figure 10) of the ratios of concentrations of Sm to

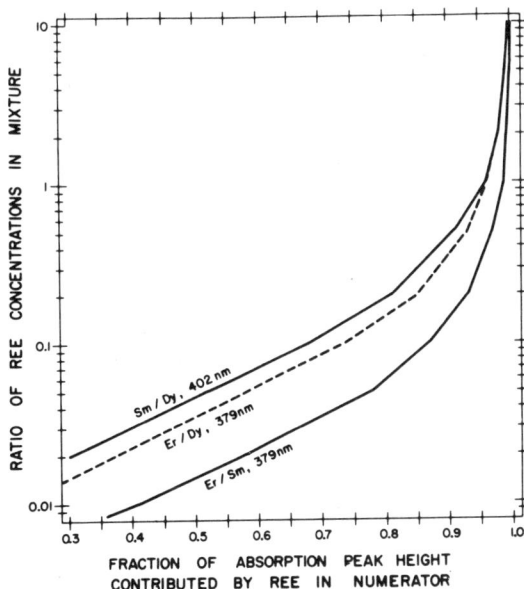

FIGURE 10

Correction factors used in hypothetical spectrophotometric determination of REE in minerals. See text for detailed explanation.

those of Dy against the fraction of the height of the 402 nanometer peak that is contributed by Sm. To use this graph, first obtain the approximate value for Sm by using the height of the 402 nanometer peak in the mixture. An approximate ratio of the concentration of Sm to that of Dy is then calculated, and the approximate fraction of the height of the 402 nanometer peak that is contributed by Sm is read off the graph. From the corrected height for the 402 nanometer peak, obtained by multiplying the fraction taken from the graph times the observed peak height, a better approximation to the actual concentration of Sm is obtained. This new value for the concentration of Sm can then be used to obtain a new ratio of concentrations of Sm to Dy, which in turn gives a better value for the fraction of the peak contributed by Sm, which leads to a better value for the concentration of Sm, and so on. Only a single cycle of calculation was needed to obtain an accurate value for the concentration of Sm for the minerals in figures 7 and 8.

The peak for Er at 379 nanometers has interferences from Dy, Sm, and Eu. That for Eu does not introduce an error exceeding 5% provided the ratio of concentration of Eu to that of Er does not exceed a value of 2, which it seldom does. Included in figure 10 are curves for the ratios of concentrations of Er to Dy and Er to Sm versus the fraction of the peak at 379 nanometers that is contributed by Er. The corrected value for Er is obtained in the same manner as was described for Sm, except that each cycle of calculation includes a correction for two interfering elements instead of only one.

The accuracy of the values obtained for the remaining REE from the estimated comparison diagrams depends on the accuracy with which the concentrations of the elements Nd, Sm, Dy, and Er were obtained and on the extent to which the interpolations and extrapolations based on the ratios for those elements match the shape of the actual comparison diagram. The uncertainty of the spectrophotometric determinations should not exceed a few percent. In the examples studied here, the smallest contribution of any element to the peak used for its measurement exceeded 75%. This kind of uncertainty is ignored in the tabulation that follows.

The uncertainties arising from the interpolations and extrapolations on the comparison diagrams are compiled in table 2, where the ratios of estimated concentrations to the actual concentrations are given. The poor-

TABLE 2

Ratios of estimated REE concentrations to measured REE concentrations

	F[*]	G	G	I
La	1.42	1.36	0.87	0.64
Ce	1.25	1.22	0.96	0.79
Pr[x]	1.12	1.08	0.99	1.19
Nd	(1.0)	(1.0)	(1.0)	(1.0)
Sm	(1.0)	(1.0)	(1.0)	(1.0)
Eu	3.4	1.40	0.98	3.8
Gd	0.96	0.83	0.85	1.29
Tb	0.96	0.91	0.92	1.08
Dy	(1.0)	(1.0)	(1.0)	(1.0)
Ho	1.00	1.00	0.93	1.01
Er	(1.0)	(1.0)	(1.0)	(1.0)
Tm[x]	1.00	1.03	1.04	0.96
Yb	0.98	1.06	1.09	0.93
Lu	0.98	1.14	1.10	1.16

[*] See table 1 for letter designations of columns

[x] "Measured" REE concentrations estimated from accurate comparison diagrams

est estimates are for Eu, as expected, because there is no way to guess the concentration of an element that behaves anomalously. The values obtained from the estimate serve only as a useful upper limit to the probable concentrations of Eu. The estimates for La are the second poorest but, despite the extrapolation, none of the values for that element is in error by as much as 50% of the correct value. No value for Ce was in error by as much as ±25% of the correct value, but there was no anomalous value for that element in any of the cases examined. Of the 14 REE, then, this approach provided concentrations accurate to ±10% for 10 elements in the Torihama apatite, for 9 in the Skaergaard apatite, for 8 in the eudialyte, for 12 in the dysanalyte. Only for Eu was any estimate in error by more than 50%. Considering that values for concentrations of 14 elements are obtained from

actual measurements on only 4 elements, the method looks surprisingly good.
We emphasize, however, that these estimates are based on our treatment of
literature data for spectrophotometry, not on experimental work. A proper
study of this method would entail calibration of a spectrophotometer, determination of extents of mutual interferences with that instrument under the
actual conditions of measurement, construction of curves like those in
figure 10 from the resulting data, and testing on carefully analysed samples.
The spectrophotometric determinations could be expanded to include some
additional elements, according to whether the minerals to be analysed are
rich in light or heavy REE.

A similar approach can be taken using flame emission spectroscopy instead of spectrophotometry as outlined above. Hingle, Kirkbright and West
[65] have tabulated the extents of interference for a number of REE by
emission lines from other REE. Based on their tabulation, which takes into
account only the most intense peaks, it appears that data of sufficient
accuracy to provide an estimated comparison diagram could be obtained for
Nd, Sm, Dy, Er, and Yb in the mixture of REE obtained from a REE mineral.
The 421.172, 400.797, and 398,798 nanometer lines for Dy, Er, and Yb, respectively, appear to be free of any serious interferences from other REE.
The 492.453 nanometer line for Nd has an interference from Sm that is as
large as 20% when the concentrations of those elements are equal; in natural minerals the concentration of Sm is nearly always less than that of Nd.
That line also suffers an interference from Pr that amounts to 9% when the
concentrations of the two elements are the same; the contribution from Pr
would not exceed 3% on any of the minerals under consideration here. Also,
because Nd is the only significant interference with the 495.936 nanometer
line for Pr, both elements could be determined and the mutual corrections
made. Data for Pr, however, will not significantly enhance the accuray of
the estimated comparison diagram.

The 488.397 nanometer of Sm has an interference from Nd amounting to
about 16% when the concentrations of the two elements are the same. Since
in most minerals the concentration of Nd exceeds that of Sm, in the case of
dysanalyte by a factor of 10, only for certain minerals such as euxenite
could concentrations of reasonable accuracy for Sm be obtained.

Procedures for correcting line intensities for the various interferences and using the estimated comparison diagrams would be the same as outlined above for the spectrophotometric method. A careful study of REE emission lines might reveal wavelengths more suitable than those listed above for this particular application.

Another instrument found in many analytical laboratories is the X-ray fluorescence spectrometer which, as mentioned previously [58,60] has been used regularly for analysis of REE in ores and minerals. With most X-ray fluorescence spectrometers L X-rays are used for analysis of the REE, especially for the heavier REE for which the K electron binding energies are approximately 60 keV. In a sample containing REE separated from a mineral, most of the L_α lines for the REE La-Sm are free from interferences, whereas the L_α line for a heavy REE has an interference from an L_β or L_γ line from some lighter REE. With careful calibration corrections can be applied and data for all REE obtained. In natural materials, accuracy is better for elements of even atomic number because of the odd-even abundance effect. Comparison diagrams could be used to assure that accurate data have been obtained for all elements, or to provide accurate concentrations for the less abundant, odd atomic numbered elements in cases where their abundances were too low to yield satisfactory results.

X-ray fluorescence, atomic emission spectroscopy, and absorption spectrophotometry have been selected for discussion because they require relatively inexpensive instruments that are commonly found in analytical laboratories. The purpose of this discussion has been to demonstrate how use can be made of the distant genetic relationships between relative REE abundances in ores and minerals and those in chondritic meteorites to test the accuracy of analytical data and to provide estimates of concentrations of REE that cannot be readily determined from data for a few that can. No attempt has been made to provide a thorough description of any analytical method. Each of the methods considered is the principal topic of a separate paper in this volume.

Acknowledgements

The REE analyses of the dysanalyte and eudialyte were supported in part by the National Science Foundation through grant GA-25626.

References

[1] L.A. Haskin, F.A. Frey, R.A. Schmitt and R.H. Smith, Physics and Chemistry of the Earth, vol. 7, p. 167-321, L.H. Ahrens, ed., Pergamon, New York, 1966.

[2] L.A. Haskin and F.A. Frey, Science 152, 299-314 (1966).

[3] A.G. Herrmann, Handbook of Geochemistry, 39, 57-71, K.H. Wedepohl, ed., Springer-Verlag, Berlin, 1970.

[4] B. Jensen and A.O. Brunfelt, Norsk Geol. Tidsskr. 45, 249-283 (1965).

[5] B. Mason, Meteorites, Wiley and Sons, Inc., New York, 1962.

[6] L.H. Aller, The Abundances of the Elements, Interscience Publishers, New York, 1961.

[7] R.A. Schmitt, R.H. Smith, J.E. Lasch, A.W. Mosen, D.A. Olehy and J. Vasilevskis, Geochim. Cosmochim. Acta, 27, 577-622 (1963).

[8] R.A. Schmitt, R.H. Smith and D.A. Olehy, Geochim. Cosmochim. Acta, 67, 67-86 (1964).

[9] L.A. Haskin, M.A. Haskin, F.A. Frey and T.R. Wildeman, Origin and Distribution of the Elements, pp. 889-912, L.H. Ahrens, ed., Pergamon, 1968.

[10] L.A. Haskin, T.R. Wildeman and M.A. Haskin, J. Radioanal. Chem. 1, 337-348 (1968).

[11] L.A. Haskin, P.A. Helmke, T.P. Paster and R.O. Allen, Activation Analysis in Geochemistry and Cosmochemistry, pp. 201-218, A.O. Brunfelt and E. Steinnes, eds., Universitetsforlaget, Oslo (1971).

[12] C.D. Coryell, J.W. Chase and J.W. Winchester, J. Geophys. Res. 68, 559-566, (1963).

[13] A. Masuda, J. Earth Sci., Nagoya Univ. 10, 173-187 (1962).

[14] L.A. Haskin and M.A. Gehl, J. Geophys. Res. 67, 2537-2541 (1962).

[15] L.A. Haskin and M.A. Haskin, Science 154, 507-509 (1966).

[16] L.A. Haskin, R.R. Wildeman, F.A. Frey, K.A. Collins, C.R. Keedy and M.A. Haskin, J. Geophys. Res. 71, 6091-6105 (1966).

[17] Yu.A. Balashov, A.B. Ronov, A.A. Midgisov and N.V. Turanskaya, Geochem. Intern. no. 5, 951-969 (1964).

[18] O. Høgdahl, S. Melsom and V.T. Bowen, Advances in Chemistry No. 73, American Chemical Society, pp. 308-325 (1968).

[19] E.D. Goldberg, M. Koide, R.A. Schmitt and R.H. Smith, J. Geophys. Res. 68, 4209-4217 (1963).

[20] T.R. Wildeman and L.A. Haskin, J. Geophys. Res. 70, 2905-2910 (1965).

[21] J.G. Schilling and J.W. Winchester, Contr. Mineral. and Petrol. 23, 27-37 (1969).

[22] A.G. Herrmann, Contr. Mineral. and Petrol. 17, 275-314 (1968).

[23] F.A. Frey, M.A. Haskin, J.A. Poetz and L.A. Haskin, J. Geophys. Res. 73, 6085-6098 (1968).

[24] F.A. Frey and L.A. Haskin, J. Geophys. Res. 69, 775-780 (1964).

[25] R. Kay, N.J. Hubbard and P.W. Gast, J. Geophys. Res. 75, 1585-1613 (1970).

[26] J.G. Schilling, Phil. Trans. Roy. Soc. Lond. A268, 663-706 (1971).

[27] A.E. Ringwood, Advances in Earth Science, pp. 357-399, P.M. Hurley, ed. MIT Press, Cambridge, 1966.

[28] F.A. Frey, L.A. Haskin and M.A. Haskin, J. Geophys. Res. 76, 2057-2070 (1971).

[29] J.G. Schilling and J.W. Winchester, Mantles of the Earth and Terrestrial Planets, pp. 267-283, S.K. Runcorn, ed., Interscience, New York 1967.

[30] P.W. Gast, Geochim. Cosmochim. Acta 32, 1057-1086 (1968).

[31] L.A. Haskin, R.O. Allen, P.A. Helmke, T.P. Paster, M.R. Anderson, R.L. Korotev and K.A. Zweifel, Proc. Apollo 11 Lunar Science Conf. Geochim. Cosmochim. Acta Suppl. No. 1, vol. 2, pp. 1213-1231, A.A. Levinson, ed., 1970.

[32] R.L. Cullers, L.G. Medaris and L.A. Haskin, Science 169, 580-583 (1969).

[33] R.L. Cullers, L.G. Medaris and L.A. Haskin, Geochim. Cosmochim. Acta, in press (1972).

[34] C.C. Schnetzler and J.A. Philpotts, Geochim. Cosmochim. Acta 34, 331-340 (1970).

[35] H. Nagasawa and C.C. Schnetzler, Geochim. Cosmochim. Acta 35, 953-968 (1971).

[36] A. Masuda and I. Kushiro, Contr. Mineral. Petrol. 26, 42-49 (1970).

[37] F.A. Frey, Geochim. Cosmochim. Acta 33, 1429-1447 (1969).

[38] P.A. Helmke and L.A. Haskin, Geochim. Cosmochim. Acta, in press (1972).

[39] T.P. Paster, D.S. Schauwecker and L.A. Haskin, A trace element study of the Skaergaard layered series, to be published (1972).

[40] P.A. Helmke, L.A. Haskin, R.L. Korotev and K.E. Ziege, Proc. Third Lunar Sci. Conf., Geochim. Cosmochim. Acta Suppl. 3, vol. 2, in press, 1972.

[41] D.G. Towell, J.W. Winchester and R.V. Spirn, J. Geophys. Res. 70, 3485-3496 (1965).

[42] J.A. Philpotts, Earth Planet. Sci. Lett. 9, 257-268 (1970).

[43] N.J. Hubbard, P.W. Gast, C. Meyer, L.E. Nyquist and C. Shih, Earth Planet. Sci. Lett. 13, 71-75 (1971).

[44] J.A. Philpotts, C.C. Schnetzler and H.H. Thomas, Nature 212, 805-806 (1966).

[45] T.H. Green, A.O. Brunfelt and K.S. Heier, Earth Planet. Sci. Lett. 7, 93-98 (1969).

[46] T.R. Wildeman and L.A. Haskin, Geochim. Cosmochim. Acta, in press (1972).

[47] A.W. Mosen, R.A. Schmitt and J. Vasilevskis, Anal. Chim. Acta 25, 10-24 (1961).

[48] K. Tomura, H. Higuchi, N. Miyaji, N. Onuma and H. Hamaguchi, Anal. Chim. Acta 41, 217-228 (1968).

[49] E.B. Denechaud, P.A. Helmke and L.A. Haskin, J. Radioanal. Chem. 6, 97-113 (1970).

[50] J.C. Cobb, Anal. Chem. 39, 127-131 (1967).

[51] G.E. Gordon, K. Randle, G.G. Goles, J.B. Corliss, M.H. Beeson and S.S. Oxley, Geochim. Cosmochim. Acta 32, 369-396 (1968).

[52] L.A. Haskin and K.E. Ziege, Instrumental Methods for Analysis of Soils and Plant Tissue, pp. 185-208, L.M. Walsh, ed., Soil Science Society of America, Inc., Madison, Wis., 1971.

[53] R.O. Allen, L.A. Haskin, M.L. Anderson and O. Muller, J. Radioanal. Chem. 6, 115-137 (1970).

[54] N.J. Hubbard, P.W. Gast, M. Rhodes and H. Wiesmann, Revised Abstracts, Third Lunar Science Conference, Abstract no. 407, C. Watkins, ed., Lunar Science Institute, Houston, Texas, 1972.

[55] C.C. Schnetzler, H.H. Thomas and J.A. Philpotts, Anal. Chem. 39, 1888-1890 (1967).

[56] P.W. Gast, N.J. Hubbard and H. Wiesmann, Proc. Apollo 11 Lunar Sci. Conf., Geochim. Cosmochim. Acta Suppl. 1, vol. 2, pp. 1143-1163, A.A. Levinson, ed., Pergamon, New York, 1970.

[57] S.R. Taylor, Geochim. Cosmochim. Acta. 29, 1243-1261 (1965).

[58] N.V. Turanskaya, Kand. dissertatsiya, GEOKhl Akad. Nauk SSSR, Moscow, 1958.

[59] A.G. Herrmann and K.H. Wedepohl, Zeit. Anal. Chem. 225, 1-13 (1967).

[60] H.J. Rose and F. Cuttitta, Appl. Spec. 22, 426 (1968).

[61] M.M. Woyski and R.E. Harris, Treatise on Analytical Chemistry, vol. 8, pp. 1-146, I.M. Koltoff and P.J. Elving, eds., Interscience, New York, 1963.

[62] K.J. Murata, H.J. Rose, M.K. Carron and J.J. Glass, Geochim. Cosmochim. Acta 11, 141-161 (1957).

[63] W.T. Carnall and P.R. Fields, Lanthanide/Actinide Chemistry, Advances in Chemistry Series, pp. 86-101, R.F. Gould, ed., American Chemical Society, Washington, 1967.

[64] D.C. Stewart and D. Kato, Anal. Chem. 30, 164-172 (1958).

[65] D.N. Hingle, G.F. Kirkbright and T.S. West, Analyst 94, 864-870 (1969).

[66] E.I. Semenov and R.L. Barinskii, Geochemistry (English translations) no. 4, 398-419 (1958).

THEORETICAL BASIS FOR THE USE OF RARE EARTHS

IN OPTICAL AND LUMINESCENT MATERIALS

by

G. Blasse

Solid State Chemistry Department, Physical Laboratory,

University of Utrecht, Utrecht, The Netherlands

ABSTRACT

In this lecture the energy level scheme of rare earth ions and possible optical transitions therein will be discussed. Broad band as well as sharp peak transitions are included. Their relevancy to luminescent materials is indicated. Further we will deal shortly with the phenomena of energy transfer and concentration quenching.

1. INTRODUCTION

Let us first consider the possible optical processes that may occur in a crystal containing rare earth ions (see figure 1). We assume that the crystal itself absorbs no radiation. In the crystal we have drawn two centres, marked A and S. A is the rare earth ion, S may be another ion. If the rare earth ion absorbs radiation it is raised to an excited state. It returns to the ground state by giving up the excitation energy as radiation (luminescence) or heat. It is also possible to excite the rare earth ion indirectly by exciting the centre S. This excitation may be followed by radiation from S, by dissipation of the excitation energy as heat or by energy transfer from S to A. It is evident that a knowledge of the energy level scheme of the ions is necessary to understand these phenomena. This is done in the first section of this paper. In the second we deal with optical transitions within these sections. The process of energy transfer will be dealth with in the third section. The last section contains a few remarks on the radiationless processes (heat dissipation).

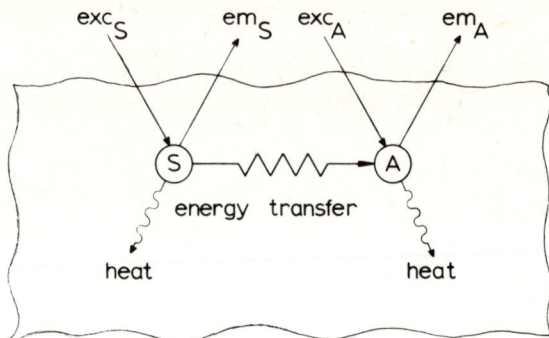

FIGURE 1

Luminescence processes. Exc means excitation
(absorption of radiation); em means emission

2. THE ENERGY LEVEL DIAGRAM OF RARE EARTH IONS

The characteristic properties of the RE ions are attributable to the presence in the ion of a deep-lying 4f shell. Electrons in this shell are screened from the environment by the outer electrons and give, therefore, rise to a number of discrete energy levels which closely resemble those of the free ion. The 4f shell may contain 14 electrons. The energy level diagrams of $Ce^{3+}(4f^1)$, $Eu^{2+}(4f^6)$, $Gd^{3+}(4f^7)$ and $Tb^{3+}(4f^8)$ are given in figure 2 as an example. The Ce^{3+} ion has only one 4f electron which gives rise to two levels ($2F_{7/2}$ and $2F_{5/2}$). As the number of electrons increases, there is in general a rapid increase in the number of possible states. This will be discussed in more detail in our paper on magnetism of the RE.

In addition to the discrete 4f levels there are other levels present. They are schematically indicated as hatched levels. These levels depend strongly on the lattice in which the RE ion is incorporated. They fall into two groups.

In the first group one of the 4f electrons is raised to the higher 5d

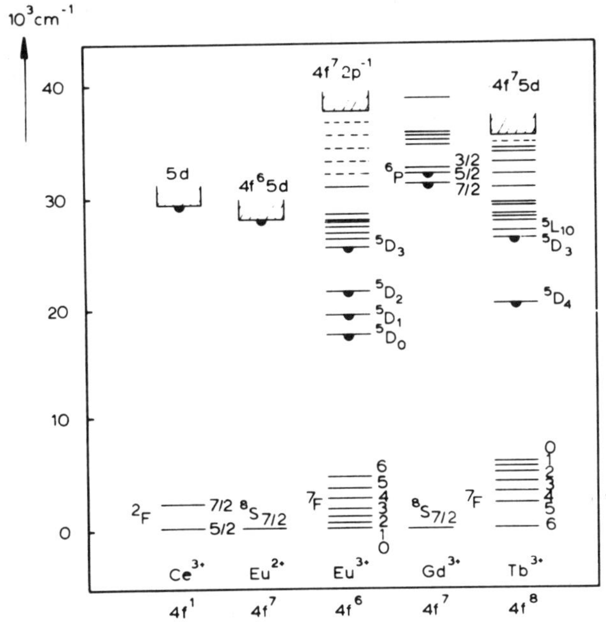

10^3 cm^{-1}

40

$4f^7 2p^{-1}$ $4f^7 5d$

30

5d $4f^6 5d$ ^6P — 3/2, 5/2, 7/2 $^5L_{10}$, $^5D_3$

$^5D_3$

20

$^5D_2$ $^5D_1$ $^5D_0$ $^5D_4$

10

^2F — 7/2 $^8S_{7/2}$ — 5/2 ^7F 6 5 4 3 2 1 0 $^8S_{7/2}$ ^7F 0 1 2 3 4 5 6

0

Ce^{3+} Eu^{2+} Eu^{3+} Gd^{3+} Tb^{3+}

$4f^1$ $4f^7$ $4f^6$ $4f^7$ $4f^8$

FIGURE 2

Energy-level diagram of some RE ions. Horizontal lines indicate narrow 4f levels. Dashed lines present levels which are not well known. Hatched levels correspond to charge-transfer or $4f^{n-1}5d$ states. Levels tabelled with half-circles have been observed in luminescence.

level: $4f^n \rightarrow 4f^{n-1} 5d$. The 5d orbit lies at the surface of the ion and is therefore strongly influenced by the lattice. In the second group one of the electrons of the surrounding anion of the lattice jumps into the 4f orbit of the central RE ion. This is called a charged-transfer transition. The question whether the lowest band corresponds to a $4f^{n-1} 5d$ state or a charge-transfer state is easily answered if one reminds the fact that completely or half-filled shells are very stable. We give two examples. In the case of Tb^{3+} ($4f^8$, half-filled plus one) the 4f shell readily releases an electron and the transition $4f^8 \rightarrow 4f^7 5d$ takes place at low energy. In the case of Eu^{3+} ($4f^6$, half-filled less one) the 4f shell readily accepts an electron and the charge-transfer state has a low energy.

3. OPTICAL TRANSITIONS AND SELECTION RULES

Transitions from the ground state to the charge-transfer state or the $4f^{n-1}5d$ state are allowed for the emission and absorption of electric dipole radiation.

The 5d state is influenced strongly by the surroundings of the ion and splits into several components depending on the site symmetry of the RE ion. Some special cases for Ce^{3+} have been drawn in figure 3. As a consequence the $4f^{n-1}5d$ state consists of several components leading to a number of absorption bands. Table 1 gives some examples for Ce^{3+} in different lattices.

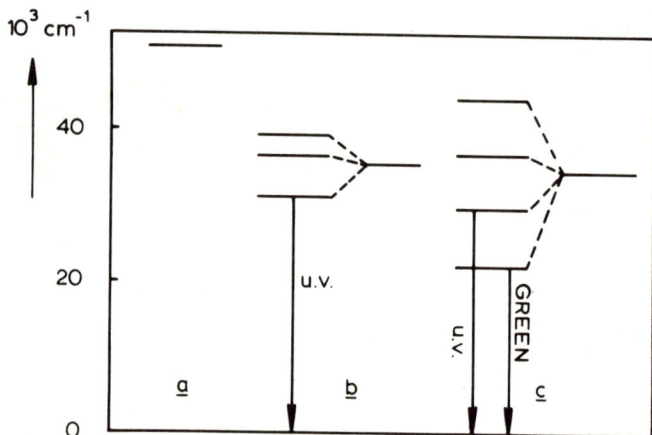

FIGURE 3

Energy-level scheme of Ce^{3+}. a. free ion; b. in $YAl_3B_4O_{12}$ (with centre of gravity); c. in $Y_3Al_5O_{12}$ (with centre of gravity). Cases b and c show the crystal-field splitting of the 5d level. The emission has also been drawn.

TABLE 1

Position of crystal-field levels of the 4f-5d transition of Ce^{3+} ion
in several host lattices (from absorption and excitation spectra
data in 10^3 cm^{-1})

$(Y,Ce)BO_3$	27.4	29	40.8	43.5
$(Y,Ce)_3Al_5O_{12}$	22.0	29.4	37	44
$(Y,Ce)PO_4$	32.8	34.2	39.6	

If an ion is excited into its $4f^{n-1}5d$ state it may return to the ground
state in two different ways. In the case of Ce^{3+} and Eu^{2+} luminescence
occurs from the $4f^{n-1}5d$ state (see figure 2). In the case of Tb^{3+} excitation into the $4f^75d$ state is followed by non-radiative transitions to the
$4f^8$ configuration in which also luminescence occurs but of a different type.

Figure 4 gives the emission of Ce^{3+} in several host lattices. The
spectral position is strongly influenced by the host lattice, because the
position of the lowest 5d component from which the luminescence occurs is
also influenced strongly by the lattice (fig. 3). It is further noteworthy

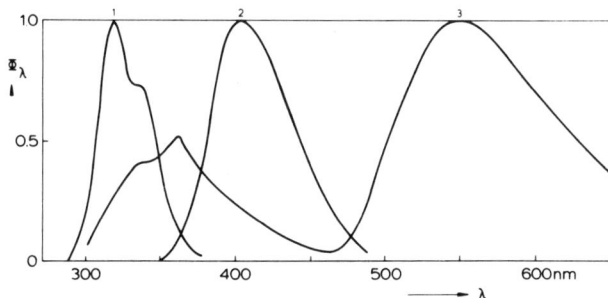

FIGURE 4

Emission spectrum of Ce^{3+} in
1. YPO_4, 2. $Ca_2Al_2SiO_7$, 3. $Y_3Al_5O_{12}$. 254 nm excitation.

that Ce^{3+} in $Y_3Al_5O_{12}$ shows two emission bands. These are ascribed to luminescence from the lowest and the one but lowest component of the 5d state (fig. 3). Note further that the emission bands are split due to the splitting of the ground state ($^2F_{7/2}$ and $^2F_{5/2}$).

Charge-transfer transitions are found for Eu^{3+} ($4f^6$), Sm^{3+} ($4f^5$) and Yb^{3+} ($4f^{13}$). In table 2 we give the position of the charge-transfer transition of Eu^{3+} in several host lattices. In isomorphous series we find a relation between the position of this absorption band and the ionic radii of the cations in the host lattice. This relation can be understood (see e.g. ref. 1).

TABLE 2

Position of charge-transfer level of Eu^{3+} in various lattices.
The two groups of data relate to isomorphous lattices.

Composition	Charge-transfer level (10^3 cm^{-1})	Ionic radii (Å)
(La,Eu)OBr	30.7	Br$^-$ 1.95, La^{3+} 1.14
(La,Eu)OCl	33.3	Cl$^-$ 1.81, La^{3+} 1.14
(Gd,Eu)OCl	35.0	Gd^{3+} 0.97
(Y, Eu)OCl	35.4	Y^{3+} 0.92
Na(La,Eu)O$_2$	36.0	Na$^+$ 0.94, La^{3+} 1.14
Na(Gd,Eu)O$_2$	41.1	Na$^+$ 0.94, Gd^{3+} 0.97
Li(Y, Eu)O$_2$	42.0	Li$^+$ 0.68, Y^{3+} 0.92
Li(Lu,Eu)O$_2$	43.0	Li$^+$ 0.68, Lu^{3+} 0.85

We now turn to transitions between 4f levels of the RE ions. These transitions are strongly forbidden as electric-dipole transitions. For this reason their optical strength is extremely low.

As an example we consider a special case, viz. transitions between the ^5D and the ^7F levels of the Eu^{3+} ion (see fig. 2). Optical transitions are forbidden for two reasons. The first is that these transitions are spin-forbidden. The selection rule is relaxed by spin-orbit coupling. As a result the ^7F states contain a slight admixture of ^5D and reversed.

The second selection rule that forbids transitions between ^7F and ^5D is the parity selection rule (Laporte's rule) which states that parity should change in electric-dipole transitions. Because the ^7F and ^5D states both originate from the $4f^6$ configuration they have the same parity. This implies that only the weak magnetic-dipole transitions (selection rule $\Delta J = 0, \pm 1$ and $J = 0 \rightarrow J = 0$ forbidden) are allowed.

The parity selection rule for electric-dipole transitions can be lifted, however, by the influence of the crystal field. Just as the spin prohibition was cancelled by the mixing of the ^7F and ^5D state by spin orbit coupling, so can the parity prohibition be cancelled by mixing the $4f^6$ configuration with a state of different parity. This occurs by the so-called odd crystal-field terms. These are the terms that change sign on inversion with respect to the RE ion. This implies that the parity selection rule can only be lifted if the RE ion occupies a lattice site that lacks inversion symmetry. Electric-dipole transitions that occur in this way are called forced electric-dipole transitions. The selection rule for these transitions is $\Delta J \leq 6$. However, if $J = 0$ for the initial or final level, then $\Delta J = 2, 4$ or 6 [2].

We now apply this knowledge to a specific transition, viz. the emission from the $^5D_0$ level of Eu^{3+} to the ^7F state consisting of the levels $^7F_{0-6}$ (see fig. 2). If the Eu^{3+} ion occupies a site with a centre of symmetry, electric-dipole transitions are forbidden. Only magnetic-dipole transitions can occur. In view of the selection rules for J we expect only $^5D_0-^7F_1$. If the Eu^{3+} ion occupies a crystallographic site lacking inversion symmetry, the following transitions are expected: $^5D_0 - F_{2,4,6}$ (forced electric-dipole) and $^5D_0-^7F_1$ (magnetic-dipole).

This has been confirmed experimentally [3]. In Ba_2GdNbO_6 where Eu^{3+} (on Gd^{3+}-sites) occupies a centre with pure cubic symmetry (figure 5) mainly $^5D_0-^7F_1$ emission is observed (figure 6). The weak and broad bands corresponding to $^5D_0-^7F_1$ emission are caused by coupling with lattice vibrations. These can give rise to a temporary deviation from pure cubic symmetry.

FIGURE 5

Crystal structure of Ba_2GdNbO_6.
Open circles: O^{2-}; black circles: Nb^{5+}; small hatched circles:
Gd^{3+}; large hatched circles: Ba^{2+}.

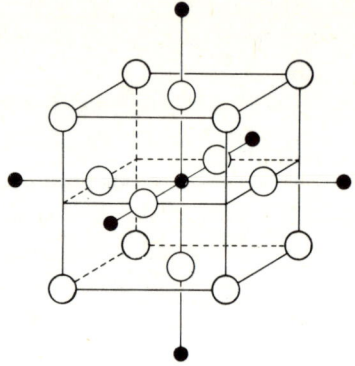

FIGURE 6

Emission spectrum of Eu^{3+} in
Ba_2GdNbO_6.

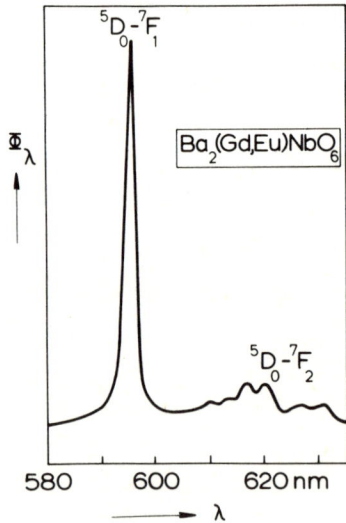

Very impressive is a comparison of the emission spectrum of Eu^{3+} in NaGdO$_2$ and NaLuO$_2$ [4]. The crystal structures of these host lattices are similar, because they are both based on the rock-salt structure (NaCl,MgO). Due to crystallographic order between the Na$^+$ and the RE^{3+} ions the structures differ in minor detail (figure 7). As a consequence Eu$^+$ in NaGdO$_2$

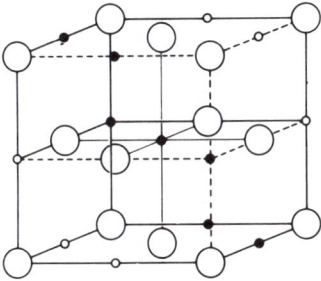 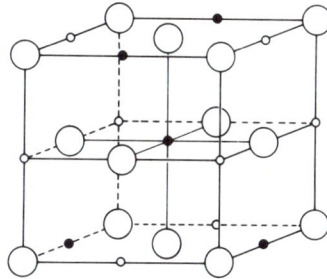

FIGURE 7a   FIGURE 7b

Crystal structure of NaLuO$_2$ (a) and NaGdO$_2$ (b) (schematic).
The rock-salt unit cell is shown. Large circles: O^{2-};
small open circles: Na$^+$; black circles: RE^{3+}

occupies a site without inversion symmetry, whereas Eu^{3+} in NaLuO$_2$ has inversion symmetry. The emission spectra are completely different (figure 8). In NaLuO$_2$-Eu^{3+} the $^5D_0$-$^7F_1$ emission dominates; in NaGdO$_2$-Eu^{3+} the $^5D_0$-$^7F_2$ emission. Note that these transitions are split into a number of components. This is due to the surroundings which split the threefold-degenerate J = 1 level and the fivefold-degenerate J = 2 level, because the site symmetry is lower than cubic. As a matter of fact the non-degenerate J = 0 level is not split. Note that these splittings are much smaller than in the case of Ce^{3+} (about 100 cm^{-1} versus about 10.000 cm^{-1}, respectively). This illustrates the fact that the 4f electrons are shielded from the environment by other electrons, whereas the 5d electrons are not.

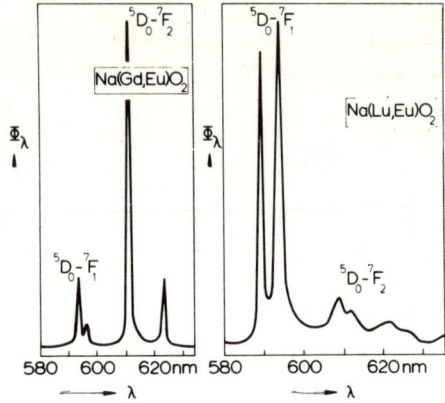

FIGURE 8

Emission spectrum of Eu^{3+} in $NaLuO_2$ and $NaGdO_2$.

The lifetime of the luminescent $^5D_0$ level is about 10^{-3} s. This is about 10^5 times longer than the lifetime of a level that luminesces via an allowed electric-dipole transition (viz. 10^{-8} s) and illustrates how strictly forbidden the 4f-4f transitions are. Short lifetimes have been found for the Ce^{3+} emission (5d-4f) which is in fact an allowed transition [5].

Finally we turn to the question which level is mixed with the $4f^6$ configuration in order to lift the parity selection rule. It seems not unreasonable to assume that this is the charge-transfer state in the case of Eu^{3+}. Strong evidence for this comes from the fact that the emission of Eu^{3+} in fluorides has mainly magnetic-dipole character, even if the Eu^{3+} ion occupies a site without inversion symmetry. As a matter of fact the charge-transfer state in fluorides is situated as very high energies due to the very high electronegativity of the F^- ion. In other cases, however, the 5d state may be admixed. This probably occurs in the case of Tb^{3+}.

In closing this section it may be stressed that the example treated, viz. the ^5D-^7F transitions of Eu^{3+} is a fairly simple one. Generally speaking the situation is more complex, but the theory necessary to describe these phenomena is the same as used above.

4. ENERGY TRANSFER

A first requirement for a luminescent material is that it emits radiation. It is trivial that this leads to the requirement that exciting radiation must be absorbed. We have seen already that the 4f-4f transitions of the RE are not very suitable for absorption of radiation, because they are strongly forbidden. Excitation may occur efficiently in either the charge-transfer state or the $4f^{n-1}5d$ state. An example of the first case is Gd_2O_3 -Eu^{3+} [6]. Exciting short-wave u.v. radiation is strongly absorbed by the Eu^{3+} ion raising this ion into its charge-transfer state. The ion then relaxes to the $^5D_0$ level from which luminescence occurs. An example of the second case is $YTaO_4$ -Tb^{3+} which can be excited efficiently by u.v. radiation into the $4f^75d$ state [7].

It is also possible, however, to excite the RE luminescence indirectly. This is done by building into the lattice another ion or group of ions that absorb strongly the exciting radiation and, subsequently, transfer this energy to the RE ion.

We give two examples. The Tb^{3+} ion in $YAl_3B_4O_{12}$ cannot be excited by 254 nm radiation (from a low-pressure mercury lamp), because Tb^{3+} in this lattice does not absorb this radiation. Ce^{3+} in $YAl_3B_4O_{12}$, however, does absorb this radiation. Energy transfer from Ce^{3+} to Tb^{3+} occurs so that excitation into the Ce^{3+} ion is followed by Tb^{3+} emission [8]. This follows from the excitation spectra (figure 9).

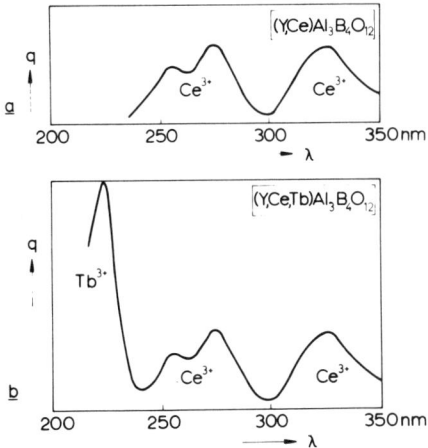

FIGURE 9

a. Excitation spectrum of the Ce^{3+} emission of $(Y,Ce)Al_3B_4O_{12}$.

b. Excitation spectrum of the Tb^{3+} emission of $(Y,Ce,Tb)Al_3B_4O_{12}$. The presence of Ce^{3+} bands in this spectrum indicates transfer from Ce^{3+} to Tb^{3+}.

Our second example concerns Eu^{3+} in YPO_4. In this lattice Eu^{3+} absorbs 254 nm only to a minor amount. If part of the phosphorus is replaced by vanadium $((Y,Eu)P_{1-x}V_xO_4)$ the radiation is absorbed by the vanadate group and transferred to the Eu^{3+}.

What is the mechanism by which this energy transfer occurs [9]? Figure 10 shows what should happen. We start with the system S(excited) + A(ground state) and end up with the system S(ground state) + A(excited). This is only possible, if one of the levels of A lies at the same height as the luminescent level of S (resonance condition). Further we need an interaction between S and A.

FIGURE 10

Energy transfer from S to A.
Transfer occurs to level 4,
followed by radiationless decay
to level 2 from which emission
occurs.

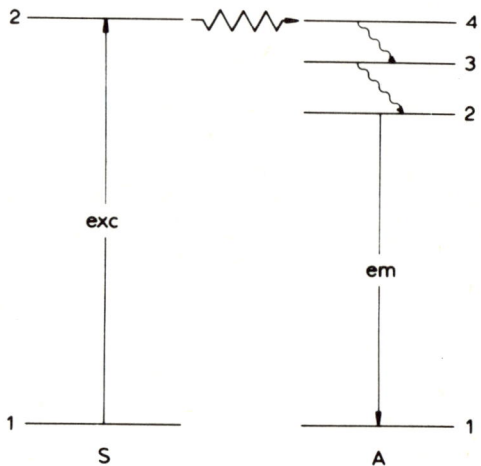

Transfer can be brought about in the first place by the Coulomb interaction between all charged particles of S and A. If S and A are so far apart that their charge clouds do not overlap, this form of energy transfer is the only one possible. If these do overlap, however, another process is possible by exchange interaction between the electrons of S and A. In this process electrons are exchanged between S and A, in the former the electrons remain with their respective ions.

This may be summarized as follows. For transfer by Coulomb interaction we write

$$P_{SA} = g_{SA} \cdot E_{SA} \qquad (1)$$

and for transfer by exchange interaction

$$P_{SA} = f_{SA} \cdot E_{SA} \qquad (2)$$

Here P_{SA} is the transfer probability. E_{SA} represents the resonance condition (in practice the spectral overlap of the emission of S and the relevant absorption of A) and occurs in both formulas. The quantity g_{SA} comprises the optical strengths of the relevant transitions and a distance-dependence of the type r_{SA}^{-n} (n = 6,8,etc.). The quantity f_{SA}, however, is proportional to the wave function overlap of S and A and therefore comprises an exponential distance-dependence.

We now apply the theory to our examples $Ce^{3+} \rightarrow Tb^{3+}$ and $VO_4 \rightarrow Eu^{3+}$. In both cases the broad band luminescence of Ce^{3+} and VO_4 shows spectral overlap with absorption levels of Tb^{3+} and Eu^{3+}, respectively. In view of the low absorption strength of the latter transitions transfer occurs by exchange interaction as can also be shown by other ways. This implies that the transfer from Ce^{3+} to Tb^{3+} and from VO_4 to Eu^{3+} can occur only over short distances in view of the necessary wave function overlap.

Since transfer to RE ions occurs efficiently even if their concentration is low we need still another process. This is SS transfer. Experimentally this is borne out by the fact that if x in $(Y,Eu)P_{1-x}V_xO_4$ is low (< 0.2) excitation into the VO_4 groups is followed by VO_4 and Eu^{3+} luminescence. Those VO_4 groups that do not have Eu^{3+} neighbours cannot transfer their energy to Eu^{3+} and show luminescence. If, however, x > 0.2 no VO_4 luminescence is observed [10]. The vanadate concentration is now so large that the excitation energy jumps from VO_4 to VO_4 group through the lattice until an Eu^{3+} ion is reached. The high efficiency of the commercial phosphor $(Y,Eu)VO_4$ is based on this phenomenon.

It is interesting to note that $(Y,Eu)NbO_2$ is not efficient [10]. This phosphor shows always blue NbO_4 luminescence and red Eu^{3+} luminescence. The reason for this is the fact that in $YNbO_4$ the resonance condition for NbO_4 - NbO_4 transfer is not fulfilled. In YVO_4, however, this condition is fulfilled (figure 11).

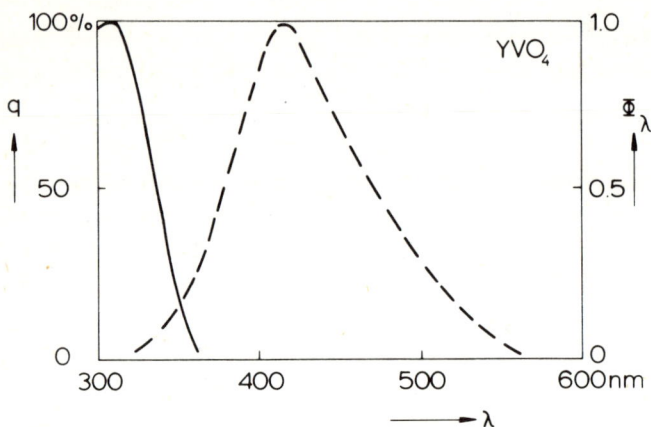

FIGURE 11a

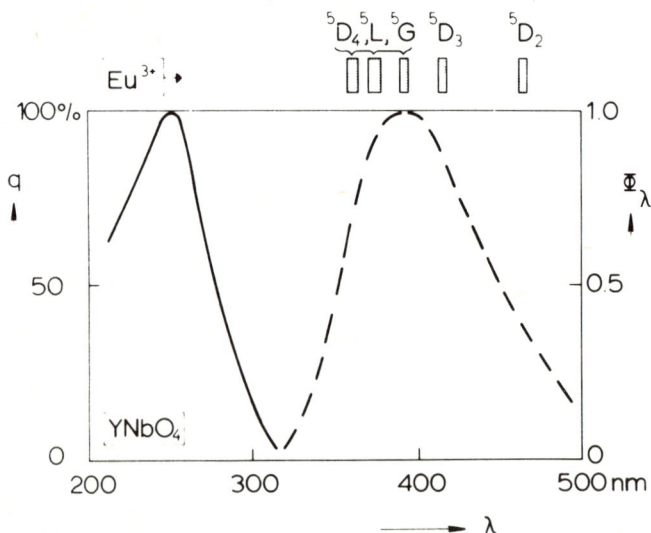

FIGURE 11b

Spectral overlap of VO_4 emission and absorption. (a) Absence of overlap in the analogous case of NbO_4. (b) Spectral overlap of both broad band emissions with Eu^{3+} absorption (levels indicated above the figure). (b) Absorption drawn in full lines, emission in broken lines.

If it would be possible to increase the RE concentration to high values without losing luminescent efficiency, we should not need SS-transfer. This is, however, not possible due to so-called concentration quenching which we will discuss now.

5. CONCENTRATION QUENCHING

Concentration quenching is the phenomenon that the luminescence efficiency of a given luminescent composition decreases above a certain concentration of the luminescent centre (the so-called critical concentration of the activator). There are a number of different processes that are responsible for this quenching. Here we can only mention these shortly.

In the case of broad-band emitters (like Ce^{3+}, Eu^{2+}) the critical concentration amounts to a few mol %. Quenching is due to transfer through the lattice from activator until a "killer" site in the lattice is reached [11]. The exact nature of these killer centres is unknown. This, by the way, is an example of energy transfer over considerable distances ($\sim$ 20 Å).

In the case of sharp-line emitters interaction between pairs (Dy^{3+}, Sm^{3+}) or at least trimers (Eu^{3+}, Tb^{3+}) is responsible for the quenching. For a short review see ref. [12].

A more elaborate review of all these problems along the same lines is given in ref. [13].

References

[1] G. Blasse and A. Bril, Z. physik. Chemie N.F. 57, 187 (1968).

[2] G.S. Ofelt, J. chem. Phys. 37, 511 (1962).

[3] G. Blasse, A. Bril and W.C. Nieuwpoort, J. Phys. Chem. Solids 27, 1587 (1966).

[4] G. Blasse and A. Bril, J. chem. Phys. 45, 3327 (1966).

[5] G. Blasse and A. Bril, J. chem. Phys. 47, 5139 (1967).

[6] A. Bril and W.L. Wanmaker, J. electrochem. Soc. 111, 1363 (1964).

[7] G. Blasse and A. Bril, J. of Luminescence 3, 109 (1970).

[8] G. Blasse and A. Bril, J. chem. Phys. 47, 1920 (1967).

[9] For a full account see D.L. Dexter, J. chem. Phys. <u>21</u>, 836 (1953).

[10] G. Blasse and A. Bril, J. electrochem. Soc. <u>115</u>, 1067 (1968).

[11] D.L. Dexter and J.H. Schulman, J. chem. Phys. <u>22</u>, 1063 (1954).

[12] G. Blasse, J. of Luminescence <u>1,2</u>, 766 (1970).

[13] G. Blasse and A. Bril, Philips Techn. Rev. <u>31</u>, 304 (1970).

APPLICATION OF RARE EARTHS IN OPTICAL MATERIALS AND DEVICES

by

P.N. Yocom

RCA/David Sarnoff Research Center, Princeton, New Jersey 08540, U.S.A.

ABSTRACT

Optical materials and devices which contain rare earth elements can be divided into the divisions of light emitting and light modifying. In the light emitting category a subdivision of incoherent and coherent materials and devices is made. The incoherent area comprises materials used in cathode ray tubes, fluorescent lamp devices, electroluminescent devices, thermoluminescent dosimeters, and upconverting phosphors or quantum counters. The coherent devices are a large number of optically pumped lasers of which the most useful are the neodymium activated materials such as yttrium aluminium garnet and glasses. Some of the experimental laser types are mentioned as possible future directions in which rare earth containing lasers may go. In the area of light modifying materials and devices the first subject treated is the materials which show photochromic behaviour or reversible coloration under suitable light sources. The other areas mentioned are electro-optic materials and devices covering the ferroelectrics and finally mentioning magneto-optic materials and some devices associated with the effect.

In surveying the field of optical materials and devices containing rare earth ions a major division can immediately be made between those materials and the devices dependent upon them which emit light and those which act upon light. In table 1 a listing of the materials to be touched upon in this lecture are grouped according to these two major classifications. In the light emitting class two major areas have now developed, namely the incoherent phosphors and the coherent materials or lasers which have been developed during only the past decade.

TABLE 1

Rare Earth Containing Optical Materials

I. Light Producing Materials
- A. Incoherent Emitters
    1. Cathodoluminescent materials
    2. Photoluminescent materials
    3. Electroluminescent materials
    4. Thermoluminescent materials
    5. Up conversion or anti-stokes materials
- B. Coherent Emitters (lasers)
    1. Optically pumped
        a. crystalline
        b. glass
        c. liquid

II. Light Modifying Materials
- A. Photochromic materials
- B. Electro-Optic materials
- C. Magneto-Optic materials

In the previous paper Professor Blasse has described the theoretical basis for rare earth containing luminescent and optical materials and in the following paper Dr. Mathers is going to discuss the more practical production problems associated with incoherent luminescent materials of present commercial importance. In light of this division I wish to discuss some properties that are needed in the phosphor area and how present rare earth materials compare to known materials.

In the area of cathodoluminescence the prime requirements are the ability of the phosphor lattice to extract the energy of the beam and transfer it to the emitting centre which in turn is able to convert the energy to emitted photons in high efficiency. The phosphor material should also be stable in the cathode ray tube environment. These basic requirements should be met by all cathodoluminescent materials. Beyond these basic requirements the particular device requires further properties, television kinescopes require the decay times of the various colour phosphors to be about equal and shorter than the frame time (about 1/30 second). In commercial television very high energy efficiency is important along with specified chromaticities.

Since Dr. Mathers will discuss the commercial red, europium activated phosphors, I will pass on to the blue and green phosphors. The presently used phosphors are based on $(Zn,Cd)S$ and have very high power efficiency, about 20%. Due to the properties of the human eye and the particular colours involved, a narrow line emitter as a rare earth f-f transition would have to show about the same energy efficiency as the broad emitters to compete with them. So far no such rare earth phosphors are known. In the area of rare earth band emitters, $CaS:Ce^{3+}$ [1] has about the necessary efficiency as a synthesized powder, but this has not yet been transferred to a screen because of the stability problems of CaS. In the case of blue emitters no reported rare earth phosphor, line or band can yet match the efficiency of ZnS:Ag.

Besides the normal TV phosphors, radar CRT's require long decay phosphors of the order of tenths of a second or more and as yet no rare earth emitter has displaced the conventional materials.

In the area of fast decay CRT phosphors for flying spot scanners or sensing tubes, cerium activated materials have always been in the forefront. Recently much work has been done in this area and many phosphors have been disclosed spanning the range of the ultraviolet to the blue and then skipping to a range in the green yellow. Table 2 shows a list of such materials [2,3]. Besides Ce^{3+} any rare earth d-f transition has the possibility of being very fast.

In the area of photoluminescent phosphors, the desired properties are

TABLE 2

Cerium Activated - Fast Decay CRT Phosphors

Phosphors	Power Efficiency	Emission bands A
YBO_3:Ce	2 %	4200, 3930
$LaPO_4$:Ce	3 %	3360, 3170
YOCl:Ce	3.5 %	3800
Y_2SiO_5:Ce	6 %	4050
$Y_2Si_2O_7$:Ce	8 %	3800
$Y_2Si_2O_7$:Ce	6.5 %	3750
$Y_3Al_5O_{12}$:Ce	3.5 %	5500, 3600

a strong absorption band which results in a high quantum efficiency for light emission. The most desired absorption bands have been those which match the mercury arc emission but other emission sources could conceivably be used. The next paper is going to be discussing this type of phosphor so I shall not devote more time to them.

The next type of incoherent emitter to be mentioned is the rare earth activated electroluminescent phosphor. In this case the passage of an electric current, either AC or DC, through the phosphor results in emission. Two types of phosphors are possible, the p-n junction single crystal type or the poly-crystalline type. In the p-n junction type emission occurs by band to band recombination of electrons and holes. So far no rare earth compounds have been reported that show this type of emission. However, in the area of poly-crystalline materials, rare earth emission has been obtained from powder phosphors dispersed in a binder and made part of a circuit and in evaporated films. In an AC powder system Goldsmith et al. [4] observed emission from a number of rare earth ions. Very recently Waite and Vecht [5] have reported emission in the DC powder case for a number of rare earth ions; Krupka [6] has reported emission of Tb^{+3} from evaporated films. In all of these cases the emission was due to f-f transitions. Up to the present all of these rare earth electroluminescent emissions are less efficient than other known materials; but it is a field of great potential if sufficient efficiency can be obtained.

Another area of recent interest is thermoluminescent phosphors. In this use the phosphor material is subjected to high energy radiation such as X-rays or gamma rays and does not show any strong emission during irradiation. However if the phosphor material is subsequently warmed, light is emitted and the total amount of light emitted is proportional to the amount of original radiation. These materials are obviously useful as radiation dosimeters and they have an advantage of being of small size and having a large dose range. Their advantages over film, the main competitor, are a better energy dependence, higher precision over a large range of dose, and an easy method of reading out information. Three of the most commonly used phosphors in this area are LiF:Mg, $Li_2B_4O_7$:Mn, and CaF_2:Dy [7].

Another area of incoherent emitters which received much interest in recent years is that of up-conversion or anti-stokes phosphors. In this case all of the materials involved contain rare earth ions. The useful property of these materials is that they absorb infrared radiation, typically from a GaAs:Si diode, and convert it to visible light. All of the most promising materials contain Yb^{+3} ions which absorb the infrared and then transfer the energy to one ion from the group of Er^{+3}, Tm^{+3}, or Ho^{+3}. In these latter ions, the absorbing, low lying energy levels become saturated and are then able to absorb the energy of a second photon in the excited state so that the overall result is the absorption of two photons before emission occurs. The observable result being visible light produced from infrared. Typically the combination Yb^{+3}, Tm^{+3} produces blue light by three-photon absorption while Yb^{+3}, Ho^{+3} produces green light and Yb^{+3}, Er^{+3} can produce green or red depending upon the lattice. Because Tm^{+3} requires three photons it is quite inefficient, but Er^{+3} can approach .1% power efficiency which puts it close to the level of GaP light emitting diodes. If these systems can equal or surpass the GaP efficiency at modest input power levels, interest in them will increase greatly [8].

The idea of saturating excited states of rare earth ions so that they will act as absorbers has resulted in much research on materials which could behave as infrared quantum counters by emitting higher energy photons for each incoming infrared photon. Unfortunately no practical device has yet resulted from this work.

At this point I should like to move onto the subject of coherent light emitters otherwise known as lasers. This is a subject in which rare earth materials have played a very great role. The basic operation of a laser depends upon obtaining a population inversion for a particular emissive transition; that is, more ions must be in the excited state than in the ground state so that one photon can "trigger" emission. Lasers are known in all three states of matter, gases, solids, and liquids. In the solid laser area there are two main types, the semiconductor or electrically pumped laser and the optically pumped. It is in this latter type that the rare earth materials play a large role. The optically pumped systems are divided into two groups: the pulsed and the continuously operating or CW (continuous wave). Both of these systems operate by focussing a strong light source on the active medium which must have high optical perfection so that little light is lost by scattering. The most used CW, optically pumped laser is yttrium aluminium garnet (YAG) doped with neodymium. In its most advanced form it can now put out some hundreds of watts of power continuously.

In a geometrical sense the largest lasers are the neodymium glass lasers. These lasers can only be operated practically in a pulsed mode but they can put out large amounts of energy over relatively large areas. The glass lasers suffer from thermal distortions which limit their repetition rate; and from this aspect an YAG laser running in a pulsed mode can give outputs at a much higher repetition rate.

As indicated, one of the principle problems in laser fabrication is to obtain as nearly a perfect optical medium as possible, with no parasitic processes occurring. To this end, the rare earth materials should be of at least phosphor grade purity and have no other foreign impurities which can cause precipitates to be present in the finished optical cavity. The solubility limit of neodymium in YAG has been a problem which has been alleviated by codoping with lutetium which causes the solubility limit of neodymium to increase [9].

One novel approach to overcoming the problem of inhomogeneities in laser materials is the use of a liquid as the active medium. One such type of laser has been a solution of rare earth chelate compounds in organic solvents. The rare earths used in these systems have been europium, terbium, and neodymium. Another type of liquid is the aprotic which have been based on either selenium oxychloride or phosphorous oxychloride. In either of these two materials neodymium chloride is dissolved by the use of tin (IV) chloride or zirconium chloride [10].

Most of the rare earth ions having f-f transitions have shown laser action in one or more of the many host materials [11]. However, beside the neodymium doped materials mentioned, erbium as the active ion is receiving attention since its emission has reduced danger to the human eye and is at a frequency which has maximum transmission in the atmosphere. Holmium is also of interest since it can show high efficiency. Other host systems which are at present under study include rare earth orthoaluminates, cubic oxide ceramics and silicate oxyapatites.

I should now like to pass from light emitting materials to light modifying materials and first treat the photochromic materials which change the spectral distribution of light. After this group of materials the more classical modulating type of materials will be discussed.

For the present purposes photochromic materials may be defined as materials which develop optical absorption bands when irradiated with certain wavelengths. If subsequently they are exposed to another critical wavelength, usually of lower energy, the absorption band is then bleached [12]. Two varieties of photochromic materials contain rare earth elements. Certain glass compositions containing cerium or europium [13,14] show the property and alkaline earth fluorides containing one element of the group of lanthanum, cerium, gadolinium or terbium can be made photochromic when treated with alkali metal vapour at high temperatures. In the glass systems it is thought that the rare earth undergoes a valence change to cause the absorption band. In the alkaline earth fluoride systems the mechanism is complex but more thoroughly studied and better understood than in the glasses [12,15]. It seems to involve two electrons trapped near the rare earth ion. In both of these systems the available contrast is limited and the effective concentration of the rare earth species is low before spurious processes set in. Also the induced colour centres bleach at room temperature after some period of storage which can vary from minutes to days.

Now I should like to pass on to the more classical type of modulator processes; the best known of which are the electro-optic process, the acousto-optic, and the magneto-optic. From the point of view of rare earths being used in device materials, the most active of these areas are the materials which make use of electro-optic effects. Two materials stand out in this area, namely the lanthanum modified lead zirconate titanate (PLZT) ceramics and the rare earth molybdate crystals. In both of these cases the basic phenomenon is a change in the birefringence of the material caused by the electric field which then is observed by its effect on the passage of polarized light.

The ceramic system is very interesting for,depending upon the lanthanum content and the zirconium to titanium ratio, materials which show the linear electropic effect, the quadratic effect or memory effect can be obtained.

Basic to these various effects, is the preparation of polycrystalline ceramic wafers which show good optical properties. This was accomplished by Land and Haertling [16,17] who have optimized a process for the fabrication of good quality material by precipitating mixed hydroxides from high purity solutions, calcining and then hot pressing under an oxygen atmosphere. After the required polishing, plates showing up to 70% transmission are obtained. This type of plate has been considered for applications such as displays, image storage, optical voltage sensors, spectral filters, shutters, gates, and memories.

One of the most developed devices using this material is the so-called "ferpic" which is the work of Meitzler [18,19,20] and coworkers. This is an image storage device which allows selective erase/write capability. It consists of a PLZT plate with two poling electrodes on opposite edges to align all the domain. On each of the faces films of a photoconductor are applied over which transparent electrodes are placed. The photoconductor films allow domain switching proportional to the light intensity and so generate an image as a spatial variation in birefringence which can be viewed by polarized transmitted light or projected onto a viewing screen. By strain basing the PLZT plate the image can be indefinitely stored with the power off.

The rare earth molybdates with orthorhombic structure are both ferroelectric and ferroelastic so that the domains can be switched by either an electric field or by a mechanical stress. The most studied material is gadolinium molybdate (GMO) which has no absorption in the visible; however, according to Barkley et al. [21], gadolinium, dysprosium molybdate is easier to grow in a higher state of perfection but shows the characteristic dysprosium absorption resulting in a yellowish material. Some of the device applications that have been investigated for these molybdates are a page composer [22] and a light scanner [21]. However the switching speed is in the millisecond range which places some difficult restrictions on their device use.

Another well recognized light modulating process is the magneto-optic effect. In this case a magnetic field applied to a material containing magnetic ions can cause a rotation of the azimuth of plane polarized light [23]. Since most of the rare earth ions are magnetic, this effect is present in most rare earth materials. However those materials which show a large rotation such as EuO, EuS, EuSe and YIG have poor transmission properties [24] so only a minor amount of conventional device work has been done with them. However I would like to point out a rather novel device based on this effect with a rare earth ion and that is a phonon spectrometer. Sabisky and Anderson [25] have demonstrated the detection of phonons at liquid helium temperatures with alkaline earth fluoride crystals containing divalent thulium. This ion has broad absorption bands in the visible showing strong paramagnetic circular dichroism which can be perturbed by weak interaction with phonons. A light beam monitors the phonon temperature by recording the fractional change in the circular dichroism.

Two other types of optical materials are of considerable interest in the current literature; they are the acousto-optic materials and second harmonic generation materials. So far no rare earth containing material has shown much promise as an acousto-optic modulator. As for second harmonic generation, gadolinium molybdate shows the effect to a considerable degree but still is not as effective as other materials such as the niobates.

References

[1] W. Lehmann, J. Electrochem. Soc. $\underline{118}$, 1164 (1971).

[2] G. Blasse and A. Bril, J. Chem. Phys. $\underline{47}$, 5139 (1967).

[3] A.H. Gomes de Mesquita and A. Bril, Mat. Res. Bull. $\underline{4}$, 643 (1969).

[4] G.J. Goldsmith, S. Larach, R.E. Shrader, and P.N. Yocom, Solid State Communications, $\underline{1}$, 25 (1963).

[5] M.S. Waite and A. Vecht, Appl. Phys. Letters $\underline{19}$, 471 (1971).

[6] D.C. Krupka, J. Appl. Phys. $\underline{43}$, 476 (1972).

[7] L.A. DeWerd and T.D. Stoebe, American Scientist $\underline{60}$, 303 (1972).

[8] T.C. Rich and D.A. Pinnow, J. Appl. Phys. $\underline{43}$, 2357 (1972).

[9] L.A. Riseberg and W.C. Holton, J. Appl. Phys. $\underline{43}$, 1876 (1972).

[10] A. Lempicki in "Handbook of Lasers", Editor R.J. Pressley, Chemical Rubber Co., 1971, p. 355.

[11] M.J. Weber in "Handbook of Lasers", Editor R.J. Pressley, Chemical Rubber Co, 1971, p. 371.

[12] Z.J. Kiss, Physics Today - 42, 1970.

[13] A.J. Cohen, U.S. Pat. 3,269,847 and U.S. Pat. 3,278,319.

[14] T.C. Shutt, M. Martin, C.J. Lewis and J. Drobnick, Canadian Ceramic Society Annual Meeting, Feb. 15-17 (1971).

[15] W. Phillips and R.C. Duncan, Jr., Metallurgical Transactions $\underline{2}$, 769 (1971).

[16] C.E. Land, Metallurgical Transactions $\underline{2}$, 781 (1971).

[17] G.H. Haertling and C.E. Land, Ferroelectrics $\underline{3}$, 269 (1972).

[18] A.H. Meitzler and J.R. Maldonado, Electronics $\underline{34}$ (1971).

[19] A.H. Meitzler, J.R. Maldonado and D.B. Fraser, The Bell System Tech. J. $\underline{49}$, 953 (1970).

[20] J.R. Maldonado and A.H. Meitzler, Proc. IEEE, $\underline{59}$, 368 (1971).

[21] J.R. Barkley, L.H. Brixner, E.M. Hogan and R.K. Waring, Jr., Ferroelectrics $\underline{3}$, 191 (1972).

[22] A. Kumada, Ferroelectrics $\underline{3}$, 115 (1972).

[23] H. Jaffe in "Handbook of Lasers", Editor R.J. Pressley, Chemical Rubber Co., 1971, p. 445.

[24] Di Chen in "Handbook of Lasers", Editor R.J. Pressley, Chemical Rubber Co., 1971, p. 460.

[25] E.S. Sabisky and C.H. Anderson, Appl. Phys. Letters <u>13</u>, 214 (1968).

PRODUCTION OF RARE EARTH RED PHOSPHORS

FOR COLOUR TELEVISION AND LIGHTING APPLICATIONS

by

J.E. Mathers

GTE SYLVANIA INCORPORATED, Precision Materials Group,

Chemical and Metallurgical Division, Towanda, Penn. 18848, U.S.A.

ABSTRACT

   Many of the problems inherent in the fabrication of cathode-ray tube screens for colour television and high pressure mercury vapour lamps for lighting are directly related to the particular phosphors used. This report considers some of these problems with emphasis on the effects that the rare earth raw materials have on the Eu^{3+} activated phosphors: YVO_4, and Y_2O_3, and Y_2O_2S. Methods for phosphor synthesis and raw material evaluation are presented. The significance of high purity yttrium and europium oxide is discussed with appropriate comments as to future quality requirements.

1. INTRODUCTION

   In recent years there has been a growth of interest in luminescent materials based on the rare earths, either as host lattice constituents or activators. Rare earth phosphors have found important commercial application in such fields as lasers, colour television, and lighting. Rare earth phosphors have also been useful in theoretical studies of luminescent mechanisms.

   The red-emitting europium-activated phosphors have been of particular commercial interest since the introduction of yttrium vanadate in 1964 as an efficient cathode-ray phosphor [1]. From 1964 to the present, europium-activated phosphors have been adopted as the standard red primary in colour television picture tubes. These phosphors include: $(Y,Eu)VO_4$, $(Y,Eu,Bi)VO_4$,

$(Y,Eu)_2O_2S$, $(Y,Eu)_2O_3$, $(Gd,Eu)_2O_3$, and $(Gd,Y,Eu)_2O_3$. The growth pattern of the red colour TV phosphor brightness is shown in figure 1.

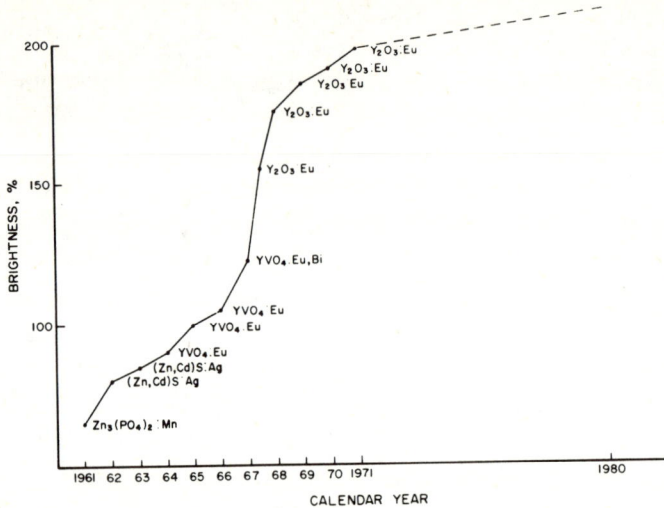

FIGURE 1

Growth pattern of the red colour TV phosphor brightness

In lighting applications, the vanadate phosphor has found commercial utility as a colour-correcting component in high pressure mercury vapour lamps [2].

This paper will deal first with some of the basic concerns of the phosphor manufacturer as well as the device manufacturer, specifically, colour television tubes and high pressure mercury vapour lamps, and how these concerns relate to the rare earth raw material.

2. COLOUR TELEVISION APPLICATIONS

Colour TV tubes commonly consist of an orderly array of dots, each dot

containing one of the three primary colours - blue, green, and red. The
dots comprising the three colours form a triad, and by suitably mixing the
colours, all colours of the visible spectrum may be formed. The phosphors
used in forming these dots are commonly put down by a slurry or dusting
technique.

In the slurry technique, the phosphor is dispersed in a polyvinyl
alcohol-bichromate-water solution. This slurry is spread on a TV panel and
partially dried. Then an aperture mask, consisting of an orderly array of
holes in a metal grid, is placed over the panel.

Intense light from a point source is focussed on the mask, striking the
panel only through holes in the mask. Where light strikes the panel, the
polyvinyl alcohol and bichromate undergo a complex cross-linking reaction
which renders the polymer insoluble. Washing the panel with water removes
the unexposed, soluble polymer, leaving behind a dot pattern of the exposed
polymer. Repeating the process two more times using different coloured
phosphors and by slightly shifting the position of the light point source
forms the triad pattern of dots.

Rigid quality specifications on the finished TV panel necessitate ex-
acting control throughout the entire fabrication process. A great deal of
the success in producing an acceptable panel can be directly related to the
phosphors. Not only the luminescent character of the phosphors, but equally
important, their screenability which can be influenced by their physical as
well as chemical properties. Such screen characteristics as texture, optimum screen weight, exposure time, adherence, and dot morphology, including
size, density, and geometry, are phosphor-related. An example of a phosphor-
related screening defect would be cross contamination. In the process of
laying down, exposing, and developing the second and third set of dots, usually some phosphor from the slurry will adhere to previously substrated
dots. The phosphor of one colour adhering to the dot of another colour is
called cross contamination and is not removable by water washing.

The order of putting down the colour dots is commonly the green first,
then blue, and the red phosphor last. Because of the high brightness of
the green, cross contamination on it is generally not noticeable nor particularly objectionable. Also, because it is substrated first, it cannot

contaminate the other two colours.

The blue phosphor, which is put down second, has a relatively low light output and contamination of it by the high brightness red phosphor is objectionable. The presence of the red phosphor in the blue dots results in loss of saturated blue colours and causes poor colour resolution in the colour TV screen. The particle size distribution of the red phosphor, specifically a high percentage of particles less than five micrometers, will quite often be associated with this type of contamination.

Another objectionable screen characteristic that can be associated with the phosphor is graininess of the solid screen. Again, phosphor particle size and distribution can be contributing factors. Likewise, dot morphology, particularly high porosity, can be caused by either large mean size particles or an excessive amount of particles greater than twelve micrometers.

In order to maintain phosphor usage efficiencies at a maximum, the tube screening engineer prefers to utilize the lowest possible optimum phosphor screen weight. The crystal density, particle size, and bulk density of the phosphor are directly related to the weight of powder required to produce optimum luminescence. While not much can be done to significantly change the crystal density of a given phosphor composition, the particle size and bulk density can usually be controlled. The ability to fix a phosphor particle to a substrate to just the right degree is quite often dependent upon a particular phosphor's chemical reactivity and related particle surface phenomena. The specific degree of phosphor adherence desired can, in fact, be controlled by special treatment of the phosphor during or after synthesis.

Another screening parameter that can be affected by the phosphor is exposure time. Since the shortest possible exposure times are desirable, the whitest body colour phosphor is preferred. Dark body colours absorb some of the ultraviolet radiation, diluting the radiation required for the proper polyvinyl alcohol-bichromate reaction; consequently, exposure times have to be increased.

Of course, there are many other factors involved in the successful screening of a commercially acceptable colour TV tube. Only the areas that are pertinent to the phosphors are within the scope of this discussion.

The intrinsic cathodoluminescent properties of the phosphors, particularly the rare earth reds, are not really a concern in the screening process; however, in the finished tube, these properties are a major concern and will be discussed in some detail in the section covering rare earth, raw material, impurity effects.

3. LIGHTING APPLICATIONS

Another application of rare earth phosphors that has been successful commercially is colour correcting in high pressure mercury vapour lamps (HPMV). HPMV lamps are good emitters of blue, green, and yellow light, but are poor emitters of red light. Consequently, unless corrected for this deficiency, they distort the true colour of many objects and are not suitable for applications in which good colour retention is necessary. HPMV lamps are also excellent emitters of ultraviolet radiation and, therefore, a number of phosphors has been developed which take advantage of this characteristic by emitting red light in response to excitation by ultraviolet radiation. These colour-correcting phosphors are generally placed on the inner surface of a transparent envelope surrounding the arc tube discharge source.

Keeping in mind the preceding general discussion pertaining to specific rare earth red phosphor applications, I would like next to consider the syntheses of the three most commercially successful rare earth phosphors: $(Y,Eu)VO_4$, $(Y,Eu)_2O_3$, and $(Y,Eu)_2O_2S$.

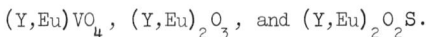

4. PHOSPHOR SYNTHESES

Actually, one does not have to be very familiar with the art to prepare any of these materials. Practically any method of combining the required elements to form the desired composition will luminesce. The real challenge is to prepare any one of the three phosphors on a large commercial scale efficiently, uniformly, and consistently to meet the customers' technical specifications, as shown in table 1, in such a way that it is profitable and competitive in the marketplace. With the rare earth phosphors you are

- 246 -

TABLE 1

Typical Phosphor Specifications

Customer	Particle Size μm		Particle Size Distribution		Cathode-Ray Brightness % of Standard	Chromaticity Coordinates		pH	Conductivity μmhos
	FSSS	Coulter Counter 50% Level Sonified	Quartile Deviation	% <5 μm		x	y		
A	-	8 ± 1.0	.30 ± .05	15 max	$\geq$ 97	.658	.338	7.5 ± 1	$\leq$ 60
B	5-6	9 - 11	-	<15	96 min	.650	.340	7 - 8	< 100
C	7-8	10 - 13	-	-	Equal	Equal to Std.		9 - 10.5	-
D	-	7 - 8	.32 max	20 max	Equal	Equal to Std.		< 9	400 max
E	6.5-7	8 - 9	-	-	Equal	.657	.337	6.5 - 75	-
F	6-7.5	-	-	-	Equal	.653	.347	7 - 9	< 150
G	-	7 - 8.5	.35 max	-	Equal	.650 min	.340 max	9 -10.5	-
H	5-6	6.5 ± 1.0	-	-	Equal	Equal to Std.		-	-

handicapped from the start due to relatively high raw material costs compared to most other commercial phosphors. Processing errors and customer rejects are costly to absorb. Because of these economic concerns, generally the most direct method of forming the phosphors is highly desirable. However, these direct methods require especially uniform and consistent rare earth oxide starting materials with very low rare earth and non-rare earth impurity levels.

A good example of a direct method is one that can be used to prepare $(Y,Eu)_2O_3$ phosphor [4]. The Y_2O_3 and Eu_2O_3 raw materials are thoroughly blended with suitable fluxes, hydrostatically pressed, and directly fired at high temperatures to form the luminescent composition. This method eliminates any wet chemical steps such as oxalate coprecipitation and costly high temperature crucibles, since the pressed forms can be directly placed on the furnace hearth. The vanadate phosphors can be directly synthesized by firing a raw material mixture of rare earth oxides and vanadium [5]. The oxysulphide phosphors are conveniently prepared by heat treating the raw material mixture in a suitable sulphurizing atmosphere or by firing the raw material mixture in the presence of sulphur and special fluxing agents [6]. All these direct methods, while basically simple, do need a great deal of optimization in order to have the necessary control for the many modifications required to meet customer specifications. Due to the dynamic nature of the processing to meet customer needs, rare earth raw materials must be carefully evaluated, as described in the following discussion.

5. RARE EARTH PHOSPHOR RAW MATERIAL TESTING

Unfortunately, phosphor chemistry has not yet progressed to a point where scientific intuition is not a major factor. With all our experience and sophisticated testing techniques, we still have to rely upon an empirical "use test" to determine the utility of any given combination of raw material. The overall process of luminescence is only qualitatively understood; predicting the outcome of a phosphor synthesis based on analytical data can be very frustrating. However, this type of information is useful and can serve as a guide.

The "use test" is a convenient and generally reliable method of evaluating raw material. The most desirable form of this test employs all the features of the actual large batch manufacturing process in that sufficient quantity of raw mix is prepared to give a minimum batch size that can be accommodated by the manufacturing equipment. Generally, this is then processed along with a regular manufacturing lot, the manufacturing material serving as a process control. In evaluating yttrium oxide, for example, the only variable would be the particular lot of Y_2O_3 being tested. The same would hold true for evaluating a lot of europium oxide. If the physical and luminescent properties of the finished test phosphor are equal to or better than the production phosphor, the raw material lot is acceptable. It should be emphasized that only materials that meet certain minimum physical and chemical specifications, based on extensive research, get to be phosphor tested. The phosphor production engineer has much more confidence in "use tested" raw materials. He is, understandably, quite reluctant to process a two-thousand-pound-batch of one-hundred-doller-per-pound phosphor based on analytical data alone.

The vast majority of "use test" failures are materials that were borderline or below on physical and chemical specifications. Infrequently, material of this nature has been waivered because of supply and demand, pricing, and other factors. Material that fails the "use test" for no apparent reason is usually assigned to the research and development engineer for comprehensive evaluation. If all the facts are at his disposal, he will have a scientific explanation in the majority of cases. Usually, existing analytical data have not been thoroughly utilized.

The characterization of Eu_2O_3 and Y_2O_3 raw materials, in relation to other rare earth and non-rare earth impurities and their effects, is a major factor in continuing phosphor progress.

6. IMPURITY EFFECTS

Efficient production of the rare earth red phosphors requires very high purity yttrium and europium oxide raw materials. Since as little as 1 ppm of certain impurities can affect luminescence [7], the Y_2O_3 and Eu_2O_3 should be as pure as possible. In fact, the standard luminescent grade of 99.9999% is desirable. To fully appreciate the need for such high purity requirements, one needs to be aware of some of the basic luminescent mechanisms.

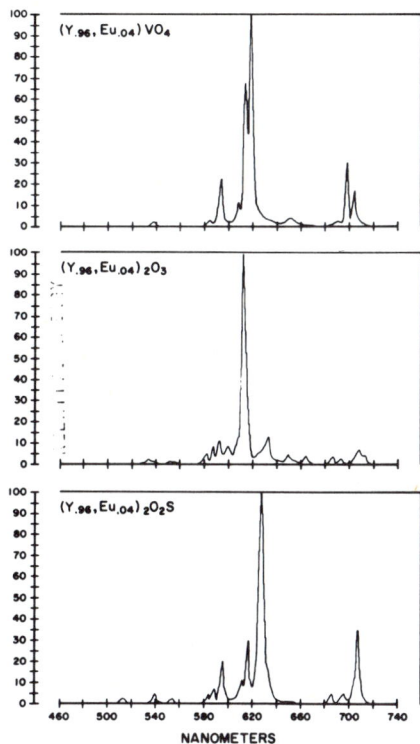

FIGURE 2

Spectral energy distribution of $(Y,Eu)_2O_3$, $(Y,Eu)VO_4$, and $(Y,Eu)_2O_2S$ under cathode-ray excitation.

Figure 2 shows the spectral energy distribution of $(Y,Eu)_2O_3$, $(Y,Eu)VO_4$, and $(Y,Eu)_2O_2S$ under cathode-ray excitation. The intensity of the emission lines will vary with change in Eu^{3+} activator concentration. This effect is more noticeable in the oxysulphide, where the intensity of the minor green emissions increase rapidly at lower activator levels (figure 3). If we compare chromaticities derived from these curves (table 2), the YVO_4 maintains

TABLE 2

Chromaticity of Vanadate, Oxide, and Oxysulphide Phosphors

as a Function of Eu Concentration

$(Y_{1-x}, Eu_x)_2 O_2 S$

Eu_x	Chromaticity Coordinates	
	x	y
0.01	.572	.420
0.02	.606	.385
0.03	.641	.354
0.04	.658	.338
0.05	.665	.334
0.06	.670	.329
0.07	.670	.330
0.08	.674	.325
0.09	.675	.325
0.10	.676	.324

$(Y_{1-x}, Eu_x)_2 O_3$

0.01	.622	.375
0.02	.635	.363
0.03	.639	.360
0.04	.648	.352
0.05	.653	.347
0.06	.655	.344
0.07	.652	.347
0.08	.652	.348
0.09	.654	.346
0.10	.656	.344

$(Y_{1-x}, Eu_x) VO_4$

0.01	.666	.333
0.02	.665	.334
0.03	.665	.334
0.04	.665	.335
0.05	.667	.332
0.06	.668	.332
0.07	.668	.332
0.08	.665	.334
0.09	.669	.331
0.10	.669	.331

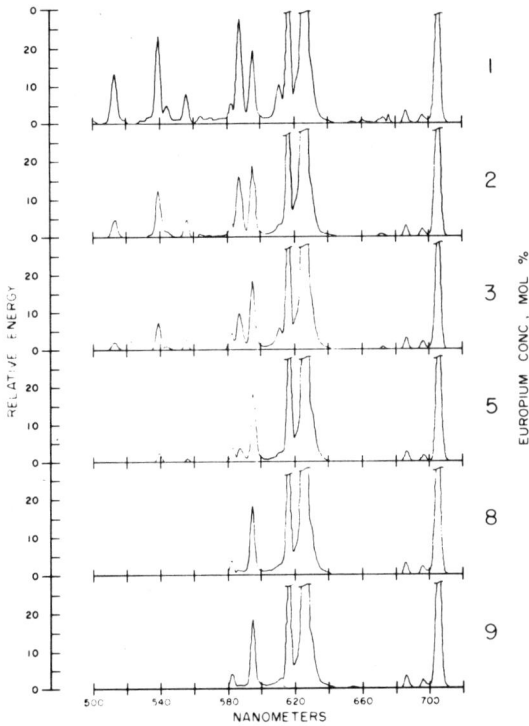

FIGURE 3

Spectral energy distri-
bution of $(Y,Eu)_2O_2S$ at
1, 3, and 9 mol percent
Eu.

almost constant chromaticity with Eu content, the Y_2O_2S shows a large change
in chromaticity with Eu content, and Y_2O_3 shows a smaller change in chromaticity with Eu content. Figure 4 shows the relative cathode-ray brightness
of YVO_4, Y_2O_3, and Y_2O_2S over an europium activator concentration range of
from 1 to 10 mol percent. The relationship with Y_2O_2S is linear, the oxide
system peaks about 3 mol percent, and the vanadate varies only slightly over
the range from 2 to 8 mol percent. The phosphor manufacturer takes advantage of these relationships by operating at the lowest possible europium
level without jeopardizing colour and brightness. This can only be done
with high purity raw materials since impurity quenching is much more critical at low activator levels. In the early days of the vanadate phosphor,

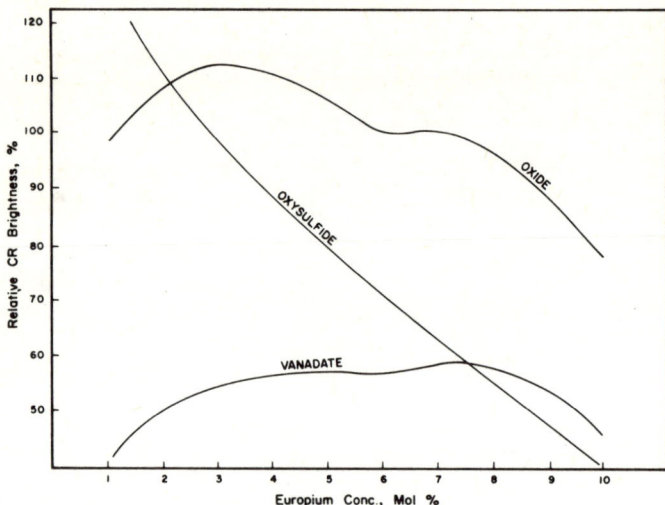

FIGURE 4

Relative cathode-ray brightness of $(Y,Eu)_2O_3$,
$(Y,Eu)VO_4$, and $(Y,Eu)_2O_2S$ as a function of europium concentration

activator levels of about 5 mol percent Eu were commonly used because commercial quantities of high purity Y_2O_3 and Eu_2O_3 were not available. As high
purity materials became more available, this level was gradually reduced to
as low as 2 mol percent. This represents a reduction of over sixty percent
and a substantial cost saving when you consider europium oxide was selling
from \$600 to \$1,200 per pound.

Cerium, without doubt, is the most dramatic brightness quencher in all
three phosphor systems. As little as 1 ppm Ce will cause significant brightness losses as shown in figure 5. It is interesting to note that similarly
charged elements (Ti, Zr, Hf, and Th) also quench the brightness of these
phosphors as shown in figure 6.

It is well known that terbium enhances the brightness in all three phos
phor systems [8,9,10]. The effective concentration will vary depending upon

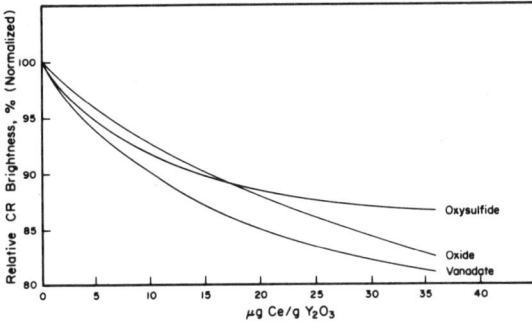

FIGURE 5

Effect of cerium impurity
on the cathode-ray brightness of $(Y,Eu)_2O_3$, $(Y,Eu)VO_4$,
and $(Y,Eu)_2O_2S$.

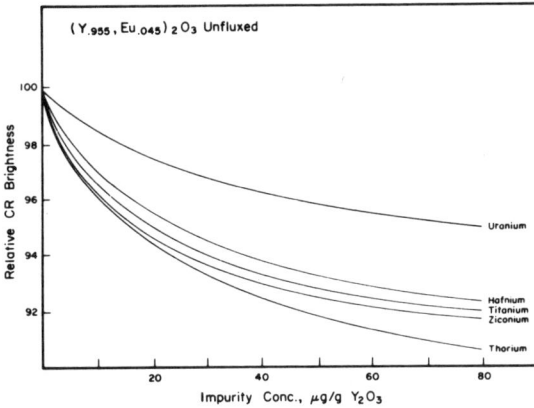

FIGURE 6

Effect of Ti, Zr, Hf, and
Th on the cathode-ray
brightness of $(Y,Eu)_2O_3$.

other impurities present and Eu activator concentration; however, the range
is usually from about 10 - 100 ppm. This does not include the bismuth modified vanadate, $(Y,Eu,Bi)VO_4$. Terbium is detrimental in this particular system, quenching the desirable broad band bismuth emission, as shown in figure
7. Similar effects are observed with praseodymium.

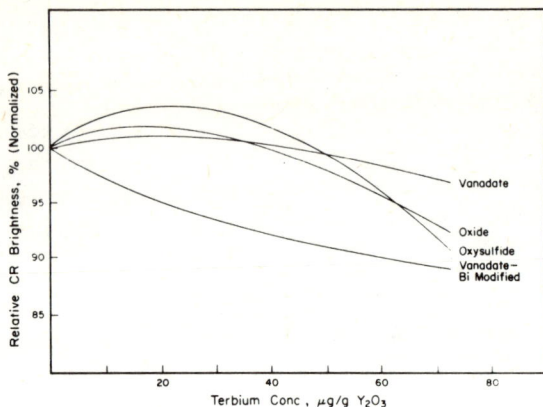

FIGURE 7

Effect of terbium on the cathode-ray brightness of

$(Y,Eu)VO_4$, $(Y,Eu,Bi)VO_4$, $(Y,Eu)_2O_3$, and $(Y,Eu)_2O_2S$

Calcium, a common impurity in Y_2O_3 and Eu_2O_3 raw materials, is generally detrimental. Figure 8 shows increased brightness loss in both the
oxide and oxysulphide systems with increasing calcium content, especially
the oxysulphide system, where calcium concentrations (on the Y_2O_3 basis)
at 15 ppm quench brightness in some phosphors [11]. The replacement of Y^{3+}
by Ca^{2+} in the zircon structure of YVO_4 results in a deficiency of positive
charge which, in order to preserve electrical neutrality, must be compensated
for by any undesirable Eu^{2+} present going to the efficient Eu^{3+}. Calcium
also tends to neutralize the strong, positive charge field produced by any
cerium impurity present, thereby decreasing its harmful effect.

At < 100 ppm, bismuth enhances the intensity of the Eu^{3+} emission in
YVO_4 without affecting chromaticity; however, greater concentrations produce the characteristic broad band bismuth emission that does shift colour
[12]. As noted earlier, the chromaticity of the oxide and especially the
oxysulphide phosphor is critically dependent upon europium concentration.

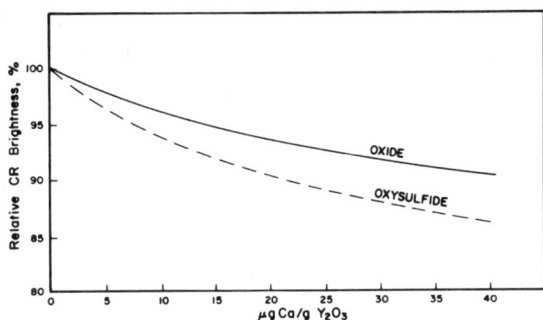

FIGURE 8
Effect of calcium impurity
on the cathode-ray brightness of $(Y,Eu)_2O_3$ and
$(Y,Eu)_2O_2S$.

This dependency is utilized in controlling colour in these phosphors. In
the vanadate system, the europium concentration has little effect on colour
over a rather broad range. Bismuth coactivation of $(Y,Eu)VO_4$ found commercial utility in colour television during a period when brightness was a particular concern, the very slight colour shift due to the yellow bismuth
emission increasing total intensity [13].

The alkali metal impurities are not usually considered detrimental to
brightness at concentrations less than 100 ppm. They can affect particle
growth and distribution, especially lithium and potassium. In the vanadate
system, practically any source of potassium ion will inhibit growth. This
effect is used commercially as one of several methods of controlling growth
and distribution [14]. Lithium, on the other hand, generally favours
growth even at very low concentrations [15].

Generally, the vanadate and oxysulphide phosphors' particle characteristics are more readily controlled than those of the oxide phosphor. This
in part is due to the relatively large excesses of vanadium and sulphur utilized in their synthesis to obtain optimum luminescence and composition.
In the oxide system this is not the case; therefore, the alkali metal
fluxes are commonly used to achieve the desired particle characteristics.
Another approach to controlling particle size and distribution in the oxide
phosphor is the utilization of yttrium oxide raw materials that possess a

discrete particle morphology and size. Most commercial sources of yttrium
oxide range in size from about one to three micrometers and lack any well
defined crystallinity. Phosphors synthesized from this type of yttrium
oxide also lack the size and crystallinity desired. We have found that a
homogeneous mixture of well defined cubic yttrium oxide crystals can be prepared, at various sizes, by resuspending thoroughly washed and dried yttrium
oxalate trihydrate in an oxalic acid solution before conversion to the oxide
[16].

Dysprosium is an efficient activator in all three red systems [17].
In the vanadate, dysprosium concentrations of one ppm show the characteristic Dy line emissions in the broad band blue matrix emission of YVO_4.

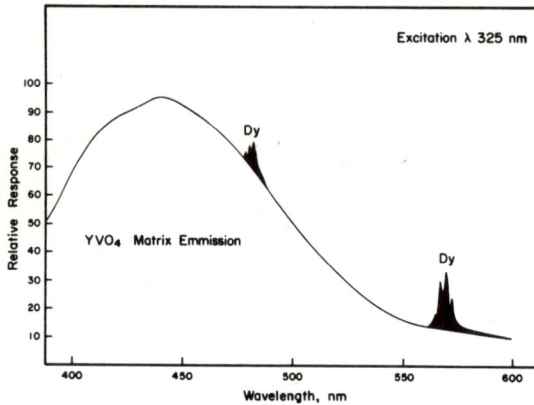

FIGURE 9

Spectral energy distribution of dysprosium contaminated YVO_4

Figure 9 shows the emission spectra of pure YVO_4 doped with one ppm Dy.
With europium activation, any dysprosium impurity present is in direct competition for energy, diluting the desirable optimum effects of europium.
In the $(Y,Eu)_2O_3$ phosphor, dysprosium impurity not only quenches luminescence
emission but also increases phosphorescence emission (decay). Figure 10

shows the increased decay time and decreased brightness of $(Y,Eu)_2O_3$ phosphor with increasing amounts of dysprosium. This phosphorescence is generally not desirable in colour television applications.

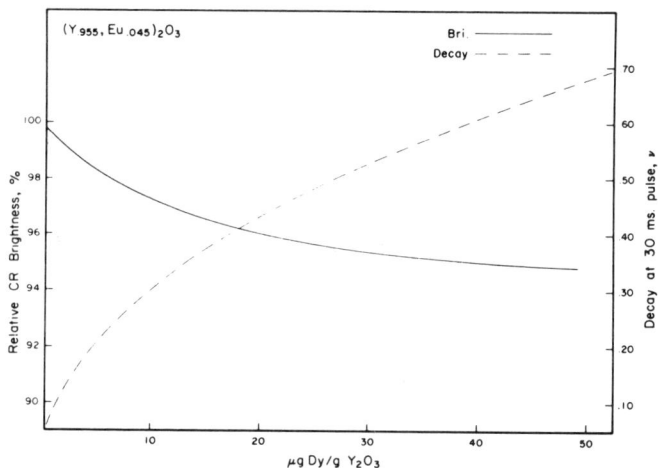

FIGURE 10

Effect of dysprosium on the cathode-ray brightness
and decay time of $(Y,Eu)_2O_3$

We have found that small additions of terbium and/or praseodymium in the presence of dysprosium impurity will effectively reduce decay in $(Y,Eu)_2O_3$ phosphor as shown in figure 11 [18].

It has been pointed out that dark body coloured phosphors are undesirable in the colour television screening process. While none of the three red systems possess a body colour equivalent to magnesium oxide, every effort is made to maintain discoloration at a minimum. Of course, other preparation parameters also affect body colour. The elements frequently found in virgin and reclaim sources of rare earth raw materials, that can contribute to poor body colour are: Zn, Bi, Cu, Ti, V, Cr, Mn, Fe, Co, Ni, Cd, and Ag. Table 3 shows the effect on body colour of several elements in the oxide

TABLE 3

Effect of Selected Elements on

the Body Colour of $(Y,Eu)_2O_3$ Phosphor

	BODY COLOUR	
Element	Unfluxed System	Fluxed System
Ti	White	Very Light Brown
V	Light Yellow	Light Yellow
Co	Brown	Green
Mo	Light Yellow	Light Yellow
W	Off White	White
Mn	Brown	Grey
Re	Yellow	Very Light Yellow
Fe	Off White	Pinkish
Ru	Brown	Grey
Os	Off White	White
Co	Grey	Grey
Rh	Grey-White	Grey-White
Ir	Very Light Orange	Very Light Yellow
Ni	Brown	Grey
Pd	Light Yellow	Grey
Cu	Grey-Green	Grey-Green
Ag	Light Purple	Off White
Zn	White	Off White
Cd	White	Light Yellow
Ge	White	Very Light Yellow
Pb	Light Yellow	Light Yellow
Sb	White	Purplish-White
Bi	Yellowish-White	Yellowish White

system. Generally, these elements are in the 10 - 100 ppm range.

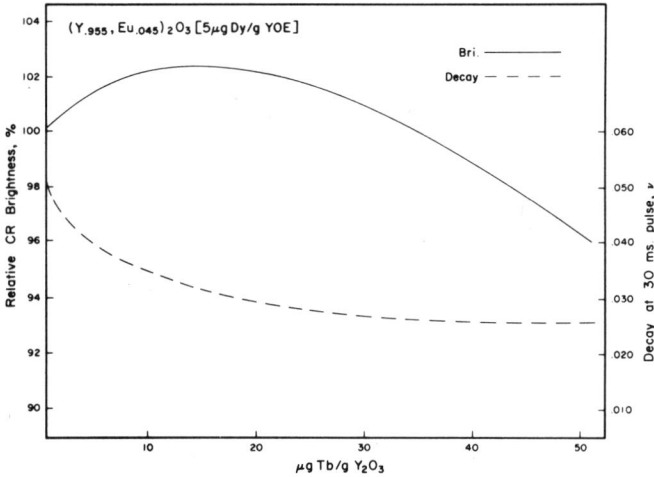

FIGURE 11

Effect of terbium on the decay time of dysprosium doped $(Y,Eu)_2O_3$

The total effect of impurities must also be considered. Often times, higher concentration of a particular impurity can be tolerated in an otherwise pure material. In a less pure material, this same impurity may be extremely detrimental. For example, we have found that the rare earth elements Sm, Tm, Yb, Er, and Ho individually have an insignificant effect on the red phosphors at concentrations less than 100 ppm. However, in combination, these impurities will affect brightness at a total concentration as low as 40 ppm as shown in table 4.

TABLE 4

Total Rare Earth Impurity Effects in (Y,Eu)$_2$O$_3$ Phosphor

Ce	Pr	Nd	Sm	Gd	Tb	Dy	Ho	Er	Tm	Yb	Lu	Total	Cathode-Ray Brightness, % Difference vs. Standard
	15		15	15	15		15					75	− 4
	15	15			15				15			60	− 18
50	15				15							80	− 30
5	5				10							20	− 3
5	10		15									30	− 7
	10	5										15	+ 3
	15	5	15		15		15					65	− 2
2				15	5							22	− 5
2			15				15	15		15		62	− 4
2				15								17	− 2
					15		15	15		15		60	+ 2
			10			10						20	=
			10	15	15	10	10					60	+ 2
					5	5	10					20	+ 2
							10	10	10	10	10	50	=
10	10			100								120	+ 5
	10	5	10		5	10			10	5		55	− 15
2	10	5										17	− 8
< 2	< 5	<0.1	< 5	< 1	<0.2	< 0.2	<10	< 2	< 2	<0.5	<15	None Det.	− 5
5			10	150	10		10	10		10	15	220	+ 4
	10		10	150								170	− 7
					15	50						=65	− 8
5	< 5	<0.1	< 5	< 1	<0.2	< 0.2	<10	< 2	< 2	<0.5	<15	> 5	− 5
15			10	100	5		10	5		10	100	255	+ 4

Rare Earth Impurity Concentration, μg/g Y$_2$O$_3$

7. FUTURE RARE EARTH RAW MATERIAL REQUIREMENTS

Very comprehensive experimental work has been completed during the past several years pertaining to the effect of both rare earth and non-rare earth impurities in the rare earth red phosphors [7,19,20]. The overall situation is extremely complex. The study of possible combinations of impurities and their effect on a particular modification of any given phosphor system (and we are dealing with at least three systems) is a never-ending process.

The critical nature of many impurities, the limitations of internal rare earth raw material control, the economics involved, and the dynamic nature of phosphor processing, collectively, create a situation that demands the highest quality of raw materials obtainable. The rare earth industry, with a few exceptions, has met this challenge. While the situation has been highly competitive, apparently it is not too late for newcomers, as evidenced by the efforts of the Metal Extractor Group of Norway.

The future quality and quantity needs of the colour television and lamp phosphor industry will depend a great deal upon the device. With the current commercial devices, the europium-activated oxide, vanadate, and oxy-sulphide phosphors apparently have no serious threat. However, I feel certain that continued progress in the device field will yield major changes in the present method for displaying an image, probably necessitating either completely different phosphor requirements or their elimination altogether.

References

[1] A.K. Levine and F.C. Palilla, Appl. Phys. Letters $\underline{5}$, 118 (1964).

[2] F.C. Palilla and A.K. Levine, Appl. Optics $\underline{5}$, 1467 (1966).

[3] S. Faria and E.J. Mehalchick, U.S. Patent 3,480,819 (1969).

[4] J.E. Mathers and R.L. Yale, U.S. Patent pending.

[5] J.E. Mathers, F.F. Mikus, and E.J. Mehalchick, U.S. Patent 3,629,131 (1971).

[6] M.R. Royce, U.S. Patent 3,418,246 (1968).

[7] E.J. Mehalchick, F.F. Mikus, and J.E. Mathers, J. Electrochem. Soc. $\underline{116}$, 1017 (1969).

J.E. Mathers and E.J. Mehalchick, U.S. Patent 3,574,130 (1971).

[8] L.G. VanUitert and R.R. Soden, J. Chem. Phys. 36, 1289 (1962).

[9] R.C. Ropp, J. Opt. Soc. Amer., 57, 213 (1967).

[10] T. Kano, Y. Otomo, and H. Yamamoto, U.S. Patent 3,655,577 (1972).

[11] F.F. Mikus, J.E. Mathers, and F.C. Palilla, U.S. Patent 3,360,675 (1967).

[12] S.Z. Toma, F.F. Mikus, and J.E. Mathers, J. Electrochem. Soc. 114, 953 (1967).

[13] F.F. Mikus and J.E. Mathers, U.S. Patent 3,360,674 (1967).

[14] S.Z. Toma, J.E. Mathers, and F.F. Mikus, U.S. Patent 3,424,692 (1969).

[15] R.L. Hickok, U.S. Patent 3,562,175 (1971).
Y. Kobayoshi, M. Masuda, S. Murayama and H. Mizyno, U.S. Patent 3,458,451 (1969).

[16] J.L. Ferri and J.E. Mathers, U.S. Patent 3,635,658 (1972).

[17] R.A. Buchanan, K.A. Wickersheim, J.L. Weaver, and E.E. Anderson, J. Appl. Phys. 39 (9), 4342 (1968).
S. Faria and D.T. Palumbo, J. Electrochem. Soc. 116, 157 (1969).
M.R. Royce and A.L. Smith, Electrochem. Soc. Meeting, Abstract no. 34, Boston, May 1968.
S. Faria and D.T. Palumbo, U.S. Patent 3,555,337 (1971).

[18] J.L. Ferri and J.E. Mathers, U.S. Patent 3,639,932 (1972).

[19] T. Kano and Y. Otomo, Electrochem. Soc. Meeting, Abstract no. 38, Boston, May 1968.

[20] G.L. Thompson, U.S. Patent 3,322,682 (1967).

THE SURFACE ACTIVITY OF SOLID RARE EARTH OXIDES

by

Menachem Steinberg

Department of Inorganic and Analytical Chemistry,

The Hebrew University, Jerusalem, Israel

ABSTRACT

Although the rare earth oxides have been used as heterogeneous catalysts since the 19th century it is only recently that the interest in them
has become widely spread, presumably because of their greater availability.
This availability has opened new horizons in their use as catalysts. In
this paper their activity in parahydrogen conversion, hydrogen deuterium
equilibration, hydrogenation and dehydrogenation of hydrocarbons, oxidation
of carbon monoxide, and their use in air pollution research will be presented and discussed. Epr results are reported on the adsorption of oxygen on
cerium (IV) oxide. The mechanism of the catalytic oxidation of carbon monoxide will be discussed according to the epr data. The rare earth oxides
are used as catalysts both as pure oxides and supported on a carrier.[*]

1. PARA-HYDROGEN CONVERSION AND HYDROGEN-DEUTERIUM EQUILIBRATION

The catalytic effects of paramagnetic solids on ortho-para hydrogen
conversion are well known. Although the magnetic properties of the rare
earth oxides are well known, not many investigations have been made on the
influence of the rare earth oxides on the para-hydrogen conversion. Moreover, comparative studies of the conversion catalyzed by these oxides are
few. Recent work by Ashmead et al. [1] showed that an apparent relationship
exists between the magnetic moment of the trivalent cations in the sesquioxides and the rate of conversion constants. They were able to demonstrate
a constancy in the value k/μ^2 where k is the rate constant and μ the magnetic moment in Bohr Magnetons (at room temperature). Their study on the para-

---

[*] There is a glossary at the end giving some definitions in catalysis.

hydrogen conversion was carried out at low temperatures on the oxides Nd_2O_3, Sm_2O_3, Gd_2O_3, Dy_2O_3 and Er_2O_3. A poorer constancy in the value of k/μ^2 was obtained by Selwood [2], who ascribes this to the uncertainty in estimating the actual specific surface taking part in the catalytic conversion. Even so, the values obtained by him are quite satisfactory taking into account the errors involved in the specific surface - BET method. On the other hand, at higher temperatures the rate of conversion is only slightly greater as compared to the hydrogen-deuterium equilibration [1].

The results may indicate that these two reactions are governed by the same mechanism, the rate controlling step being the dissociative adsorption of hydrogen. The somewhat higher rate for the conversion reaction may be due to a contribution from a magnetic mechanism.

Recent work carried out by Selwood [2] shows a very interesting relationship between the effects of an extrinsic magnetic field on the para-hydrogen conversion. It was shown that the conversion rate at a high magnetic field (18 kOe) and room temperature is significantly higher (positive magneto-catalytic effect). If must be emphasized, however, that the magnetic properties of the surface may play a decisive role in the catalytic activity. These magnetic properties may be caused by free surface valences.

2. CONVERSION OF HYDROCARBONS

The rare earth oxides are active in the hydrogenation and dehydrogenation of hydrocarbons [3]. Some data have been published by Komarevsky [4] about the selectivity of these oxides in dehydrogenation reactions. It was found that at 525° neodymium and samarium oxides are active in dehydrogenation of paraffins but not in cycloparaffines, differing in this respect from chromia and vanadia which are active for both types of compounds. In the dehydrogenation of cyclohexane to benzene it was found that at 545° the activity of the oxides increased with the formula weight of the oxide [5]. However, it was also realized that the specific surface area of the oxides monotonously increases with the formula weight of the rare earth oxides. A very interesting observation in Minachev's [5] work was that the apparent activation energies of the dehydrogenation were inversely proportional to the changes in the effective magnetic moments of the trivalent rare earth ions. No explanation was given. The oxides La_2O_3, Er_2O_3 and Ho_2O_3 were reported to be good catalysts for the hydrogenation of unsaturated hydrocarbons. In reactions of dehydrocyclization [5], it was realized that the reactivity follows the same trend as in the dehydrogenation of cyclohexane.

Reports appeared in the literature, mainly by Russian workers, on the catalytic dehydrogenation by rare earth oxides supported on carriers. The work carried out on carbon as a carrier in the dehydrogenation of cyclohexane showed that the activity of the catalyst is not substantially affected by the nature of the oxide. The results differ from those obtained with the pure rare earth oxides. It is very interesting to notice that in reactions where hydrogen is present, such as in cyclohexane dehydrogenation, hydrogen deuterium equilibration and high temperature parahydrogen conversion, the rare earth oxides show similarly or regularly changing activities.

The discovery of the effect of rare earths on the catalytic properties of the "molecular sieves" in the early sixties [6] has resulted in improved refinery efficiencies and greater market flexibility in the production of petroleum products. These rare earth molecular sieves have already improved the yield of gasoline products by 87%.

3. THE ADSORPTION OF OXYGEN ON CERIUM (IV) OXIDE

The cerium (IV) oxide used in our work was of stated purity 99.9999% with respect to rare earths. The oxide which was prepared by the conventional method of decomposition of cerium (III) oxalate at 900°C in air, showed an epr signal at $g = 1.96$ [7]. The signal is not expected from the electronic configuration of the quadrivalent cerium cation $[Ce] \, 4f^o \, 5d^o \, 6s^o$. The intensity of this signal changes by the treatment of the oxide; when the time of heating is longer, the intensity of the signal is higher until it reaches its maximum. Also, when the temperature of the heating is higher, the number of spins increases. On the other hand, when the oxide is prepared by decomposition in air of cerium (III) oxalate at lower temperatures (375°) a very small signal is observed at $g = 1.96$ [8]. The results from these experiments are in a very good agreement with the well known fact that

CeO_2 is a non-stoichiometric compound, CeO_{2-x} This oxide is a semiconductor of the n-type. The epr signal may be ascribed to quasi-free electrons, e^-, in the conduction band:

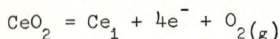

$$CeO_2 = Ce_i + 4e^- + O_{2(g)}$$

where Ce_i stands for an interstitial quadrivalent cerium cation. The intensity of the signal is a function of e^-. In the low temperature preparation x is smaller and hence the concentration of e^- is low which results in a small signal. By heating the low temperature preparation in vacuum at higher temperatures it was found that the epr signal increases. Moreover, oxygen is evolved by this heating, thus increasing x in CeO_{2-x}. Oxygen is adsorbed on cerium (IV) oxide [8]. Simultaneously, with the adsorption, a new signal appears at g = 2.020. The intensity of the signal at low oxygen pressures is a function of the oxygen equilibrium pressure. When oxygen is desorbed the intensity of the signal decreases. It is suggested that the g = 2.020 signal is from the anion-molecule O_2^-:

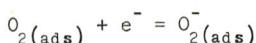

$$O_{2(g)} = O_{2(ads)}$$

$$O_{2(ads)} + e^- = O_{2(ads)}^-$$

Recent results obtained in our laboratory [9] by using $^{17}O_2$ support the suggestion given above.

4.   THE OXIDATION OF CARBON MONOXIDE

The rare earth oxides catalyze the oxidation:

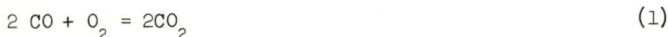

$$2 CO + O_2 = 2CO_2 \tag{1}$$

The mechanism of this oxidation reaction with ceria as the catalyst may be explained by the results mentioned above:

$$CO_{(g)} = CO_{(ads)} \tag{2}$$

$$CO_{(ads)} + O_{2(ads)}^- = CO_{3(ads)}^- \tag{3}$$

$$CO_{3(ads)}^- = CO_{2(g)} + O^- \tag{4}$$

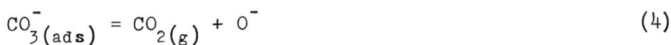

The $O^-_{(ads)}$ species may take part in the carbon monoxide oxidation reaction. However, beyond a critical concentration of $O^-_{(ads)}$ on the surface, which depends on the specific surface of the oxide, the following reaction may take place:

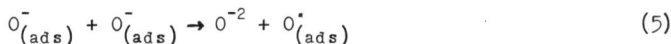

$$O^-_{(ads)} + O^-_{(ads)} \rightarrow O^{-2} + O^{\cdot}_{(ads)} \tag{5}$$

O^{-2} is the oxidic anion. By this reaction $O^{\cdot}$ radicals are provided for the reaction. These radicals may react with carbon monoxide:

$$CO_{(ads)} + O^{\cdot}_{(ads)} = CO_{2(ads)} \tag{6}$$

The $CO_{2(ads)}$ can also react with $O^-_{(ads)}$

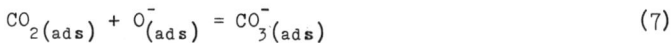

$$CO_{2(ads)} + O^-_{(ads)} = CO_{3(ads)}^- \tag{7}$$

This reaction is followed by (4).

The concentration of quasi-free electrons is lowered by (5) and thus the reaction rate decreases with time. At the last stage of the reaction the concentration of electrons on the surface is very low as was shown from esr measurements[*]. A support of the suggested mechanism of CO_3^- formation is found in the results of a study on CO_2 adsorption on CeO_2. When oxygen is not present the rate of desorption is high, whereas in the presence of oxygen it becomes lower.

In a recently published work by Claudel and Veron [10] it was shown that the rate of oxidation of carbon monoxide on a catalyst of cerium, supported on thoria as a carrier, depends on the electrical conductivity of the solid. The conductivity, on the other hand, depends on the thermal

---

[*] Unpublished results

treatment of the catalyst; the longer the treatment time, the higher the conductivity. These results support the mechanism of the carbon monoxide oxidation proposed above.

5. THE USE OF RARE EARTH OXIDES IN AIR POLLUTION RESEARCH

Libby [11] has recently reported that the d-banded solid $LaCoO_3$ is active as a catalyst in the conversion of cis-2-butene to n-butane. This compound shows great promise in the petroleum industry and may be useful as a catalyst for automobile pollution control. Libby suggests that $LaCoO_3$, which he calls "the poor man's catalyst", should be tested as a potential automobile exhaust catalyst. In a later report [12] it is suggested that there is little difference between various preparations of $XCoO_3$ (X = rare earth cation). A very important and interesting result from this work is the finding that a naturally unseparated mixture of rare earths (RE) can be used in the preparation of $RECoO_3$, giving a striking similarity in the reactivity. A communication by Vickery [13] reports on results obtained by the author in 1958 in which Pr_6O_{11} and Tb_4O_7 were treated as catalysts for automobile exhaust gases and found to be catalytically active even in the presence of lead. Vickery suggests that rare earths, and particularly those exhibiting paramagnetism, possess a high potential in this area. It seems, however, that this is not the only property that affects the catalytic activity. Results from the Bell Laboratories recently announced, show that $Nd_{1-x}Pb_xMnO_3$, $PrCoO_3$ and similar compounds are promising as exhaust catalysts [14]. There is yet another topic which awaits active research on the activity of rare earth oxides, namely the conversion of NO that might be useful in the control of oxides of nitrogen, NO_x, found in the product gases from high temperature combustion.

References
[1] D.R. Ashmead, D.D. Eley and R. Rudham, J. Catal. 3, 280 (1964);
     Trans. Faraday Soc. 59, 207 (1963).
[2] P.W. Selwood, J. Catal. 19, 353 (1970); ibid 22, 123 (1971).

[3] C.B. McGaugh and G. Houghton, J. Phys. Chem. 65, 1887 (1961);
     K.V. Topichieva and F.M. Ibragimova, Russ. J. Phys. Chem. 41, 812
     (1967).

[4] V.I. Komarevsky, Industr. Engng. Chem. 49, 264 (1957).

[5] Kh. Minachev and M.A. Markov, Problem of Kinetics and Catalysis 11,
     248 (1966) (Translated by Israel Program for Scientific Translations
     Ltd. - 1968, into English).

[6] (a) C.J. Plank and E.T. Rosinsky, U.S. Pat. 3,140,253 (July 7, 1964)
     and U.S. Pat. 3,271,322 (Sept. 6, 1966).
     (b) R.L. Koffler, Preprint 11b AIChE, Materials Conference on Rare
     Earth Applications, March 31 - April 4, 1968.

[7] M. Steinberg, Israel J. Chem. 8, 877 (1970).

[8] M. Gideoni and M. Steinberg, J. Solid State Chem. 4, 370 (1972).

[9] N. Kauffher, C. Naccache and M. Steinberg, Israel J. Chem. (Proceedings),
     (in press).

[10] B. Claudel and J. Veron, Compt. Rend. Acad. Sci. Paris C 267, 1195
     (1968).

[11] W.F. Libby, Science 171, 499 (1971).

[12] L.A. Pedersen and W.F. Libby, ibid. 176, 1357 (1972).

[13] R.C. Vickery, Science 172, 86 (1971).

[14] RIC News (K.A. Gschneidner Jr., Ed.) 7, 3 (1972).

GLOSSARY

Absorption - The transfer of a substance through a phase boundary
        followed by solution or a form of combination with the receiving
        phase.

Activity - The ability of a catalyst to alter the rate of chemical
        reaction.

Adsorption - The adhesion of molecules of gases or of dissolved substances
        or liquids to the surface of solid bodies with which they are in con-
        tact.

BET-method - A method of estimating surface area from physical adsorp-
        tion isotherms.

Carrier - A physical support for a catalyst deposited in or on it.

Catalysis - A phenomenon relating to a change in the rate of a chemical reaction, either an increase or decrease, which is brought about by a substance not appearing in a compound resulting from the reaction.

Extrinsic Field - The magnetic field coming from an external field.

Intrinsic Field - The magnetic field produced by a paramagnetic species in the solid.

Selectivity - The preferential activity of a catalyst for a specified reaction.

Specific Surface - The surface area per unit weight of a carrier or catalyst as measured by gas adsorption ($m^2 \cdot g^{-1}$).

MISCELLANEOUS USES OF RE MATERIALS

(CERAMICS, GLASSES, NUCLEAR APPLICATIONS, ETC.)

by

Karl A. Gschneidner, Jr.

Ames Laboratory USAEC and Department of Metallurgy

Iowa State University, Ames, Iowa 50010, U.S.A.

ABSTRACT

The more important commercial uses of the rare earth elements in ceramics, glasses and the nuclear area are described. Several other (miscellaneous) applications, which are not included in the various categories reviewed at the Conference, have been noted and discussed. A few unusual uses
are also mentioned in the four categories reviewed in this paper.

1. INTRODUCTION

One difficulty in preparing this paper is knowing exactly what the
other authors are going to include or exclude in their papers. Hopefully,
there are no serious omissions and only a few minor overlaps, if any.

Although glasses can quite properly be included with ceramics, there
are enough important uses involving glasses that it is appropriate to make
a special category of them. Under the term ceramics I will consider only
crystalline inorganic materials, inorganic substances which have amorphous
structures are considered glasses.

2. CERAMICS

Most of the ceramic uses are minor in nature, but since they use separated rare earths in many of the applications, the dollar value is not insignificant.

The rare earths are added to ZrO_2 to stabilize the cubic modification, preventing the monoclinic and tetragonal phases from forming between room temperature and the melting point. In general Y_2O_3 is the rare earth which is usually added (5 to 12 mol%) to ZrO_2. Although CaO and MgO also stabilize ZrO_2, there is a tendency for the calcia- or magnesia-stabilized zirconia to deteriorate during prolonged high temperature usage. Thus, yttria-stabilized zirconia is used in high temperature portions of furnaces, heaters, etc. The chemical stability of zirconia is also reported to be enhanced by yttria. The yttria-stabilized zirconia is available as felt, yarn, tape, roving, woven cloth, and fibrous board for a large variety of high temperature applications [1,2].

The rare earth oxides, primarily CeO_2, La_2O_3 and Nd_2O_3 are added to $BaTiO_3$ to improve its capacitance properties. A variety of different effects can be achieved by adding from 0.5 to 50 wt.% rare earth oxide. These include reduced aging, increased capacitance and changing of the temperature coefficient of capacitance from positive through zero to negative. Furthermore, rare earth additions lower the resistivity by as much as 10^{10} ohm-cm and account for the use of rare earth-doped $BaTiO_3$ as thermistors (temperature-sensitive resistors) and thermistor switches [3].

Minor uses involving yttria in a wide variety of applications include: crucibles for molten metals [4], thermocouple insulators [5], transparent ceramics [2,6] and thin film capacitors [7]. The transparent ceramic can be prepared by high pressure techniques [6] or by addition of 10% ThO_2 followed by normal sintering methods [2]. The greatest drawback for this material is that under high temperature reducing conditions, it forms a black sub-oxide, Y_2O_{3-x} where $x \simeq 0.01$ and the material is no longer transparent.

An important use for praseodymium is the addition of 3 to 5 mol% Pr_6O_{11} to zirconia to give a bright yellow glaze to ceramic tiles [8]. One of the more important electron gun cathode materials is LaB_6, and it is finding use in electron microscopes and other equipment which require electron emissive devices [9].

There are many other potential uses of rare earths in or as ceramic materials. Three of the most exciting ones are: lanthanum (7 at.%)-modified lead zirconate-lead titanate (PLZT), an electro-optic ceramic* [10,11]; gadolinium molybdate, a ferro-elastic compound* [12]; and the

___

* This use is discussed in more detail in the chapter by P.N. Yocom.

rare earth cobaltites as ceramic high temperature cathodes [13] and auto exhaust catalysts* [14]. The lanthanum addition to PLZT makes it transparent.

3. GLASSES

The largest uses of the rare earths in the glass industry are as decolourizers and polishing compounds. Their uses as colouring agents and in fibre optics, lenses and laser cladding glasses are minor in comparison to the first two.

In the last two years there has been a very large acceptance of cerium as decolourizer in the container class industry in the United States. The cerium decolourizes glass (primarily flint glass) by oxidizing Fe^{+2} impurities, which impart a bluish colour to glass, to Fe^{+3} which has a faint yellow colour. Cerium, in addition to being an effective oxidizing agent (1) is colourless in both of its ionic states, +3 and +4, (2) is an efficient absorber of ultraviolet light, and (3) reduces the amount of complementary colourants which need to be added to "neutralize" the colour due to Fe^{+3} and other impurities in the glass. About 0.25 wt.% CeO_2 is sufficient to decolourize all iron levels up to 0.05% Fe_2O_3. At the same time, the complementary colourants Se and/or Co concentrations can be reduced by one-half by the use of cerium as the oxidizing agent instead of arsenic which had been previously used. In Europe a large amount of neodymium is used as a complementary colourant to "neutralize" the yellow colour of Fe^{+3}. The absorption of ultraviolet light is beneficial since it reduces the rate of deterioration of the product in the bottle; this is especially important for vitamins, drugs and foods [8,15]. Cerium is also used in glasses subjected to γ-ray or electron radiation which would turn brown due to iron impurities in the glass. This is especially important for TV tubes and space applications [16].

The rare earths have been used for many years as polishing compounds in the glass industry. Most polishing materials have the same composition as the parent ore, containing 45 - 50% CeO_2. Specialized high-speed polishing materials require higher CeO_2 concentrations up to 90%. The rare earths are used over cheaper polishing compounds, such as rouge, because they are

---

* This use is discussed in more detail in the chapter by M. Steinberg.

cleaner, faster, have a longer life and give a superior finish. They are
used to polish plate glass, mirrors, lenses (optical, camera and precision),
TV tubes, marble, onyx, granite, gems and semiconductor wafers.

Lanthanum oxide is added to optical glasses to increase the index of
refraction, accounting for its use in camera lenses and fibre optics. Rare
earth oxides are added to colour decorative glassware. Neodymium oxide
gives a reddish-purple colour, Nd_2O_3/Se combination pink and CeO_2/TiO_2 combination yellow [8]. Neodymium-praseodymium oxide mixture in the ratio as
found in their common ores is used in glassblower's goggles to filter out
the intense yellow colour due to the presence of sodium. A Pr/Ce oxide combination is used in welders' goggles to filter out the blue, violet and
ultraviolet radiation [6]. The addition of 1.5% cerium hydrate to glass
produces the pink tint which is so popular in today's fashion glasses.

4. NUCLEAR

Major uses of the rare earth elements in the nuclear field are concerned
with their use as control rod materials and burnable poisons. The nuclear
cross-sections of selected rare earth elements are shown in Table 1. The
four rare earth elements, which have cross-sections greater than 1100 barns,
are the highest known for naturally occurring mixtures of isotopes. Although
Gd and Sm have the two highest cross-sections, Eu and Dy are of greater inter
est for control rod use because they have several large cross-section isotopes which are formed successively one after another (see Table 2). In contrast the Sm and Gd isotopes, which have high cross-sections, form by neutron capture isotopes which are essentially transparent to neutrons. And
thus, the chain is broken, since the higher mass isotopes with large cross-
sections cannot be formed at least in sufficient quantities to sustain the
chain. For ^{151}Eu and ^{160}Dy isotopes, a maximum of five neutron captures is
possible before they become worthless as absorbers. For naturally occurring
Eu, 3.96 neutrons are absorbed per atom and for Dy 2.41, which compares to
0.4 for Sm and 0.3 for Gd. The large capture cross-sections for Gd and Sm,
however, make these two materials quite attractive in reactor shutdown and
safety devices [17,18,19].

TABLE 1

Nuclear Properties of Large Cross-Section Rare Earths

Element	Absorption Cross-Section of Naturally Occurring Elements (barns/atom)
Sm	5600
Eu	4300
Gd	40,000
Dy	1100
Er	170
Tm	125
Lu	108

TABLE 2

Nuclear Properties of Sm, Eu, Gd and Dy Isotopes

Element	Isotope mass no.	Abundance (%) or (half-life)	Cross-Section (barn/atom)	Half-life of New Isotope After Neutron Capture [a]
Sm	144	3.16	< 2	400 d
	147	15.07	87 ± 60	---
	148	11.27	---	---
	149	13.84	40,800 ± 900	---
	150	7.47	---	---
	151	(73y)	10,000 ± 200	---
	152	26.63	224 ± 7	47 h
	154	22.53	6 ± 1	24 m
Eu	151	47.77	9,000	9.2 h
	152	(13y)	5,000	---
	153	52.23	420 ± 10	16 y
	154	(16y)	5,000 ± 400	1.7 y
	155	(1.7y)	14,000 ± 4000	15.4 d
Gd	152	0.020	125	230 d
	154	2.15	---	---
	155	14.73	61,000 ± 5000	---
	156	20.47	---	---
	157	15.68	240,000 ± 12,000	---
	158	24.87	4	180 d
	160	21.90	7	3.6 m
Dy	160	2.298	130 ± 130	---
	161	18.88	680 ± 40	---
	162	25.53	240 ± 30	---
	163	24.77	220 ± 50	---
	164	28.18	2780 ± 150	---

[a] y = years, d = days, h = hours, and m = minutes.

The use of Gd and Sm as burnable poisons has been described in the literature. Because of the need for enriched fuel in power production re- actors, fissionable isotopes are added in an amount in excess of their cri- tical mass, and thus a neutron absorber (the burnable poison) is added to compensate for the excess core reactivity. Ideally, the depletion of the fuel and the poison occurs in such a manner that the reactivity remains constant throughout the core lifetime. Usually, this is not realized, but it can be compensated for by moving the control rods [17,18,19].

The individual rare earths in these two applications are used primarily in the form of mixed oxides, e.g., R_2O_3/Al_2O_3 or as oxide dispersions in stainless steel [18,19]. In the case of Eu control rod materials, one must be careful to be sure that Eu is in the trivalent state. The reason for this is that all the final daughter products of the neutron capture pro- cesses are various Gd isotopes, and if Eu were in the divalent form in the control rod material, there would be a volume contraction of as much as 30% as the trivalent Gd atoms are formed, leading to dimensional instabil- ity of the control rods [20]. Fortunately, all the reactors using Eu con- trol rods today involve Eu_2O_3.

Minor uses will include the rare earths as diluents in fuels, primari- ly Ce and Y because of their low cross-sections [17], in neutron radiography imaging devices [21] and as radioisotopes. The high neutron cross-section of Gd and Dy, along with the fact that after the capture of a neutron they emit β- and γ-rays, makes these materials useful in neutron radiography. The emitted β- and γ-rays are allowed to strike an x-ray film to record the image. Gd is used in situ because of the short half-life of the emitting isotopes, but Dy, which continues to emit the rays after the neutron beam is shut off, can be transported to another site to expose the x-ray film.

Some of the rare earth radioisotopes being used or considered for vari- ous devices or applications will be found listed in Table 3.

5. MISCELLANEOUS

Miscellaneous uses include rare earth core carbon arcs, yttrium alumin- ium garnets (YAG) as substitute diamonds, cerium hydrate in silicone rubber

gaskets, lanthanum trichloride in blood analysis and all the rare earths
in scientific research.

TABLE 3

Applications of Rare Earth Isotopes

Isotope	Radiation	Uses
Y^{90}	β/γ	Cancer therapy [22,23], treatment of epilepsy [23]
Ce^{144}	β/γ	Thermoelectric generators, industrial gauges [22]
Pm^{147}	β/γ	Thermoelectric generators [22], industrial gauges [22], industrial x-ray fluorescence [22], lighting dials and switches [24], radioisotope thrustors [25], vacuum gauges [25], power sources [26], cardiac pacemakers [27], detector for Sn and Ba in ores [28]
Gd^{153}	γ	Industrial x-ray fluorescence [22], detector for W and Pb in ores [28]
Tm^{170}	β/γ	Industrial radiography [22], medical teletherapy [22], power sources [29]
Tm^{171}	β/γ	Power sources [30]
Yb^{169}	γ	Portable x-ray sources [31], radiography [32]

Carbon arc electrodes have been a major application for the mixed rare
earths for about 70 years. Rare earth fluorides or oxides added to the core
of a carbon electrode increase the arc intensity by a factor of ten. This
is due to the electrons, which have been excited into higher energy levels
in the arcing process, emitting light as they return to their lower energy
levels. The emitted light is essentially identical to natural sunlight.
The amounts of the various rare earths present is important to obtain the
proper balance, and thus, essentially all the carbon arcs contain a rare
earth mixture based on the distribution of rare earths as found in monazite.
Other ores such as bastnasite are not acceptable unless the rare earth composition is modified [33].

The hardness (8.5 on Mohs' scale compared to 10 for diamond), colour
and refractive index of YAG make it an excellent substitute for diamond.
The YAG's can be cut and faceted into the same gem styles as diamonds, and

they sell at the retail level for about $40 per carat or about 1/100th of real diamonds [34].

Cerium hydrate is added (0.1 to 1.0%) to silicone rubber gaskets which are used in aircraft windows and laboratory environmental chambers. The cerium improves the heat stability and helps maintain the colour and elastic properties of the silicone rubber under extreme temperature variations.

Lanthanum chloride is used in blood tests for determining Ca concentrations by atomic absorption spectrophotometry. The presence of Na^+, $PO_4^=$ or $SO_4^=$ ions in the blood lead to erroneous results. The presence of Na^+ gives results which indicate that the calcium concentration is higher than the true value, while the $PO_4^=$ or $SO_4^=$ ions have the opposite effect. The addition of 0.1 to 0.5% (wt./vol. basis) $LaCl_3$ eliminates these interferences [35].

A major use of rare earths in terms of dollar sales is in basic and applied research. In general the rare earths are being used as separated high purity materials in the form of metals and compounds. There is, however, an appreciable fraction of research, mostly applied, concerned with the behaviour of the mixed rare earths. It is amazing, considering the volume of the market, that there is no basic research being carried out on the fundamental chemical, metallurgical and physical properties of the mixed rare earths of natural formulations. Thousands of tons of mixed rare earths are being sold yearly, and many properties of this pseudo-element are unknown, or at best estimated from the property of the pure elements and their composition in the mixture. I am sure if these data were known many more markets would be opened to the rare earths.

There are many more miscellaneous and minor uses which have not been or will not be discussed at the Conference. But because of time and space limits nothing more will be included in this paper. Additional information can be found in the Overview series published by Molybdenum Corporation of America,* the RIC News published by the Rare-Earth Information Center,* and the two extensive review articles by Mandle and Mandle [36,37].

---

* The addresses of these two institutions are given in ref.[1] for the Rare-Earth Information Center and ref.[2] for Molybdenum Corporation of America.

References

[1] RIC News 2 [2] 5 (June 1967); published quarterly by the Rare-Earth
    Information Center, Institute for Atomic Research, Iowa State Univer-
    sity, Ames, Iowa 50010, U.S.A.

[2] Overview [9]; published irregularly by Molybdenum Corporation of
    America, 280 Park Avenue, New York, New York 10017, U.S.A.

[3] Overview [21].

[4] RIC News 2, [1], 5 (March 1967).

[5] RIC News 4, [3], 8 (September 1969).

[6] R.A. Lefever and J. Matsko, Mater. Res. Bull. 2, 865 (1967).

[7] RIC News 3, [3], 4 (September 1968) and 6, [4], 4 (December 1971).

[8] Overview [32].

[9] RIC News 4 [1], 3 (March 1969), 4, [4], 4 (December 1969) and 6, [3],
    2 (September 1971).

[10] Chem. Eng. News. 49, [20], 28 (May 17, 1971).

[11] RIC News 6, [2], 3 (June 1971) and 7, [2], 3 (June 1972).

[12] RIC News 5, [3], 4 (September 1970), 6, [1], 4 (March 1971) and 7,
     [1], 2 (March 1972).

[13] RIC News 5, [3], 3 (September 1970).

[14] RIC News 6, [2], 2 (June 1971).

[15] Overview [13].

[16] Overview [6].

[17] W.K. Anderson, p. 522 in The Rare Earths, F.H. Spedding and A.H. Daane
     eds., Wiley, New York (1961).

[18] W.E. Ray, Nucl. Eng. 17, 377 (1971).

[19] Overview [11].

[20] K.A. Gschneidner, Jr., Nuclear Met. X, 115 (1964).

[21] RIC News 3, [4], 3 (December 1968) and 6, [1], 3 (March 1971).

[22] J. Silverman and E. Thro, Nucl. News. 11, [6], 49 (1968).

[23] RIC News 2, [1], 2 (June 1967).

[24] RIC News 1, [3], 2 (September 1966).

[25] RIC News 1, [4], 5 (December 1966).

[26] RIC News 2, [3], 4 (September 1967).

[27] RIC News 5, [4], 1 (December 1970).

[28] RIC News 7, [1], 2 (March 1972).

[29] RIC News 3, [2], 4 (June 1968).

[30] RIC News 2, [3], 4 (September 1967).

[31] RIC News 1, [3], 4 (September 1966).

[32] RIC News 2, [2], 10 (June 1967).

[33] "Trends in Usage of Rare Earths", NMAB-266 (October 1970); a report of the National Materials Advisory Board, National Research Council, National Academy of Sciences -- National Academy of Engineering, Washington D.C.

[34] Overview [22].

[35] Overview [14].

[36] R.M. Mandle and H.H. Mandle, p.416 in Prog. Sci. Tech. Rare Earths, Vol. 1, L. Eyring, ed., Pergamon Press, New York (1964).

[37] H.H. Mandle and R.M. Mandle, p.190 in Prog. Sci. Tech. Rare Earths, Vol. 2, L. Eyring, ed., Pergamon Press, New York (1966).

THEORETICAL BASIS FOR THE USE OF RARE EARTHS

IN MAGNETIC MATERIALS

by

G. Blasse

Solid State Department, Physical Laboratory,

University of Utrecht, Utrecht, The Netherlands.

ABSTRACT

   This paper deals with the magnetic properties of rare earth (RE) compounds. The origin of the magnetic moment of RE is described first. Then
we consider magnetic interactions between these moments. The mechanisms in
insulators are outlined and illustrated for some special cases (EuO, garnets,
perovskites). In metallic conductors the mechanism of the magnetic interaction is different because of the presence of conduction electrons.

   Our examples here are (Eu,Gd)X, RE metals, compounds of RE and aluminium,
and compounds of RE and the 3d-transition metals. Analogies and differences
between the magnetic properties of RE and transition-metal compunds are outlined where necessary.

1. INTRODUCTION

   The occurrence of spontaneous magnetic moments is restricted to the
presence of incompletely-filled d or f electron shells. In this paper we
restrict ourselves to incompletely filled 4f shells, i.e., the RE elements.
We will first discuss the value of the magnetic moment for a $4f^n$ configuration. Then we will consider the interaction between the moments. This is a
quite complicated problem and, therefore our approach will be nonmathemati-
cal and approximative. First insulators will be discussed. A comparison
with the interaction between 3d-metal ions is inevitable. We have chosen
examples in the field of the rocksalt, the garnet and the perovskite structure.

A different mechanism of magnetic interaction occurs if conduction electrons are present. Here we will also make the comparison with transition-metal compounds. From the great variety of available examples we have chosen the (Eu,Gd)X system, the rare earth metals, and compounds between the RE and aluminium and the 3d-metals, respectively. For more examples and a more elaborate treatment of this subject the reader is referred to the literature (see e.g. refs. 1 to 5).

2. THE MAGNETIC MOMENT OF RE

In our previous paper at this meeting we have introduced the energy-level scheme of the RE ions. For the value of the magnetic moment we are only interested in the ground level and levels very near to this ground level. The ground state for a $4f^n$ configuration can be found from Hund's rules. Take, for example, $Eu^{3+}(4f^6)$ and $Er^{3+}(4f^{11})$. Now the ground state should have maximum spin-multiplicity $2S + 1$ and maximum L consistent with this multiplicity. If the shell is less than half-full $J = L - S$, otherwise $J = L + S$ (S and L are the total spin and total orbital angular momentum and J is composed from L and S). This gives for Eu^{3+} ${}^7F_0$ and for Er^{3+} ${}^4I_{15/2}$. The one but lowest level of Er^{3+} is some 7000 cm^{-1} above the ground level. This level comes also from the ${}^4I$ multiplet but has a different J value, viz. 13/2. It is designated as ${}^4I_{13/2}$. In the case of Eu^{3+} the one but lowest level is only some 400 cm^{-1} above the ${}^7F_0$ level. This is the ${}^7F_1$ level.

The magnetic moment results from the motion of the electrons around the nucleus and their own axis (spin) and is proportional to the angular moments. Its value μ is given by the formula

$$\mu = g\sqrt{J(J+1)}\ \mu_B$$

where μ_B is the Bohr magneton and g is the Landé splitting factor given by

$$g = 1 + \frac{J(J+1) + S(S+1) - L(L-1)}{2J(J+1)}$$

For Er^{3+} this gives $\mu = 9.59\ \mu_B$ to be compared with the experimental value

of 9.5 μ_B. The agreement is very good. The fact that a free-ion theory
works so well for ions in compounds is due to the screening of the 4f electrons by the outer 5s and 5p electrons. For the greater part of the RE
such a good agreement has been observed. This is not the case for the transition-metal ions where the orbital angular momentum is at least partly
quenched due to the influence of the surroundings.

Let us now return to the case of Eu^{3+}. For the ground level $(^7F_0)$
$J = 0$, so that we expect $\mu = 0$ μ_B. Experimentally 3.4 μ_B is observed at
room temperature. This value is temperature dependent. This does not mean
that our simple theory fails completely. The magnetic $^7F_1$ level is at so
low energy above the nonmagnetic $^7F_0$ level, however, that it becomes therm-
ally accessible. A similar situation occurs for Sm^{3+} $(4f^6)$. A complete description of this phenomenon becomes rather complicated.

We have here derived the magnetic moment of ions with $4f^n$ configuration.
We now turn to the interaction that may occur between these species in a
solid.

3. MAGNETIC INTERACTIONS IN INSULATORS

The magnetic interaction between ions in insulating solids is now reasonably well understood. A great contribution has been made by Anderson [6].
The interaction between magnetic ions may lead to parallel or antiparallel
alignment of their magnetic moments. It is essential in the theory that the
cation orbitals are mixed with the anion orbitals. In the case of 3d-metal
ions, for example, the "magnetic" 3d orbital has a certain 2p anion-orbital
character (in the case of oxides with O^{2-} ions with $2p^6$ configuration). Due
to this mixing the 3d-orbitals of different cations may overlap giving rise
to magnetic interaction.

Two cases may be distinguished:

a.   If the 3d-orbitals are orthogonal (i.e. if they do not overlap) a
parallel alignment of the magnetic moments results (ferromagnetic interaction). This is due to the repulsion between the electrons in the relevant orbitals (Hund's rule of maximum spin multiplicity is based on this
effect too). This situation is not often encountered and depends strongly

on the symmetry of the system. We present here a clear-cut example, viz. the magnetic interaction between Ni^{2+} and Mn^{4+} in La_2NiMnO_6 [7]. In this compound there is only one important interaction, viz. that between Ni^{2+} and Mn^{4+} in the colinear array $Ni^{2+} - O^{2-} - Mn^{4+}$. Figure 1 shows the relevant orbitals. The d-orbitals of Ni^{2+} and Mn^{4+} do not overlap for symmetry reasons. Ferromagnetic interaction occurs ($T_{Curie} \sim 300$ K).

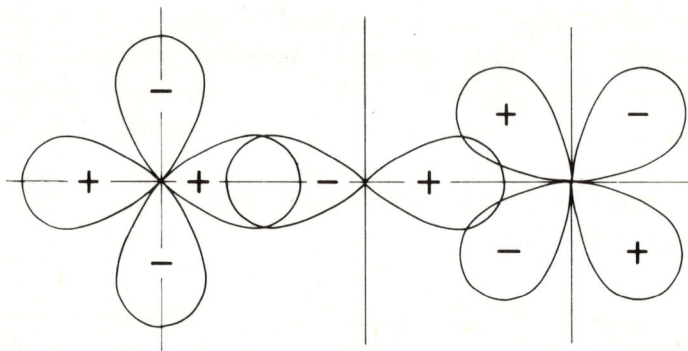

FIGURE 1

Orthogonality of the half-filled t_{2g} orbitals of Mn^{4+} (left) and the half-filled e_g orbitals of Ni^{2+} (right) in the colinear array $Mn^{4+} - O^{2-} - Ni^{2+}$.

b.   If the 3d-orbitals do overlap, antiferromagnetic interaction results, i.e., the magnetic moments are orientated antiparallel. This is because antiparallel electrons can gain energy by spreading into nonorthogonal, overlapping orbitals, whereas parallel electrons cannot (compare the chemical bonding in H_2). A clear example is $LaCrO_3$ [8]. The dominant interaction is that between two Cr^{3+} ions in the colinear array $Cr^{3+} - O^{2-} - Cr^{3+}$ (fig. 2). The d-orbitals overlap, antoferromagnetic interaction results ($T_{Néel} = 320$ K).

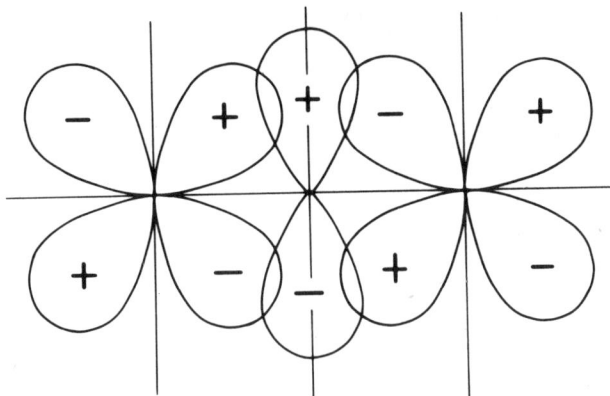

FIGURE 2

Overlap of the half-filled t_{2g} orbitals of Cr^{3+} (left and right)
in the colinear array $Cr^{3+} - O^{2-} - Cr^{3+}$.

This theory is of great importance for the interpretation of magnetic
interactions in transition-metal compounds (see e.g. ref. 9). Since it depends strongly on the admixture of anion orbitals to the cation orbitals we
may expect that the interaction between $4f^n$ ions in insulators will be extremely weak in view of the fact that the overlap of the 4f-orbitals and the
anion orbitals may be negligible. From optical work, however, it follows
that this overlap may, nevertheless, be of importance. Jørgensen et al. [10]
have attributed the crystal field splitting in RE ions to covalent bonding.
With reference to our paper on the optical properties of RE I may also
point to work by Van Uitert [11] and myself [12] that clearly shows that
covalency between 4f-orbitals and anion orbitals plays an important role
in energy transfer processes in which RE are involved. Nevertheless the
magnetic interactions between RE in insulators are usually weak. This is
borne out by magnetic investigations of the RE sesquioxides [1]. Magnetic
susceptibility measurements indicate very weak antiferromagnetic interaction.
Spin ordering occurs at extremely low temperatures only. It is without
doubt, however, that there is some interaction.

The fact that EuO becomes ferromagnetic below 70 K was reported some 10 years ago and was very surprising [13]. EuO is an insulating compound with rocksalt structure. The magnetic moment corresponds with the $4f^7$ configuration of Eu^{2+}. Later the isomorphous Eu^{2+} chalcogenides were also studied [14]. In this structure (fig. 3) we have two different magnetic interactions, viz. those between nearest Eu-neighbours (in a 90° Eu - O - Eu array) and those between next-nearest Eu-neighbours (in a 180° Eu - O - Eu array). The former is positive and decreases in the sequence EuO, EuS, EuSe, EuTe; the latter is negative and increases in this sequence. As a result EuTe is antiferromagnetic. The 180° interaction can be understood from the theory mentioned above. Since covalency increases going from O to Te, this interaction should also become stronger. The 90° interaction decreases with increasing cation distances. A number of proposals have been made to explain this interaction. The most reasonable of all seems to be that of Goodenough [15]. He assumes overlap between 4f (occupied) and 5d

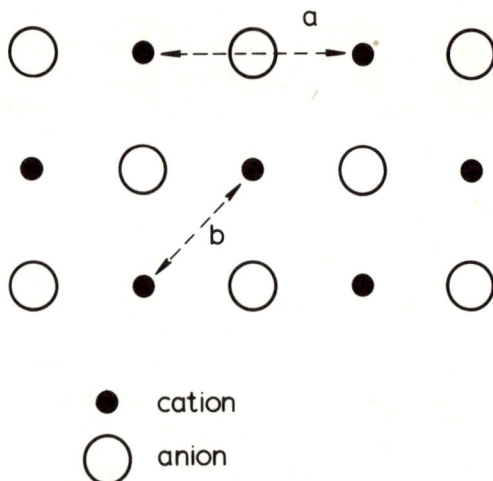

FIGURE 3

Magnetic interactions in the rocksalt structure ((100) face), for example in EuO. (a) via an intervening O^{2-} ion (b) directly between two Eu^{2+} ions.

(unoccupied) orbitals of different Eu^{2+} ions. This means that roughly speaking the 5d-orbital plays the role of the 2p-orbital in the theory outlined above: 4f of Eu(1) mixes with 5d of Eu(2); this 5d is orthogonal with 4f of Eu(2) and ferromagnetic interaction results.

Finally we mention two groups of compounds in which RE and 3d-metal ions are present simultaneously.

The first of these are compounds with garnet structure. This is a complicated structure. An oxide garnet has the general formula $\{A_3\}\,[B_2](C_3)O_{12}$, where $\{\ \}$ represents a dodecahedral, $[\ ]$ an octahedral and $(\ )$ a tetrahedral cation site. In yttrium iron garnet (YIG) with formula $\{Y_3\}\,[Fe_2](Fe_3)O_{12}$ the dodecahedral sites are nonmagnetic (Y^{3+}). The Fe^{3+} ions on the tetrahedral sites are coupled antiferromagnetically with those on the octahedral sites. The responsible mechanism is again due to overlap of Fe^{3+} - 3d-orbitals. As a result YIG is a ferrimagnet with a total magnetic moment equal to that of one Fe^{3+} ion per formula unit: $\{Y_3\}\,[\overrightarrow{Fe}_2](\overleftarrow{Fe}_3)O_{12}$. The magnetic interaction is rather strong $(T_{Curie} = 545$ K$)$.

It is possible, however, to replace the nonmagnetic Y^{3+} ion in $Y_3Fe_5O_{12}$ by magnetic RE ions, e.g. $Gd_3Fe_5O_{12}$. At low temperatures the following magnetic ordering is found: $\{\overrightarrow{Gd}_3\}\,[\overrightarrow{Fe}_2](\overleftarrow{Fe}_3)O_{12}$. The Gd^{3+} ions couple antiferromagnetically with the tetrahedral Fe^{3+} ions. The total magnetic moment is $(3\times7) + (2\times5) - (3\times5) = 16\ \mu_B$ (the moment of Gd^{3+} is $7\ \mu_B$, that of Fe^{3+} $5\ \mu_B$). This is also found experimentally. The coupling between the Gd^{3+} and the Fe^{3+} ions is much weaker than between the Fe^{3+} ions mutually. The Gd^{3+} moment therefore drops quickly with increasing temperature and at a certain temperature the tetrahedral Fe^{3+} moment becomes dominant (as it is in YIG). Therefore we expect a compensation temperature in the magnetization-temperature curve (see fig. 4). This has also been observed. For further work on the garnets, see ref.[16].

The other groups of compounds are those with perovskite structure [17]. Their general formula is $RE.FeO_3$ if the transition-metal ion is Fe^{3+}. These compounds are antiferromagnetic or tend to exhibit parasitic ferromagnetism. The Fe^{3+} - Fe^{3+} interactions dominate and determine the Curie temperature. The smaller the RE, the higher the Curie temperature (smaller Fe-O distances). As in the garnets the interaction between RE and Fe is only weak. The antiferromagnetic ordering temperature of the RE sublattice

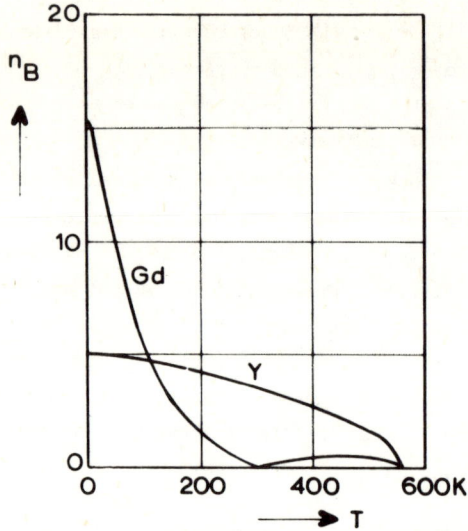

FIGURE 4

Magnetic moment of $Y_3Fe_5O_{12}$ and $Gd_3Fe_5O_{12}$ as a function of temperature. Note the compensation temperature for $Gd_3Fe_5O_{12}$.

is less than 10 K. That of the Fe^{3+} sublattice varies from about 600 to 750 K.

4. MAGNETIC INTERACTIONS VIA CONDUCTION ELECTRONS

Magnetic interactions of this type are known among other things as indirect exchange, s - d and s - f exchange, Rudermann-Kittel interaction, etc. Not all of these are identical, but we will not enter into this point here.

The type of exchange under consideration generally operates in the following way. A magnetic atom or ion polarizes the spins of conduction electrons or holes by a localized exchange interaction. As a matter of fact these conduction electrons are mobile. The exchange interaction may be ferromagnetic or antiferromagnetic. In this way a spin density is formed around the magnetic atom under consideration. This density decreases exponentially or in an oscillating way at larger distances from the magnetic atom. Another magnetic atom can interact with the first one through the induced spin density (indirect interaction).

The exchange interaction between the conduction and f-electrons is of the form

$$H_{sf} = - A(\vec{r}-\vec{R})\vec{s}\cdot\vec{S}$$

where $\vec{s}$ and $\vec{S}$ are the spins of the conduction electron and the ion at $\vec{r}$ and $\vec{R}$, respectively. $A(\vec{r})$ is the exchange integral. The Rudermann-Kittel approximation is

$$A(\vec{r}-\vec{R}) = \Gamma \, \delta(\vec{r}-\vec{R})$$

where Γ is constant.

H_{sf} is spin dependent. If $A(\vec{r}) < 0$, the spin-up electrons have minimum energy in the vicinity of $\vec{S}$ and those of spin-down do not. This causes the polarization $P(r)$ of the conduction electrons, defined as the density difference of spin-up and spin-down electrons.

The theory leads for metals to

$$\vec{P}(r) \sim \vec{S} \cdot \Gamma \cdot F(2k_F \cdot r)$$

Here $F(x) = (\sin x - x \cos x)\cdot x^{-4}$ and k_F the wave vector at the Fermi surface. F is the oscillatory function of Rudermann and Kittel. In semiconductors a nonoscillatory function is found (similar to $F(2k_F \, r)$ for $k_F \to 0$).

The polarization $\vec{P}(r)$ produced by an ionic spin $\vec{S}_i$ at $\vec{R}_i$ interacts with a second spin $\vec{S}_j$ at $\vec{R}_j$ through H_{sf}. The exchange energy between the spins is

$$H_{ij} = - J(\vec{R}_i - \vec{R}_j)\vec{S}_i \cdot \vec{S}_j$$

where $J(\vec{R}_i - \vec{R}_j) \sim \Gamma^2 \cdot F(2k_F |\vec{R}_i - \vec{R}_j|)$. This is for interaction between ionic spins $\vec{S}$. For the RE, however, J is a good quantum number and $\vec{S}$ must be replaced by $(g - 1)\vec{J}$, i.e.,

$$H_{ij} \sim - \Gamma^2 \cdot (g-1)^2 \cdot J(J+1) \cdot F(x)$$

This has indeed been found. The paramagnetic Curie temperatures of the heavier RE metals, for example, contain the factor $(g-1)^2 \cdot J(J+1)$ (fig. 5).

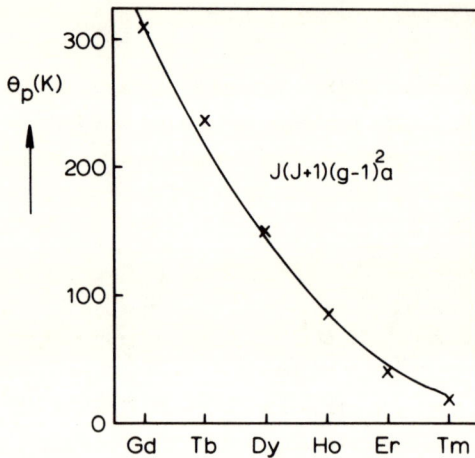

FIGURE 5

Paramagnetic Curie points of the RE metals. The drawn line is the function $J(J+1)(g-1)^2 a$. The constant a does not depend on the atomic number. It has been chosen such that the curve fits with the experimental points for Gd (original curve from J.H. van Vleck, Progr.Sci.Technology of RE,2, Pergamon Press, 1966).

We may stress that due to its oscillatory character the Rudermann-Kittel interaction in metals may result in ferromagnetism or all sorts of antiferromagnetic interactions leading to a large variety of magnetic spin structures. In semiconductors, however, only ferromagnetism is expected.

Finally it may be remembered that this interaction is only of importance if the direct interaction between the magnetic ions (atoms) is small. This is applicable in the case of the RE since the mean radius of the 4f shell is small compared to the interionic spacing. In the case of the transition metals, however, direct interaction between the d electrons dominates the magnetic properties. The d levels broaden into (relatively) narrow bands. Due to exchange interaction we have two subbands, one for electrons with spin-up, another for electrons with spin-down. Below the Curie temperature these are at slightly different energy. Because the number of electrons in both bands is therefore not equal, a (non-integral) magnetic moment results.

For indirect interaction the magnetic moment (in the case of ferromagnetism) equals that of the ions with a small additional increase due to the conduction electron polarization. In fact the moments observed for the pure RE metals are about $1 \mu_B$ larger than the ionic values.

We now turn to some specific examples.

5. EXAMPLES

Magnetic interaction via semiconducting carriers can be very clearly identified in the Eu chalcogenides [1]. Take, for example, the system $Eu_{1-x}Gd_xSe$. Each Gd contributes in principle an extra conduction electron. For small values of x the resistivity at 300 K decreases from about $10^8 \Omega$ cm for pure EuSe to 1Ω cm ($x \simeq 0.05$). At the same time the paramagnetic Curie temperature (θ_p) increases steeply from 9 K for pure EuSe to about 45 K for $x \simeq 0.05$. We note that GdSe is antiferromagnetic with $\theta_p = -60$ K. Similar effects are observed for the system $Eu_{1-x}La_xSe$, although LaSe is a diamagnetic metal. It is quite obvious that the ferromagnetic increase of θ is due to indirect exchange via conduction electrons. The impurity atoms (like Gd and La) play only one important role, viz. they provide conduction electrons that are delocalized enough to produce additional ferromagnetic interaction between the Eu^{2+} ions.

Back now to metallic materials. In the RE metals the magnetic interactions are assumed to be of the Rudermann-Kittel type as described above, although the situation is more complicated than indicated. Moreover, the magnetic ordering phenomena in the lighter RE metals, which do not become ferromagnetic at low temperatures, are only incompletely understood. For the RE metals usually different, complicated magnetic structures have been observed. This can be understood from the oscillatory character of the interaction, because this leads to competing magnetic interactions.

The RE can form alloys with other metals. Consider first nonmagnetic metals. We take as an example Al. Several alloys are known. Here we consider only those with formula $REAl_2$. These alloys are ferromagnetic with Curie temperatures up to 170 K $(GdAl_2)$, see e.g. ref.[4]. The variation of the Curie temperature of the compounds with the heavy RE follows the prediction of the Rudermann-Kittel theory (see e.g. ref. 4). This series of alloys seems therefore also a good example of indirect exchange. Dilution of the RE concentration has also been studied [18]. θ_p decreases linearly with x in the systems $Gd_{1-x}Y_xAl_2$ and $Gd_{1-x}La_xAl_2$, where nonmagnetic Y (or La) replaces Gd. In the system $Gd_{1-x}Th_xAl_2$, where the number of conduction electrons changes with increasing value of x (Th is tetravalent and Gd trivalent), the decrease of θ_p is much more pronounced than in the former systems with Y and La.

Finally we come to alloys of RE with transition-metal elements (for reviews see refs. 1 and 4). These compounds open the possibility to combine the high Curie temperatures of the transition metals and the large moments of the RE, but this has not been realized. Only Gd_6Mn_{23} shows ferromagnetic coupling between 3d- and 4f-spins [19]. The Curie temperature (478 K) seems to be due to Mn-spin coupling only.

Detailed studies have been made on compounds $RE.M_2$ and $RE.M_5$ (M = 3d-element). The magnetic moments of the RE as well as of M are often lower than in the elements. The 3d-levels are filled by 5d (or 6s) electrons of RE. The RE moments are partly quenched. Generally the Curie temperature is determined by 3d-spin ordering. The RE spins order only at lower temperatures and often antiparallel with the 3d-moment. This situation resembles those in the insulating garnets and in fact compensation temperatures have been found.

In the lighter RE the total magnetic moment is determined by $J = L - S$. Since the 3d-spins couple antiferromagnetically with the RE spins, the result is a ferromagnetic alignment of the RE moment and the 3d-metal moment. In figure 6 this is shown schematically.

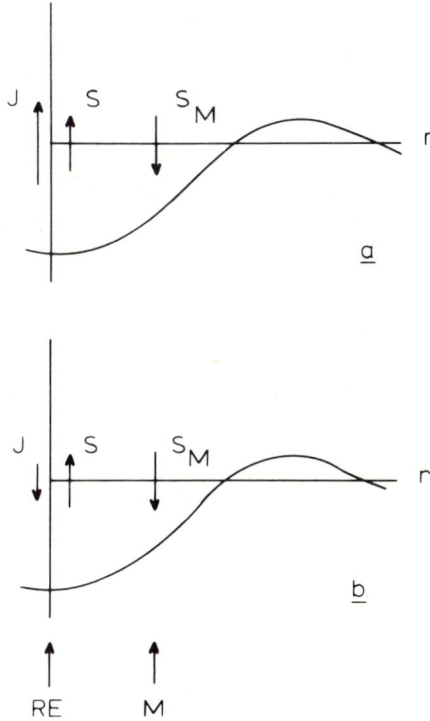

FIGURE 6

Schematic representation of the RE - 3d-metal interaction. (a) heavy RE, (b) light RE. The curve shows the oscillatory indirect exchange interaction. The position of the ions on the distance-axis (r) is indicated.

Finally we remark that it is usually assumed that the RE - 3d-metal interaction is due to some type of indirect exchange via conduction electrons.

64000- 294 -

Our knowledge of this field is at the moment still increasing.

REFERENCES

[1] S. Methfessel and D.C. Mattis, Handbuch der Physik, Band XVIII/1, Springer Verlag, 1968, p. 389.

[2] S.V. Vonsovsky and B.V. Karpenko, ibid, p. 265.

[3] T. Kasuya, in Magnetism IIB (G.T. Rado and H. Suhl, Eds.), Academic Press, (1966).

[4] K.N.R. Taylor, Adv. in Physics $\underline{20}$ (87), 551 (1971).

[5] B.R. Cooper, Solid St. Physics $\underline{21}$, 393 (1968).

[6] P.W. Andersen, Solid State Physics $\underline{14}$, 99 (1963) and in Magnetism I (G.T. Rado and H. Suhl, Eds.), Academic Press, 1963.

[7] G. Blasse, J. Phys. Chem. Solids $\underline{26}$, 1969 (1965).

[8] G.H. Jonker, Physica $\underline{22}$, 707 (1956).

[9] G. Blasse, Progress in Ceramic Science $\underline{4}$, 133 (1966).

[10] C.K. Jørgensen, R. Pappalardo and H. Schmidtke, J. Chem. Phys. $\underline{39}$, 1422 (1963).

[11] L.G. van Uitert, R.C. Linares, R.R. Soden and A.A. Ballman, J. Chem. Phys. $\underline{36}$, 702 (1962).

[12] G. Blasse and A. Bril, J. Chem. Phys. $\underline{45}$, 2350 (1966); G. Blasse, J. Chem. Phys. $\underline{46}$, 2583 (1967).

[13] B.T. Matthias, R.M. Bozorth and J.H. van Vleck, Phys. Rev. Letters $\underline{7}$, 160 (1961).

[14] See e.g. U. Enz, J.F. Fast, S. van Houten and J. Smit, Philips Res. Repts $\underline{17}$, 1451 (1962).

[15] J.B. Goodenough, Magnetism and the Chemical Bond, J. Wiley, 1963, p. 149.

[16] E.F. Bertaut and R. Pauthenet, Proc. I.E.E. $\underline{B104}$, 261 (1957); S. Geller, J. Appl. Phys. $\underline{31}$, 30S (1960); S. Geller, H.J. Williams, G.P. Espinosa and R.C. Sherwood, Bell Syst. Techn. J. $\underline{43}$, 565 (1964).

[17] J.B. Goodenough and J.M. Longo, Landolt-Börnstein (Ed. K.H. Hellwege) III, Vol. 4a, Springer Verlag, 1970.

[18] K.H.J. Buschow, J.F. Fast, A.M. van Diepen and H.W. de Wijn, Phys. Stat. Sol. 24, 715 (1967).

[19] B.F. de Savage, R.M. Bozorth and F.E. Wang, J. Appl. Phys. 36, 992 (1965).

PRODUCTION AND APPLICATION OF RARE EARTH FERRIMAGNETIC GARNETS

PART 1: APPLICATION IN MICROWAVE COMPONENTS

by

C.S. Brown

The General Electric Company Limited, Central Research Laboratories,

Hirst Research Centre, Wembley, England.

ABSTRACT

This paper deals with the application of rare earth ferrimagnetic garnets in microwave devices. The basic physics of the interaction of micro-
waves and magnetic insulators is outlined and related to the application of
ferrites in microwave isolators, circulators and phase shifters. To obtain
the required performance many material parameters must be controlled and it
is shown that the garnet structure is ideally suited in that a wide range of
ions may be substituted into the basic yttrium iron garnet structure to modify the properties in the desired manner. Ceramic production technology is
outlined and an indication of the application of ferrite devices in microwave systems is given.

1. INTRODUCTION

The electrical engineer's interest in ferrimagnetic oxides arose initially from their non-metallic nature which for the first time enabled magnetic
components to be made at other than audio frequencies. Their relatively high
resistivity and magnetic permeability has led to a large industrial market
for ferrite cored inductors capable of operating up to 100 MHz. The first
available ferrimagnetic materials were ceramic oxides analogous to the mineral spinel $(MgAl_2O_4)$. The general formula for a spinel ferrite can be written
as $M^{2+}Fe_2^{3+}O_4$ where M is typically Zn, Mg, Mn, Fe, Ni or Co or combinations of
these. As will be shown later, a key requirement for applications at microwave frequencies is that the dielectric loss of the material should be low.

This can be difficult to achieve in spinel ferrites because of the multi-valent nature of iron which, in association with the divalent cations, can give rise to "electron-hopping" conductivity.

The discovery of the garnet-structured oxides based on yttrium-iron garnet (YIG), $Y_3^{3+}Fe_5^{3+}O_{12}$ provided a major breakthrough [1]. High resistivity is more readily achieved in these ferrites as all the cations are trivalent. Since the original work other advantages of the garnet ferrites have been found, many of which arise from our ability to vary the properties of the material by controlled substitution on the dodecahedral, octahedral and tetra-hedral lattice sites.

2. MICROWAVE PROPERTIES OF FERRITES

Consider an infinite magnetic medium in a field H_o sufficient to saturate it, i.e., the magnetization can be considered as a single dipole parallel to H_o as shown in figure 1. If this dipole is perturbed, it will precess

FIGURE 1

Gyromagnetic resonance

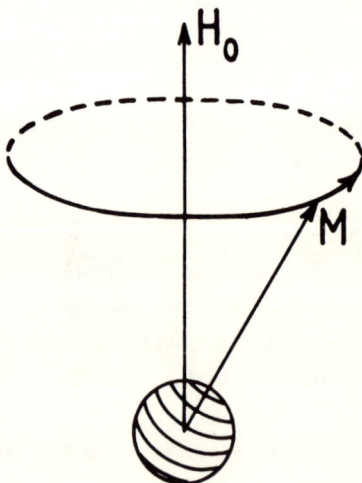

about H_o like a gyroscope with a frequency of precision given by

$$\nu = \frac{\gamma}{2\pi} H_o \qquad \text{where} \quad \frac{\gamma}{2\pi} \simeq 2.8 \text{ MHz/Oe} \qquad (1)$$

Because the system is not lossless, the magnetization will return to its equilibrium position parallel to H_o. However, if a small circularly polarized r.f. magnetic field is applied perpendicular to H_o it will couple to the precession. The coupling is a maximum when the hand of polarization and frequency of the r.f. field are the same as that of the natural precession and in these circumstances energy is absorbed from the r.f. field; there is no coupling if the polarization is reversed. In a finite medium the situation is complicated by the demagnetizing fields caused by the magnetic poles at the boundary of the material, so that the internal field is not the same as the applied external field.

In figure 2, the real and imaginary components of the permeability are shown for the two hands of polarization. It will be seen that gyromagnetic resonance does not occur at a unique field. The width of the resonance (ΔH) usually measured at half-height is an important parameter particularly in resonant-mode devices. In a single crystal, ΔH is determined by the intrinsic damping of the precession but in a ceramic two other factors must be taken

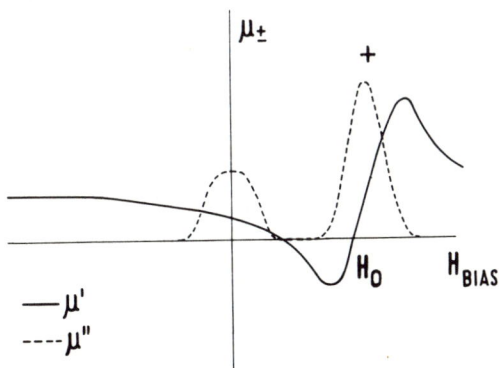

FIGURE 2

Real and imaginary components
of the permeability

into account. Magneto-crystalline anisotropy, which measures the energy required to move the magnetization vector from its preferred crystallographic
direction, broadens the linewidth since the effective field at any individual
crystallite of the ceramic is a function of the relative orientation of its
crystallographic axes and the applied field. The magnitude of the effect
increases with anisotropy and becomes larger as the magnetization is reduced.
The second factor is porosity which can be intra-grain air pores or magnetic
inhomogeneities. This gives rise to variable demagnetizing fields within the
material. For low porosities this effect is directly proportional to the
porosity and the saturation magnetization. Figure 2 also shows losses that
occur at low fields. To minimize these losses the material must either be
magnetically saturated or the saturation magnetization ($4\pi M_s$) must be sufficiently low so that at a frequency, ν

$$4\pi M_s < n \left(\frac{2\pi\nu}{\gamma} \right) \qquad\qquad (2)$$

where n is a factor of the order of 0.5. Hence there is a need for materials
with saturation magnetizations about 500 Gs for use at the lower microwave
frequencies around 2 GHz.

3. MICROWAVE FERRITE DEVICES

From figure 2, the two principal modes of operation of microwave ferrite
devices can be seen. The simplest are resonant devices where the bias field
is such that the internal field (taking into account demagnetizing fields)
corresponds to the peak of the gyromagnetic resonance. In an empty waveguide
the magnetic field components of the fundamental TE mode are in quadrature in
time and space and their magnitude varies across the width of the guide as
shown in figure 3a. At points x_1, x_2 (near L/4) the components are of equal
amplitude and hence the wave is circularly polarized. Furthermore the hand
of polarization is reversed for a wave travelling in the opposite direction.

Ferrite placed at x_1 (or x_2) with a bias field as shown in figure 3
will absorb a wave travelling in one direction and transmit a wave travelling
in the opposite. The performance of the resonant isolator as measured by the

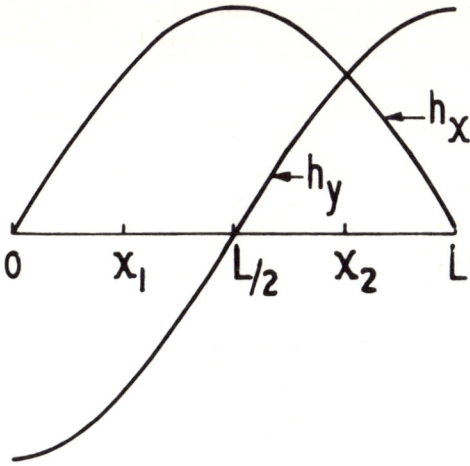

FIGURE 3

Field distribution in a resonance waveguide resonance isolator

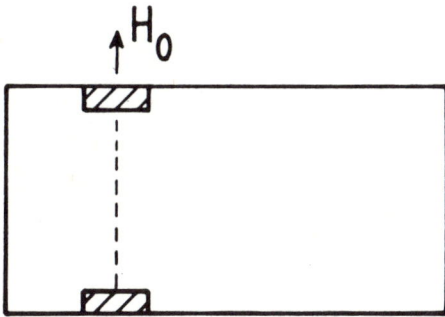

ratio of reverse to forward loss is proportional to the inverse square of
the linewidth.

The second mode of operation is non-resonant and depends on the fact
that away from resonance the permeabilities μ'_+ and μ'_- are different although
the loss represented by μ'' is small. Non-resonant devices are inherently
broad-band and are increasingly used in microwave systems. One device is
the field displacement isolator in which the ferrite is used to modify the
field distribution in the waveguide. As shown in figure 4, the electric
field is concentrated near one edge of the ferrite for the reverse direction

FIGURE 4
Electric field distri-
bution in a field dis-
placement isolator.

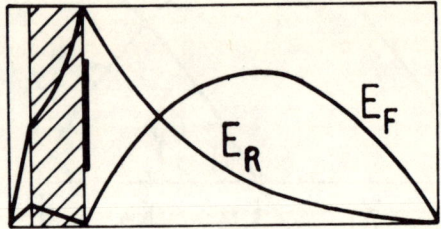

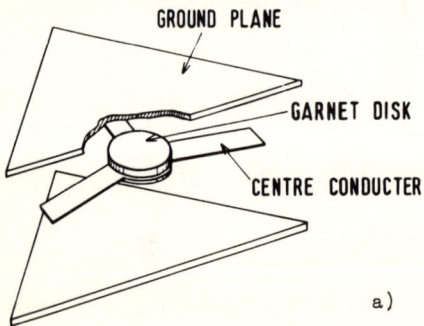

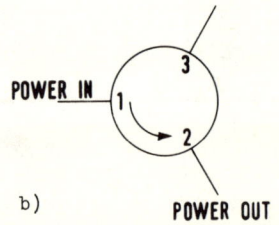

FIGURE 5
a) and b) Junction circulator

of propagation in which the permeability is positive but is near zero for
the forward wave. By placing a resistive load at this edge, the reverse
wave is attenuated whilst the forward wave propagates with little loss. The
limitations on the performance of this device are the magnetic and dielectric
loss of the ferrite at the operating field. The most important ferrite component now employed in microwave systems is the circulator. Figure 5 shows
schematically a 3 port circulator where energy entering a given port is only
coupled to the next port, i.e., 1 → 2, 2 → 3, 3 → 1 but not the reverse.
Circulators have been realized in a number of ways using the non-reciprocal
properties of the permeability but the most commonly used device is the junction circulator, which is shown in triplate stripline form in figure 5; it
can also be realized in waveguide and microstrip. Fay and Comstock [2] have
shown that the electric field distribution in the ferrite disc of a junction
circulator can be resolved into 3 modes. Two of these modes (n = ± 1) are
resonant modes of the disc and produce the field distribution as shown in
figure 6a. When a field is applied, the field pattern is rotated through
30° so that port 3 is at a null of the electric field and transmission only
occurs between ports 1 and 2 (figure 6b). The two modes can be considered as
waves travelling with opposite hand around the periphery of the disc so that
when a bias field is applied their velocity of propagation is no longer de-
generate because the effective permeability of the ferrite is different for
the two polarizations. If the third port of a 3 port junction circulator is

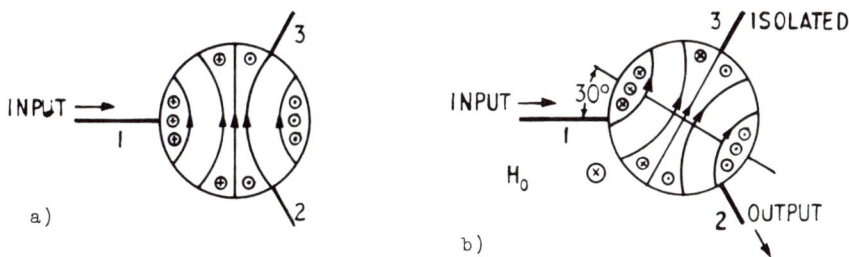

FIGURE 6
Mode field patterns in a circulator[*]

[*] From J. Helszajn, Principles of Microwave Ferrite Engineering, Wiley
Interscience.

terminated with a matched 50Ω load, it forms an isolator. This isocircula-
tor is now commonly used in microstrip form because it is compatible with
other active and passive components and allows the construction of microwave
integrated circuits.

The most recent development in microstrip ferrite devices is the edge
mode isolator [3] which can be considered to be a microstrip field displacement isolator and in common with that device has a large bandwidth but without restrictions due to waveguide cut-off.

Many other devices using the non-reciprocal properties of ferrite ceramics at microwave frequencies have been constructed, two of the more important being Faraday rotators and phase shifters. Single crystals have found
less widespread application but can be used as tunable microwave resonators,
particularly for the control of solid-state oscillators, and as power limiters
using the non-linear properties of gyromagnetic resonance [4] which occurs
when the angle of precession reaches a well-defined limit.

4. FACTORS AFFECTING THE CHOICE OF GARNET CERAMICS

From the brief outline in the previous sections, it will be seen that
to optimize a ferrite for a particular device application, a number of properties must be controlled. These will be considered in relation to the
ferrimagnetic garnets but since they are inter-related, it is preferable not
to discuss them separately. The properties to be considered are:

   i)   saturation magnetization ($4\pi M_s$)
  ii)   Curie temperature (T_c)
 iii)   temperature coefficient of magnetization (dM_s/dT)
  iv)   linewidth (ΔH)
   v)   anisotropy field (H_a)
  vi)   magnetic loss ($\tan \delta_\mu = \mu''/\mu'$)
 vii)   dielectric loss ($\tan \delta_\epsilon = \epsilon''/\epsilon'$)

Some of these properties are not intrinsic and depend on the preparative
techniques employed.

Yttrium iron garnet (YIG) is the key material from which all other garnets are derived. Its molecular formula can be written $\{Y_3\}$ $[Fe_2]$ (Fe_3) O_{12} where $\{\}$, [], and () represent the atoms located on the dodecahedral (c), octahedral (a) and tetrahedral (d) sites. The room-temperature properties of YIG are given in table 1. For device applications at frequencies below about 8 GHz, $4\pi M_s$ is too high. It can readily be reduced by substitution of Al or Ga for Fe since both show a preference for the d site and in YIG the magnetization results from the antiparallel moments on the d and a sites.

TABLE 1
Properties of Ceramic YIG

$4\pi M_s$ (at 20°C)	1750 Gs
T_c	275°C
dM_s/dT (0 - 60°C)	3.5 Gs/$^\circ$C
ΔH	45 Oe
H_a	40 Oe
$\tan \delta_\mu$ (at H = 0.5 H_{res})	< 0.001
$\tan \delta_\epsilon$	< 0.0005

The room-temperature saturation magnetization as a function of Al and Ga content is shown in figure 7 [5]. At high levels, particularly in the case of Al, some of the substituting ions enter the a site [6]. The distribution between the d and a sites is dependent on the method of preparation and quenching from a high temperature will increase the magnetization. It will be shown in the second part of this paper that this effect may be of value when it is necessary to control magnetization to very close limits.

However, since one is reducing the number of magnetic exchange interactions, substitution of non-magnetic ions also reduces T_c which results in an increase in dM_s/dT. To overcome this difficulty, a magnetic rare earth can be substituted for yttrium. Gadolinium is used since being an S-state

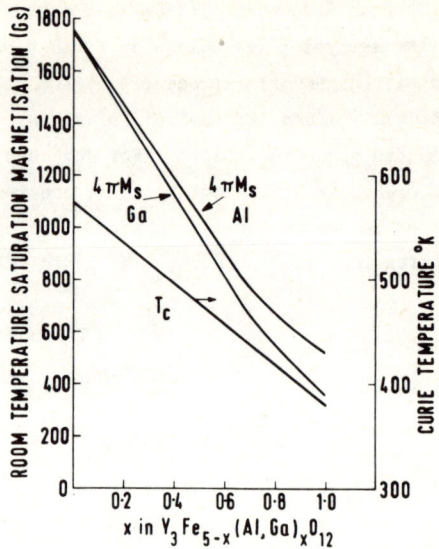

FIGURE 7
Saturation magnetization as
a function of Ga and Al substitution in YIG

ion, it does not couple to the lattice and increase the intrinsic losses.
At $0°K$, Gd on the c sublattice couples antiparallel to the resultant of the
a and d sublattices giving a moment of 16 Bohr magnetons (compared to 5 for
YIG). Although T_c for $Gd_3Fe_5O_{12}$ (i.e. 100% substitution) is similar to that
of YIG, the magnetization of the c sublattice decreases far more rapidly
with temperature giving rise to a magnetization characteristic as in figure
8 with a compensation point T_{cp} where the magnetization reverses. As the Gd
substitution is reduced T_{cp} decreases and the optimum temperature performance
is achieved when the compensation and Curie temperatures are approximately
equally spaced about ambient temperature [7]. For zero temperature coefficient the composition required is $Y_{1.8}Gd_{1.2}Fe_5O_{12}$ but with this high Gd substitution, the magnetic losses and linewidth are considerably higher than
with YIG (ΔH = 150 Oe cf 45 Oe). It may also be necessary to reduce $4\pi M_s$
so that in practice a compromise is used between c site and d site substitution. Figure 9 illustrates the trade-off between linewidth and temperature
coefficient of magnetization with saturation magnetization as a parameter [8].

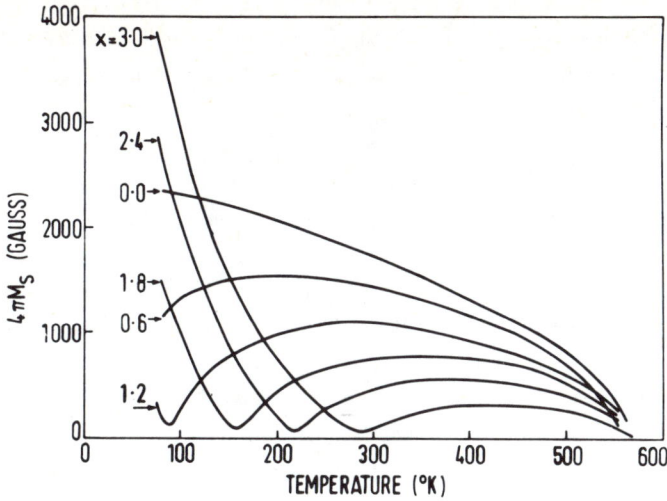

FIGURE 8

Variation of saturation magnetization as a function of temperature
in system $Y_{3-x} Gd_x Fe_5 O_{12}$ *

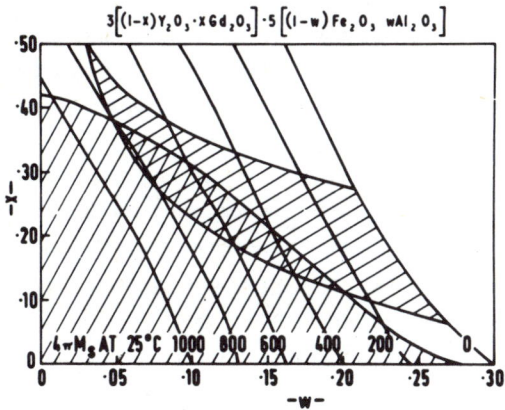

FIGURE 9

Temperature stability as a function of Gd and Al substitutions in YIG

LINEWIDTH LESS THAN 120 OERSTEDS AT 25°C.

MAX. VARIATION IN $4\pi M_s$ OF ± 50 GAUSS OVER 120°C INCLUDING THE TEMPERATURES OF −40° TO +40°C.

* From L.R. Hodges and G.R. Harrison [8].

It will be seen that a region exists where acceptable performance can be achieved. A typical low-loss, stable garnet with $4\pi M_s$ = 500 Gs would have the composition $Y_{2.78}Gd_{0.22}Fe_{4.85}Al_{0.15}O_{12}$.

All rare-earth elements other than Gd give rise to rapidly increasing magnetic losses [9], Dy, Ho and Tb having the largest effect. For low-loss materials it is essential that unwanted rare-earth elements should be at a low level and commonly 99.99% purity yttria and gadolinia are used. However, small quantities of these "fast-relaxing" ions can be used to improve the high-power properties of garnets. The same physical process which increases the loss also delays the onset of non-linearity. Above the non-linear threshold the losses rise catastrophically so that for high power devices, for example in radar systems, it may be preferable to accept a somewhat increased loss at low power. Typical substitution levels would be about 0.02 atoms per mole.

As discussed in section 2, the resonance linewidth of a ceramic is increased by anisotropy. It has been shown that In, which enters the a sites, reduces the effective anisotropy field so that linewidths as low as 3 Oe have been obtained in nearly 100% dense ceramics [10]. Recent work has shown, however, that care must be taken in relating linewidth to magnetic loss. Away from resonance, the effective linewidth may be only 5% of ΔH and is related to the spin-wave linewidth (ΔH_k) which is a property of the magnetic ions present and the ceramic microstructure but is independent of anisotropy. Hence the effect of In may not be so effective in non-resonant devices.

For phase shifters operating at remanence, the ferrite is usually made to fill one dimension of the waveguide. In these circumstances magnetostriction is an important secondary parameter, since the remanent magnetization which controls the differential phase between the two states of magnetization, will be a function of stress unless the magnetostriction is zero. Two alternative approaches have been shown to reduce magnetostriction

    i)   substitution of cerium for yttrium on the dodecahedral sites

   ii)   substitution of manganese for iron on the octahedral sites.

The flexibility of the garnet structure is such that Geller has made a

very wide range of materials by substitution on the three sites [11]. Most
of these are of academic interest only, but one material which could find
application is $\left\{ Bi_{3-2x}^{3+} Ca_{2x}^{2+} \right\} [Fe_2^{3+}] (Fe_{3-x}^{3+} V_x^{5+}) O_{12}$. The saturation magnetization can be varied in the range 200 - 800 Gs and has a relatively small mag-
netization temperature coefficient because of the surprisingly high Curie
temperature [12]. This is the only practical ferrimagnetic garnet system
which does not contain rare-earth elements.

5. PRODUCTION OF GARNET CERAMICS

The technology employed is determined by the need to obtain close stoichiometry and to control the microstructure of the ceramic whilst achieving
near 100% theoretical density. Stoichiometry is essential if dielectric loss
is to be minimized. Excess iron gives rise to Fe^{2+} ions and deficient iron
will produce a two-phase structure containing the orthoferrite phase $(RFeO_3)$.
The various stages of the technology are outlined in relation to the above
requirements.

i) Raw materials

Oxides are normally used but oxalate co-precipitation techniques are
being increasingly adopted to obtain a higher reactivity through particle
size control and to ensure complete mixing of the components. If co-precipitation is used, close control is essential to ensure that the desired stoichiometry is achieved.

ii) Mixing

If oxides are used, it is necessary to ensure adequate mixing which is
normally achieved by ball-milling.

iii) Pre-sinter reaction

In order to obtain the required microstructure with minimum porosity and
uniform grain size, it is necessary to pre-sinter the mixed oxides. A compromise must be chosen between inadequate reaction which will result in poor
microstructure due to excessive shrinkage on final sintering, and over-reaction

- 310 -

which will result in a material which is difficult to sinter. Table 2 illustrates the reaction of Y_2O_3-Fe_2O_3 as a function of temperature. A typical
pre-sinter temperature is $1050^{\circ}C$.

TABLE 2

Reaction Sintering of Y_2O_3-Fe_2O_3

Sintering Temperature (°C)	Constitution (%)	
	$YFeO_3$	$Y_3Fe_5O_{12}$
850	0	0
950	5	1
1050	10	40
1150	10	80
1250	1	99
1350	0	100

iv) Milling

The reacted material must be reduced to a grain size of the order of
1 μm. This is conventionally carried out in a stainless steel mill but, because of the hardness of the garnet phase, it can lead to difficulty in stoichiometry control by pick-up of iron from the mill. This is allowed for by
producing an iron-deficient starting mix or, where very close control is required, blending of batches may be used. Alternative milling techniques,
such as air milling, which do not contaminate the material, are attractive.

v) Pressing

This is a critical part of the processing to ensure near-theoretical
density of the fired ceramic. Die pressing is not adequate and isostatic
pressing is usually employed.

vi) Final sinter

The garnets are not highly reactive and sintering must be carried out at
relatively high temperature, say 1450 C, in order to achieve minimum porosity.

However, care must be taken to avoid excessive grain growth and oxygen loss, which will again cause Fe^{2+} ions to be formed resulting in increased dielectric loss. Oxygen loss can be minimized to some extent by firing in an oxygen atmosphere.

vii) Machining

Since it is difficult to press and sinter to the mechanical tolerances required for microwave devices, the material is usually prepared as a cylinder or rectangular block and the parts are machined using diamond-loaded tools.

6. SYSTEMS APPLICATIONS

As has been described in section 3, microwave ferrite devices can be made to act as isolators, circulators and phase shifters. Isolators are used to prevent two parts of a circuit from interacting with one another. For example, the performance of a microwave generator would depend on any reflected signal from the load into which it is feeding. An isolator inserted between the generator and the load will absorb the unwanted reflected signal whilst causing very little attenuation of the required signal.

Circulators enable a component to be shared. For example a radar system will use a circulator with transmitter connected to port 1, aerial to port 2 and receiver to port 3 (see figure 5a). The transmitted signal is radiated from the aerial and the incoming signal will be directed to the receiver. In telecommunications, a number of adjoining frequency bands are transmitted from the one aerial and must then be separated at the receiving station. This is carried out by feeding the received signal through a bank of circulators and pass-band filters; the signals in the stop band of any filter will be reflected to the next circulator.

Remanent phase shifters are finding increasing application in phase-array radar systems. By changing the phase across the wavefront of a signal fed to a large number of aerials in parallel, the direction of the radiated signal can be changed without having to physically move the aerial. This provides a much quicker scanning response but the application depends on the extent to which cost can be reduced.

Ferrite devices are now an essential part of all microwave systems and for many applications the rare earth ferrimagnetic garnets are the only materials that can be used to obtain the desired performance.

Aknowledgement

The author wishes to recognize the assistance provided by Mr. E.E. Riches and his colleagues of the Magnetism Group at the Hirst Research Centre, whose work on microwave ferrite materials and devices has provided the basis for this review.

References
General

W.H. Van Aulock, Handbook of Microwave Ferrite Materials (Academic Press 1965).

B. Lax and K.J. Button, Microwave Ferrites and Ferrimagnetics (McGraw-Hill Book Co. 1962).

J. Helszajn, Principles of Microwave Ferrite Engineering (Wiley Interscience 1969).

[1] F. Bertaut and F. Forrat, C.R. Acad. Sci. Paris, 244, 96 (1957).
S. Geller and M.A. Gilleo, Acta Cryst. 10, 239 (1957).

[2] C.E. Fay and R.L. Comstock, Trans. I.E.E.E., MTT-13, 15 (1965).

[3] M.E. Hines, Trans. I.E.E.E., MTT-19, 442 (1971).

[4] H. Suhl, Proc. I.R.E., 44, 1270 (1956).

[5] G.R. Harrison and L.R. Hodges Jr., J. Amer. Ceram. Soc. 44, 214 (1961).

[6] S. Geller, H.J. Williams, G.P. Espinosa and R.C. Sherwood, Bell System Technical Jour. 43, 565 (1964).

[7] A. Vassiliev, J. Nicholas and M. Hildebrandt, C.R. Acad. Sci. Paris, 252, 2681 (1961).

G.R. Harrison and L.R. Hodges Jr., J. Appl. Phys. 33 (Suppl.) 1375 (1962).

[8] L.R. Hodges Jr. and G.R. Harrison, J. Amer. Ceram. Soc. $\underline{48}$, 516 (1965).

[9] E. Schlomann, Trans. I.E.E.E., $\underline{MAG-1}$, 168 (1965).

[10] C.E. Patton and H.J. Van Hook, J. Appl. Phys. $\underline{43}$, 2872 (1972).

[11] References in [6].

[12] P. Mossman, Trans. I.E.E.E., $\underline{MAG-5}$, 614 (1969).

PRODUCTION AND APPLICATION OF RARE EARTH FERRIMAGNETIC GARNETS

PART 2: APPLICATION IN BUBBLE DOMAIN STORES

by

C.S. Brown

The General Electric Company Limited, Central Research Laboratories

Hirst Research Centre, Wembley, England.

ABSTRACT

The application of magnetization reversal for the representation of
the two binary states has been used for many years in digital stores. The
discovery of the magnetic bubble domain provides the first opportunity of
producing a high-density backing store without moving parts.

Following a discussion of the physics of bubble domains and the conditions under which they exist, the materials which can be used are considered.
It is shown why the rare earth garnets are suited to this application and
techniques for the epitaxial deposition of the thin films required are described. Finally a short review is given of the manner in which bubble
domains can be used as a serial shift register store as a potential replacement for the mechanical systems.

1. INTRODUCTION

Increasingly information is stored, processed and transmitted as binary
digits. Storage may be required at various levels, for example, in a computer. Small-capacity, fast stores with access times of less than 100 ns
will use semiconductor integrated circuits. The main store with a capacity
up to 10^5 bits would currently use magnetic cores with an access time of the
order of 1 μsec. The backing store, which might have a capacity of perhaps
10^8 bits, would be a pack of magnetic discs with an average access time of
several milliseconds. In general, cost per bit rises as the access time decreases and in the design of a computer, the manner in which the information

is shared between the various levels of storage is a vital consideration in optimizing performance and cost.

Magnetism has played a key role in stores since the two binary states "O" and "1" can readily be represented by the two remanent magnetization states of the hysteresis loop. A magnetic store is passive and comsumes no power, except for writing and reading, and is non-volatile, i.e., the information is not lost if the power is switched off. In the core store, the binary states are represented by the two directions of circumferential magnetization in a ceramic ferrite toroid, the state being written by a current passing along a wire which threads the core. Even with considerable miniaturization, the packing density is limited since every bit of information must be stored on a discrete core. The disc store uses a thin layer of γ-Fe_2O_3 or an evaporated cobalt film as the magnetic medium, the information being stored on circular tracks as reversals of magnetization induced by the field at the poles of a recording head as the disc rotates. There has been a demand for sometime to replace the disc store by a non-mechanical system with at least the same packing density and low cost per bit [1]. The magnetic bubble domain store may fulfil this need.

2. PHYSICS OF BUBBLE DOMAINS

In zero external field, a low-coercivity magnetic material will appear to be un-magnetized but closer examination shows that although the net magnetization is zero, individual regions are magnetized parallel to the easy axes of magnetization defined by the magnetocrystalline anisotropy. If there are boundaries, the demagnetizing field will also affect the direction of magnetization of these domains. For a thin plate, the demagnetizing field at the surfaces is such that the minimum energy configuration will usually be such that the magnetization is forced to lie in the plane of the sample. It has long been recognized that magnetic domains in a plate could be used to store binary information in the form of magnetization reversals, but the interaction of the magnetization with the edges of the plate render it difficult to move the domains reversibly.

The breakthrough was announced by Bobeck in 1967 [2], when he showed

how a plate of single crystal magnetic material with the easy direction of
magnetization perpendicular to the plate and with sufficiently high anisotropy to overcome the demagnetizing field, would support cylindrical domains
magnetized as shown in figure 1. In a uniformly magnetized plate with the
applied field as shown, cylindrical domains are stable under two counter-
acting forces. The forces generated by the applied field (H_0) and the domain wall energy (H_w), dependent on the anisotropy, act to reduce the volume
and wall area respectively whilst the demagnetizing field (H_D) tends to increase the surface area [3]. The variation of $H_0 + H_w$ and of H_D as a function of domain size are shown schematically in figure 2, from which it will
be seen that there are two equilibrium radii, but the smaller is unstable.

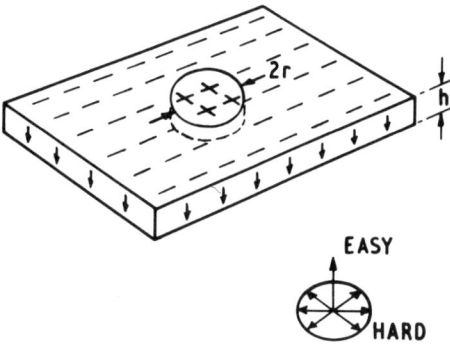

FIGURE 1
Basic geometry of bubble
domain.

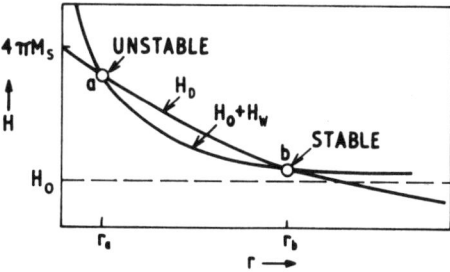

FIGURE 2
Graphical solution of bubble
domain stability[*].

* From A.H. Bobeck et al. [3]

Clearly there is a minimum radius as H_O is increased to a value where the
two curves are tangential. More complete analysis [4] shows that there is
a minimum field below which the domain runs out into a serpentine strip
configuration. Figure 3 shows the diameter of a cylindrical domain as a
function of plate thickness with bias field as parameter, indicating that
there is an optimum plate thickness to achieve a stable domain of minimum
diameter. The dimensions in figure 3 are plotted in terms of a material
parameter, the characteristic length (ℓ), which is defined by Thiele [4]

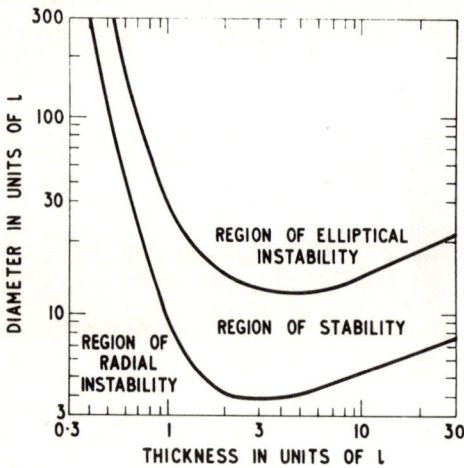

FIGURE 3

Bubble domain diameter as a
function of plate thickness[*]

as $\ell = \sigma_w/4\pi M_s^2$ where σ_w, the domain wall energy is proportional to $K_u^{\frac{1}{2}}$,
K_u being the uniaxial anisotropy constant. At optimum plate thickness the
necessary bias field is of the order of 0.3 $(4\pi M_s)$. A further restriction
is that the anisotropy field must be greater than the saturation magnetization for magnetization perpendicular to the plate to be the stable configuration which leads to the requirement that $K_u > 2\pi M_s^2$. Hence we see that to
obtain small bubble diameters for high packing densities requires low magnetization, adequate but not excessive anisotropy and plate thickness of the
order of the required bubble diameter. Added to these requirements, the
temperature coefficient of magnetization should be small, and the domain

---

[*] From A.H. Bobeck et al. [3]

wall mobility must be adequate to enable the device to operate at a high
bit rate. Unfortunately, at present no accurate means has been found to
predict wall mobility.

3. MATERIALS FOR USE IN BUBBLE DOMAIN DEVICES

The first class of materials in which bubble domains were extensively
studied were the rare earth orthoferrites, $RFeO_3$. These materials have an
orthorhombic structure with a canted antiferromagnetic spin system [5,6] as
illustrated in figure 4. The resultant uniaxial anisotropy axis at room
temperature is along [001] except for $SmFeO_3$ where it is along [100] and
the angle of cant is about $0.5°$ giving rise to a magnetization of the order
of 100 G. The materials are transparent in thin section and the domain
pattern can be observed visually by means of Faraday rotation. However,
the anisotropy is high so that the smallest domain diameter, observed in
$TbFeO_3$, is about 40 µm. This can be reduced by exploiting the fact that
in $SmFeO_3$ the easy axis is perpendicular to that in $TbFeO_3$ so that a mixed
$(Sm,Tb)FeO_3$ will have a reduced anisotropy. However, these solid solutions
are very temperature sensitive and the minimum diameter is still greater
than 15 µm.

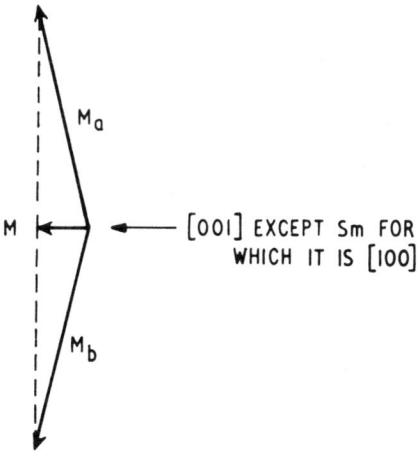

FIGURE 4
Canted antiferromagnetic ordering
in rare earth orthoferrites

The aim was to find a material which would support bubble domains of about 5 μm and be temperature insensitive. The breakthrough came with the discovery that single crystal mixed rare earth garnets could be grown to be sufficiently uniaxial to support bubble domains [7] although previously it had been assumed that the cubic structure of garnets ruled them out. Several theories for the anisotropy have been suggested but it is likely that it results from pair ordering of the rare earth elements on the dode-cahedral sites.

Once it had been shown that the garnets would support bubble domains, the wide range of possible substitutions increased the hope that an opti-mized material would be found. Bell Telephone Laboratories, who have been responsible for most of the basic work, grew many compositions with up to four rare earth elements initially attempting to minimize magnetostriction, which might increase coercivity, and to obtain a low temperature coefficient of magnetization when the saturation magnetization was reduced to about 100 Gs by Al or Ga substitution [8] (see Part 1). A typical composition is $Eu_2Er_1Ga_{0.7}Fe_{4.3}O_{12}$ for which the minimum domain diameter at the field which causes the bubbles to collapse is about 5 μm. All the initial work was carried out on specimens cut from bulk single crystals and it was found that the desired anisotropy is only achieved in particular growth zones and that the required cut varies with composition. Since the plate thickness must be of the same order as the domain diameter, it is clearly necessary to use an epitaxial growth technique to reduce material costs. Epitaxy provides an alternative means of achieving the desired anisotropy by means of magnetostriction and controlled lattice mismatch [9,10]. Negative magnetostriction along the film normal with the film in tension will cause the easy axis to be perpendicular to the substrate. Employing strain-induced anisotropy, simple rare earth garnets have been grown which support bubble domains, the main requirement now being to control the lattice con-stants of the deposit and the substrate to achieve the desired anisotropy without causing crazing of the deposit. The most commonly used substrate is $Gd_3Ga_5O_{12}$ [11] and the currently favoured garnet composition is nominally $Y_{3-z}Gd_zFe_{5-x}Ga_xO_{12}$ where z is in the range 0.2 - 0.5 and x is about 1.0. The range of compositions with the required saturation magnetization and anisotropy appears to be narrow and there may be alternative systems that will give the required anisotropy for a smaller mismatch and hence reduce the risk of crazing.

4. CRYSTAL GROWTH TECHNIQUES

As mentioned in the previous section, the early work on garnets was carried out using slices cut from bulk single crystals. Since the ferrimagnetic garnets are incongruently melting, bulk crystals cannot be grown direct from the melt and are grown from fluxed melt solution. The usual flux is $PbO.PbF_2.B_2O_3$ with a ratio by weight of lead compounds to B_2O_3 of about 50 [12]. Crystals up to 200 g have been grown in platinum crucibles 20 cm high x 20 cm diameter by cooling the melt from $1300^\circ C$ to $950^\circ C$ at $\frac{1}{2}^\circ C$ per hour. At the lower temperature the melt is drained from the crucible which is then allowed to cool slowly to room temperature.

Two epitaxial techniques, liquid phase epitaxy and chemical vapour deposition, are being developed and it is too early to state which will prove superior in producing films with the required control of properties and low defect density.

4.1 Liquid Phase Epitaxy (LPE) is an extension of the fluxed melt technique for the growth of bulk crystals; the substrate being immersed in a supersaturated solution of the garnet. Two basic modifications have been used:

1) The solution is brought to equilibrium at about $900^\circ C$ and the substrate is immersed either by using a tilting boat technique or by lowering it into the solution. The solution is then cooled by 10 - $15^\circ C$ to deposit the required thickness of garnet [13].

2) The solution is supersaturated by cooling below the equilibrium temperature [14].

It has been claimed that when supercooled up to $40^\circ C$, the melt will remain stable for periods as long as 48 hours. The substrate is immersed into the supercooled melt and a 4 μm thick film grown in about 10 minutes. It has been claimed that films having defect densities of less than 10 per cm^2 with a thickness uniformity of ± 2.5% over an area of 0.6 cm^2 can be grown by this technique. Since the Ga distribution coefficient is temperature dependent, the composition of the deposited garnet and hence the saturation magnetization is a function of the growth temperature. This would appear to favour the use of the isothermal system.

LPE is a simple technique to set up although some workers have found difficulty in obtaining the low defect density quoted above. It is particularly valuable for research into the effect of compositional changes particularly where multi-component systems are being studied.

4.2 Chemical Vapour Deposition (CVD) uses techniques commonly employed in the semiconductor industry. The materials are transported in the gas phase by a carrier gas (e.g. He) using the relatively volatile chlorides and reacted by hydrolysis in the vicinity of the substrate to produce the required garnet. For example to produce YIG, the required reaction could be

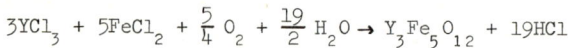

$$3YCl_3 + 5FeCl_2 + \frac{5}{4} O_2 + \frac{19}{2} H_2O \rightarrow Y_3Fe_5O_{12} + 19HCl$$

The precise reaction deposition mechanism is not known and allowance must be made for the different reactivities of the components. Furthermore the reactivity of YCl_3 is so high that HCl must be added to the reaction chamber to prevent gas phase reaction. Two forms of reactor have been described. The T-reactor introduced by Autonetics [15] is shown schematically in figure 5, and the laminar flow reactor [16] in figure 6. Although in principle the latter should more closely produce the conditions for uniform deposition of the epitaxial garnet layer, the T-reactor has been successfully used at the Hirst Research Centre and will be used to illustrate the CVD technique. The products which can occur in the reactor are Y_2O_3, $YFeO_3$, $Y_3Fe_5O_{12}$, Fe_2O_3 and Fe_3O_4 and because of the high reactivity of YCl_3, the reaction proceed along the chamber in that order. The required garnet deposition zone can be shifted and expanded by changing the relatively propor-

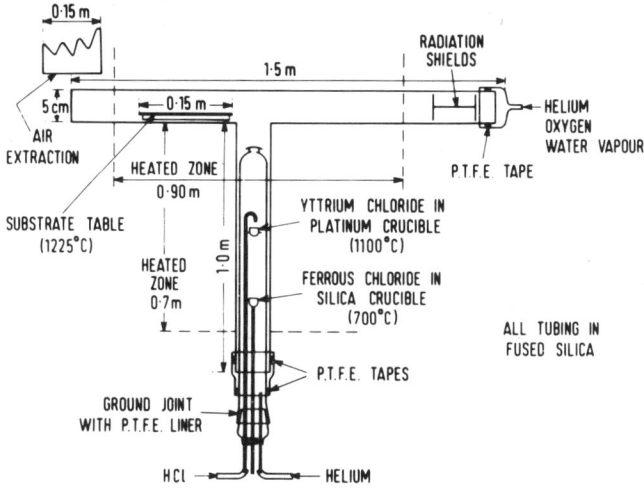

FIGURE 5

Reactor for deposition of yttrium iron garnet

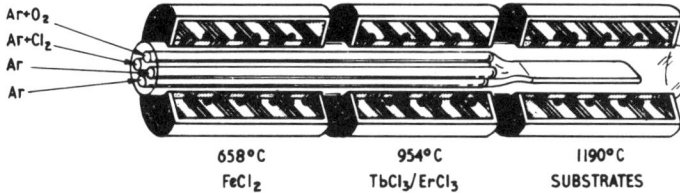

FIGURE 6

Laminar flow CVD reactor[*]

tions of the reacting gases. For bubble domain applications, gallium is
introduced into the vertical arm by in-situ reaction of the metal. Typical
conditions for the deposition of YIG:Ga with a saturation magnetization of
100 Gs are given in table 1 for which growth rates of the order of 5 μm per

[*] From M. Robinson et al.[16].

hour are achieved. Mixed rare earth garnets can be obtained either by
combining the rare earth chlorides in the required proportions to allow for
their different volatility at a fixed temperature or, probably with better
control, by the addition of further temperature controlled zones. In the
latter manner (Gd,Y)IG:Ga has been grown at the Hirst Research Centre. The
chlorides can be readily purified by vacuum distillation but care must be
taken to prevent hydrolysis of the rare earth chlorides forming the oxychloride which are too volatile at the reactor temperature.

TABLE 1
Growth Parameters for Gallium-doped YIG Films

Substrate temperature:	1175°C
Vertical helium flow rate:	10.5 l/min
Horizontal helium flow rate (dry):	4.5 l/min
(wet):	0.25 l/min
Oxygen flow rate:	0.06 l/min
HCl flow rate:	0.35 l/min
YCl_3 source temperature:	1040°C
$FeCl_2$ source temperature:	870°C
$GaCl_3$ source temperature:	135°C
Growth rate:	$\sim$ 5 μm/h

It has been found that the anisotropy of epitaxial layers deposited by
CVD is invariably strain-induced even when the composition is such that the
bulk crystal exhibits ordering anisotropy, probably due to the high deposition temperature. In contrast, epitaxial layers deposited by LPE usually
show growth-order anisotropy. The nature of the anisotropy can be determined by annealing the film at about 1200°C; this has no effect on strain

anisotropy but removes order anisotropy. Most work has been done using gadolinium gallium garnet (GGG) as the substrate and adjusting composition in the (Y,Gd)(Fe,Ga) system to obtain the optimum lattice mismatch. It is possible to adjust the lattice constant of the substrate using mixed rare earth gallium garnets but this approach is likely to be more difficult since the tolerance on mismatch is small. For some mixed rare earth garnet epitaxial layers, GGG is not suitable and samarium and mixed samarium-gadolinium gallium garnets have been used.

Work on the tolerancing of bubble domain circuits has indicated that the control on saturation magnetization must be to better than 5%. This is exceedingly difficult to achieve as grown but advantage can be taken of the effect of annealing in the range 600 - 1000°C to redistribute Ga between the a and d sites. Low temperature annealing following quench from a higher temperature should make this target feasible and allow for batch to batch and within batch variation.

5. APPLICATION OF BUBBLE DOMAINS IN STORES

In section 2, the basic physics of bubble domains and the conditions under which they occur have been outlined. In a uniform bias field an array of bubble domains will be the equilibrium state, the spacing between the bubbles being determined by the magnetic interaction between neighbouring domains. To exploit the domains in a store it is necessary to be able to generate, propagate and detect them. Initially this was done using current-carrying loops which produced a local magnetic field gradient but this would not be feasible at the packing densities required. The major breakthrough was the invention of the permalloy "T-bar structure" [17] in conjunction with a rotating in-plane field. Figure 7 shows a T-bar structure and indicates how this enables bubble domains to be propagated in a controlled manner. Bubble domains are attracted to + magnetic poles that appear when the in-plane field is directed along a long dimension of a T or bar. As the field rotates clockwise the + poles always appear immediately to the right of the domains causing them to propagate toward the right. For each revolution of the in-plane field the domain is shifted by one unit of the

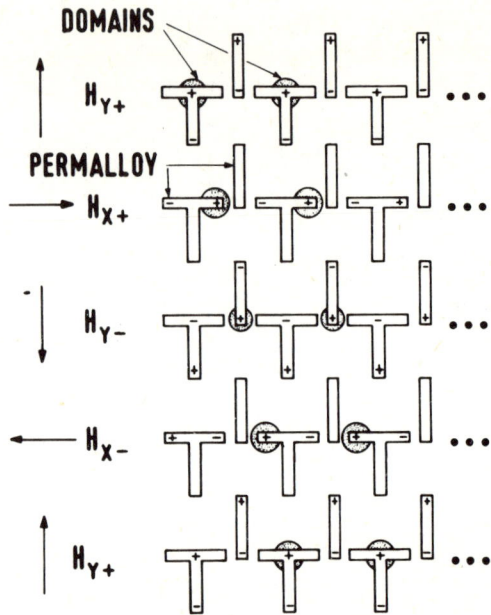

FIGURE 7
T-bar propagation[*]

structure, this being of the order of five times the bubble diameter, i.e.,
25 μm for a typical garnet. The domains can be propagated round corners so
that a folded loop structure can be built to obtain the optimum packing
density. For experimental purposes the permalloy pattern (ca. 5000 Å thick)
is defined by photolithographic techniques on a glass substrate and brought
into close proximity to the film but for systems it will probably be deposited directly onto the film. 1000 bit loops have been shown to operate
reliably and packing densities of about 2×10^5 cm^{-2} are predicted.

To use such a loop in a store, data must be written in by a generator,
the absence or presence of a bubble representing "0" and "1", respectively.
The same permalloy and rotating field scheme can form a generator [17] as
shown in figure 8. A permanent domain associated with the rotating + pole
configuration of the generator disc is forced to stretch when one end

---

[*] From A.J. Perneski [17].

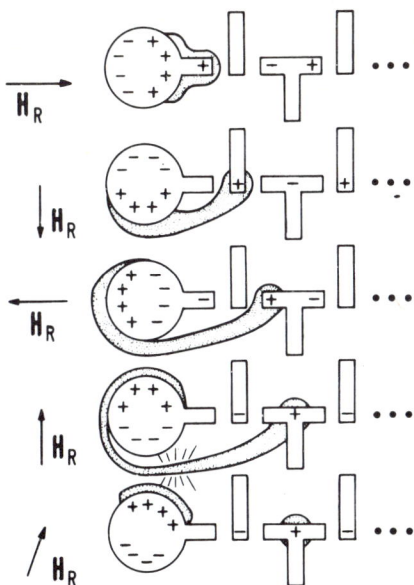

FIGURE 8
Domain generation*

becomes trapped in the T-bar propagation channel. When the in-plane rotating field is directed upward, the - poles near the stretched portion of the
domain cause it to sever creating a newly formed domain in the propagation
channel. To control the writing operation from the generator a number of
approaches has been suggested but the most elegant is to use the switch
shown in figure 9 [18]. A current pulse along the conductor biases the
position of the bubble by means of the local magnetic field to cause it to
move either into track A or B, one of which feeds into the storage loop.

The final component required to complete the store is a means for detecting bubble domains for read-out. As the size of domains has been reduced to increase packing density, this has become more difficult since the
magnetic flux change to be detected is proportional to the volume of the
domain. The most satisfactory technique at present is a deposited permalloy
magnetoresistive element incorporated in a bridge to detect the change in
resistance as the domain passes under the element. The signal to noise

---

* From A.J. Perneski [17].

FIGURE 9
Bubble domain switch[*]

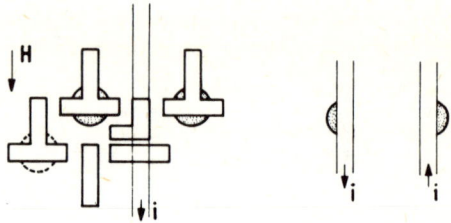

ratio is only about 5 with an output voltage of 100 μV which, although not as good as systems engineers would like, is probably adequate. For experimental work, detection is usually carried out using optical Faraday rotation.

In the design of a store, a key parameter is the access time. Since the store as described is serially accessed, the access time is limited by the length of the storage loop and the maximum velocity of propagation. Propagation is controlled by the effective field gradient produced by the rotating in-plane field and the mobility of the domains. Measured mobilities particularly in epitaxial films are lower than had been expected and are typically 100 cm^2/sec Oe. In addition some garnets are non-linear and show a limiting velocity regardless of driving field gradient. However, the close packing achievable with garnets suggests that a data rate of 1 MHz should be obtained, although at the present time systems are being built which may not exceed 100 kHz. However, it is possible to organize the store using a number of small loops which can be read in parallel so that the access time is limited by the size of the loop and not by the total size of the store.

Prototype stores of up to 5000 bits have been constructed and shown to operate as predicted. 10^6 bit stores are now under development but, in parallel, there is also potential interest in smaller stores. One simple requirement is for an electronic repertory dialler [19] which will store frequently-used telephone numbers and feed the information on to a telephone line when required.

---

[*] From G.S. Almasi et al. [18].

The commercial success of the bubble domain store will depend on the ultimate cost per bit of stored information and calculations show that this will be satisfactory if the yield of epitaxial garnet films with the required close control of properties and low defect density can be achieved in production. Any defect, whether it is caused by imperfection on the substrate surface caused during polishing or introduced during growth, is likely to prevent the propagation of bubble domains by increasing the coercivity in the neighbourhood of the defect. Because of the small area required, perhaps 1 mm^2, it is considered that a low defect density, say 10 per cm^2, should provide an adequate yield of circuits which propagate bubble domains. By 1977 at the latest, we will know whether the bubble domain store is going to play an important role in the storage hierarchy of new digital systems.

Acknowledgement

The author wishes to acknowledge Dr. P. Baxter and his colleagues of the Thin Film Group at the Hirst Research Centre for information provided on the growth and assessment of epitaxial garnets for application in bubble domain devices, and the Ministry of Defence (Procurement Executive) who have provided support for this work.

References
[1] R.E. Matick, Proc. I.E.E.E., 60, 266 (1972).
[2] A.H. Bobeck, Bell System Technical Jour., 46, 1901, (1967).
[3] A.H. Bobeck, R.F. Fischer, A.J. Perneski, J.P. Remeika and L.G. Van Uitert, Trans. I.E.E.E., MAG-5, 544 (1969).
[4] A.A. Thiele, J. Appl. Phys. 41, 1139 (1970).
[5] D. Treveo, J. Appl. Phys. 36, 1033 (1965).
[6] S. Geller, J. Chem. Phys. 24, 1236 (1956).
[7] A.H. Bobeck et al., Appl. Phys. Letters, 17, 131 (1970).
[8] A.H. Bobeck, D.H. Smith, E.G. Spencer, L.G. Van Uitert and E.M. Walters, Trans. I.E.E.E., MAG-7, 461 (1971).

[9] J.E. Mee, G.R. Pulliam, D.M. Heinz, J.M. Owens and P.J. Besser, Appl. Phys. Letters, $\underline{18}$, 60 (1971).

[10] E.A. Giess, B.A. Calhoun, E. Klokholm, T.R. McGuire and L.L. Rosier, Mat. Res. Bull $\underline{6}$, 317 (1971).

[11] P.J. Besser et al., AIP Conference Proceedings No. 5, Magnetism and Magnetic Materials, 125 (1971).

[12] L.G. Van Uitert, W.A. Bonner, W.H. Grodkiewicz, Miss L. Pictroski and G.J. Zydzik, Mat. Res. Bull. $\underline{5}$, 825 (1970).

[13] L.K. Shick et al., Appl. Phys. Letters, $\underline{18}$, 89 (1971).

[14] H.J. Levinstein, S. Licht, R.W. Landorf and S.L. Blank, Appl. Phys. Letters, $\underline{19}$, 486 (1971).

[15] J.E. Mee, G.R. Pulliam, J.L. Archer and P.J. Besser, Trans. I.E.E.E., $\underline{MAG-5}$, 544 (1969).

[16] M. Robinson, A.H. Bobeck and J.W. Nielsen, Trans. I.E.E.E., $\underline{MAG-7}$, 464 (1971).

[17] A.J. Perneski, Trans. I.E.E.E., $\underline{MAG-5}$, 554 (1969).

[18] G.S. Almasi et al., AIP Conference Proceedings No. 5, Magnetism and Magnetic Materials, 220 (1971).

[19] P.C. Michaelis and I. Danylchuc, Trans. I.E.E.E., $\underline{MAG-7}$, 737 (1971).

PERMANENT MAGNETS BASED ON RE MATERIALS

by

Helmut Stäblein

Krupp Forschungsinstitut, D 43 Essen, BRD.

ABSTRACT

In the last five years a large progress has been made in developing RCo_5-type materials into high-grade permanent magnets. They have outstanding energy products as well as resistances against demagnetizing fields, which are described in comparison to common materials. Despite these new alloys have been known since 15 years ago or so, the starting shot for development activities was given by the discovery of K. Strnat et al. of their very high anisotropy fields in 1966. Now $SmCo_5$- and $(Sm,Pr)Co_5$ magnets are commercially available from several companies in Western Europe, USA and Japan. Development trends now under way are aimed at further increasing the characteristics by utilizing materials with higher saturation polarizations, e.g., materials of the type $R_2(Co_{1-x}T_x)_{17}$ with $T = Fe$ or Mn and x somewhere between 0 and 0.5.

1. INTRODUCTION

A permanent magnet, once magnetized, shall usually retain a magnetic flux as large as possible without any further input of energy. However, the permanent magnet is exposed at least to its own demagnetizing field and in case to the fields of external circuits (motors, generators). Furthermore, its function must be ensured over a more or less extensive range of temperatures.

In recent years, a new category of permanent magnet materials based on RCo_5 has been developed. R may mean one atom or a mixture of atoms of certain rare earth metals. The magnet materials in this category are, in some respect, superior to the magnet materials known hitherto, and this is why so many laboratories are engaged in their development. In the following, we indicate the governing magnetic properties of these materials, the manufacturing processes, the characteristics obtained and their temperature dependence, and point out the possible directions of further development. A comparison with long-known materials is to demonstrate the advantages and drawbacks of RCo_5-compounds.

2. SATURATION POLARIZATION AND ANISOTROPY FIELD

The saturation polarizations J_s at room temperature are compiled in table 1 [1-3]. Values between 0.77 and 1.23 T are only obtained with Y and the "light" RE-metals against 0.24 to 0.73 T with the "heavy" RE-metals and Th [2]. Since the Ce mischmetal (MM) contains, in addition to approx. 55% Ce, approx. 26% La, 13% Nd, 5% Pr, balance other rare earth metals, J_s of $MMCo_5$ is about 15% higher than that of $CeCo_5$. Generally speaking, in RCo_5 solid solutions containing different rare earth metals, J_s appears to vary linearly with the mole fraction [4,5]. If Co is replaced successively by Cu, compounds of the type $RCo_{5-x}Cu_x$ are obtained. As x increases, the saturation polarization decreases. This is partly compensated for by adding some Fe [3]. J_s of Alnico and barium ferrite are above and below, respectively, the data of the new materials.

Apart from the value at room temperature, the dependence on temperature is of importance. For R = Y, Ce, Pr, Nd, and Sm, J_s decreases steadily from $0^{\circ}K$ to the Curie temperature T_c, while for R = Gd, Tb, Dy, Ho, and Er a minimum is found at around $120^{\circ}K$ or less, which corresponds to a compensation point [6-8]. The T_c's are roughly in the range of those of the conventional permanent magnet materials.

The anisotropy field H_A is of particular importance. In most of the RCo_5-compounds mentioned in table 1, the hexagonal c-axis is the preferred direction at room temperature. Only in the case of $NdCo_5$ the incipient transition to an easy plane at lower temperatures makes itself felt. Hence this material, taken alone, appears to be unsuitable as a permanent magnet. The anisotropy fields of the RCo_5-compounds are comparable to those of $PtCo$, but range by about one order of magnitude above those of Alnico and barium ferrite.

TABLE 1

Saturation polarization J_s, Curie temperature T_c, anisotropy field H_A and
X-ray density ρ_o of permanent magnet materials

Material	J_s (20°C) (T)	T_c (°C)	H_A (20°C) (MA/m)	ρ_o (g/cm³)
YCo_5	1.06 [1]	648 [1]	~ 10.4 [1]	7.60 [1]
$LaCo_5$	0.91 [1]	567 [1]	~ 14.0 [1]	8.03 [1]
$CeCo_5$	0.77 [1]	374 [1]	13.6-16.8 [1]	8.55 [1]
$PrCo_5$	1.20 [1]	612 [1]	11.6-16.8 [1]	8.34 [1]
$NdCo_5$	1.23 [2]	630 [55]	~ 2.4 [2]	8.39 [2]
$SmCo_5$	0.97 [1]	724 [1]	16.8-23.2 [1]	8.60 [1]
$SmCo_5$	1.10 [73]	-	-	-
$CeMMCo_5$	~ 0.89 [1]	~ 518 [1]	14.4-15.6 [1]	~ 8.35 [1]
$CeMMCo_3Cu_2$	0.54 [15]	-	-	-
$CeCo_{3,5}Cu_{1,5}$	~ 0.35 [3]	-	-	
$CeCo_{3,5}CuFe_{0,5}$	-	> 720 [56]	-	
$CeCo_{3,5}Cu_{1,25}Fe_{0,5}$	~ 0.60 [3]	-	-	~ 8.50 [1]
$SmCo_{3,5}Cu_{1,5}$	~ 0.60 [3]	-	-	
$SmCo_{3,5}Cu_{1,35}Fe_{0,5}$	~ 0.69 [3]	-	-	
Alnico	$\leq$ 1.5	$\lesssim$ 900	($\leq$ 0.5) [57]	$\leq$ 7.3
$BaO.6Fe_2O_3$	~ 0.48	450	1.4	5.3
PtCo	(0.72) [54]	~ 530 [54]	5.6 [54]	15.7 [54]

The RCo_5-compounds have densities which are considerably smaller than
the PtCo, roughly comparable to the Alnico, and much larger than the barium
ferrite values.

- 334 -

3. PRODUCING HIGH COERCIVITY VALUES

A high anisotropy field does not necessarily imply a high coercivity $_J H_c$, since magnetic reversal takes place by nucleation and movement of walls [9-12]. The ingot, mostly produced in induction or arc furnaces, has a relatively low $_J H_c$ value. The necessary high coercivity must be produced in subsequent processing steps so as to prevent nucleation and easy movement of the walls. Basically, there are two ways to achieve it: Powder-metallurgical processing of RCo_5-alloys or heat treatment with precipitation hardening of compact $RCo_{5-x}Cu_x$-alloys.

By crushing and grinding, the highest values can be obtained with $SmCo_5$: $_J H_c \approx 1.2$ MA/m [2]. $_J H_c$ can be further increased by etching in dilute acid [13], by a diffusion treatment with zinc [14], by grinding at low temperature, e.g.,under liquid N_2 [15], and by using the smallest particles. The drawback of these powders is, however, that the coercivity decreases appreciably in the course of time [13].

The second possibility to produce high $_J H_c$ is to apply precipitation hardening in the system $RCo_{5-x}Cu_x$ with R = Ce or Sm and x > 0.5 or 1, respectively. Already in the as-cast state, $_J H_c$-values are found of the order of some 100 kA/m. A further increase can be achieved by tempering at around $400^\circ C$, for instance, up to $_J H_c = 2.2$ MA/m with $SmCo_3Cu_2$ [3]. $RCo_{5-x}Cu_x$-magnets do not show such a gradual decrease of $_J H_c$ as do the RCo_5-powders.

4. PRODUCTION OF HIGH-GRADE PERMANENT MAGNETS

The term "high-grade" shall refer to a material in which all the known means of optimizing the magnetic properties were exploited, particularly by forming a c-axis fibre structure and by thoroughly avoiding weak or non-magnetic constituents such as pores, oxides, RCu_5, etc.

Compact RCo_5-magnets can be produced from powder essentially just like hard ferrites: Using a non-stoichiometrically composed powder of single-domain size, the particles are oriented in a magnetic field, compacted by pressing, and sintered to form a dense solid.

The composition of the powder is of considerable importance. It must be richer in rare earth metals than RCo_5. Powders poorer in rare earth metals gave inferior magnetic properties when R = Sm [16]. In the case of $SmCo_5$, for instance, the stoichiometric content is 33.8% by weight of Sm, and the best sintered magnets are obtained with a composition of about 37.5% [16-18]. This relatively large deviation is thought to be necessary because of the presence of several weight per cent of Sm_2O_3 [72], and to the necessary presence of Co-site vacancies which enables Sm-diffusion [76]. The optimum composition is either produced by casting [16,18] or by mixing the powders from two separate casts in the correct ratio with the composition of the main cast approaching RCo_5 while that of the "sintering aid" being much richer in R [19-21].

The particle size of the powder normally used ranges around 10 μm. Orientation requires magnetic fields preferably of at least 2 MA/m. The powder should be compacted as much as possible by pressing so that the temperature during sintering can be kept sufficiently low and an excessive grain growth and thus a considerable drop of $_JH_c$ is avoided. Sintering takes place at around 1100°C under inert conditions. Commonly the temperature is critical, particularly if the aim is a hysteresis loop as rectangular as possible [16]. Sintering can increase the coercivity to a value several times that of the compact; in the case of $SmCo_5$ from 0.9 to 3 MA/m [22] and in the case of $PrCo_5$ from 80 to 530 kA/m [23]. Magnets made under optimum conditions have a relative density of more than 90% and enclosed pores, so that the $_JH_c$-values do not decrease in the course of time. After the sintering treatment the magnets have to be cooled rather quickly [33]. $SmCo_5$ seems to be metastable below 750°C because of an eutectoid decomposition into the adjacent phases Sm_2Co_7 and Sm_2Co_{17} [75]. Dense magnets can also be made without sintering, simply by pressing them under a high isostatic pressure of about 20 kbar during uniaxial deformation [24].

The $RCo_{1-x}Cu_x$-compounds offer basically two possibilities for aligning the c-axis: Directional solidification of the melt [25,26] or powder-metallurgical processing as described above for the Cu-free alloys [27,74]. The method of directional solidification has also been used since long to produce highest grade Alnico materials [28-30].

5. MAGNETIC CHARACTERISTICS

The characteristics of different alloys are compiled in tables 2 to 5. The compositions given are nominal only. The figures represent a selection of peak values.

TABLE 2

Magnetic characteristics of binary RCo_5-compounds

SH = sintering aid

Composition, production	B_r (T)	$(BH)_{max}$ (kJ/m^3)	$_B H_c$ (kA/m)	$_J H_c$ (kA/m)	
$SmCo_5$, sintering	0.77	120	600	2200	[16]
$SmCo_5$, sintering with Sm-SH	0.88 0.93	140 160	640 500	1830 730	[39]
$SmCo_5$, sintering with Sm-SH	0.98	191	690	950	[73]
$SmCo_5$, high pressure	0.87 0.92	147 160	670 610	1260 800	[24]
$PrCo_5$, sintering with Pr-SH	0.85	134	390	440	[20]
YCo_5, pressed, 64% dense	0.29	10	120	215	[58]
YCo_5, 0.1 mm crystallite	1.06	220	365	365	[59]

Among the binary compounds (table 2), the by far best magnetic values are obtained with $SmCo_5$, regardless whether sintering or high pressure techniques were employed. B_r, $(BH)_{max}$ and $_B H_c$ closely approach the theoretically possible limits. Depending on manufacturing conditions, a smaller B_r can be combined with a larger $_J H_c$ and vice versa.

TABLE 3

Magnetic characteristics of quasi-binary, Cu-free RCo_5-compounds

SH = sintering aid

Composition, production	B_r (T)	$(BH)_{max}$ (kJ/m^3)	$_BH_c$ (kA/m)	$_JH_c$ (kA/m)	
$Sm_{0,5}Pr_{0,5}Co_5$, sintering with Sm-SH	0.89 1.00	147 183	600 540	1500 620	[60]
$Sm_{0,5}Pr_{0,5}Co_5$, sintering with Sm-SH	1.03	207	805	1320	[61]
$PrCo_5$, sintering with Sm-SH	0.94 0.90	168 160	560 710	1000 1100	[62]
$Sm_{0,5}La_{0,5}Co_5$, sintering with Sm-SH	0.73	105	560	2200	[60]
$Sm_{0,5}La_{0,5}Co_5$, pressing with binder	0.52	46	270	-	[63]
$Sm_{0,5}Ce_{0,5}Co_5$, sintering with Sm-SH	0.77	107	300	320	[60]
$Pr_{0,87}Nd_{0,13}Co_5$, pressing with 9 kbar	0.56	44	210	270	[64]

On account of the good properties of the $SmCo_5$-compound, the idea suggested itself to substitute other rare earth metals for at least part of the Sm, cf. table 3. This can be done with good success by using Pr, in which magnets with the highest $(BH)_{max}$ and the highest $_BH_c$ at room temperature are obtained through an optimum compromise between the increase of B_r and the loss of $_JH_c$.

TABLE 4

Magnetic characteristics of quasi-ternary and higher, Cu-free RCo_5-compounds
SH = sintering aid

Composition, production	B_r (T)	$(BH)_{max}$ (kJ/m^3)	$_BH_c$ (kA/m)	$_JH_c$ (kA/m)	
$Pr_{0,87}Nd_{0,13}Co_5$, sintering with Sm-SH	0.78	103	450	650	[64]
$Sm_{0,25}Ce_{0,25}La_{0,5}Co_5$, pressing with binder	0.50	40	240	-	[63]
$MMCo_5$, high pressure 82% dense	0.52	38	160	170	[65]
$MMCo_5$, 89% dense sintering with MM-SH	0.50	34	210	270	[66]
$Sm_{0,5}MM_{0,5}Co_5$, sintering with Sm-SH	0.90	159	605	1200	[31]

TABLE 5

Magnetic characteristics of $RCo_{1-x}Cu_x$-type compounds

Composition, production	B_r (T)	$(BH)_{max}$ (kJ/m^3)	$_BH_c$ (kA/m)	$_JH_c$ (kA/m)	
$Ce_{1,05}Co_{4,25}Cu_{0,75}$	0.60	48	260	295	[67]
$SmCo_{3,3}Cu_{1,7}$	0.65	61	270	-	[25]
$CeCo_{3,6}Cu_{0,7}Fe_{0,7}$ sintered	-	83	-	400	[27]
$CeCo_{3,8}Cu_{0,9}Fe_{0,5}$ sintered	0.63	74	425	560	[74]
$CeCo_{3,8}Cu_{0,9}Fe_{0,5}$	0.63	78	410	525	[56]
$CeCo_{2,9}Cu_{1,7}Fe_{0,5}$	0.43	33	300	910	[56]
Ce-Co-Cu-Fe-Leg. isotropic	0.30	17	220	-	[68]
anisotropic	0.56	56	380	-	
$SmCo_{3,5}Cu_{1,35}Fe_{0,4}$	0.65	72	320	480	[3]
$Sm_{0,75}Ce_{0,25}Co_{3,3}Cu_{1,2}Fe_{0,5}$	0.71	98	400	480	[69]

- 339 -

From the aspect of economy in use, tests conducted with the relatively inexpensive Ce-mischmetal (MM) are of particular importance, cf. table 4. With "pure" MMCo$_5$-compounds, no particularly good magnetic properties have been achieved as yet. In Sm-MM-Co-alloys, however, values can be achieved which are equivalent to those of pure SmCo$_5$-alloys [31].

On account of the lower J$_s$-values, the Cu-containing alloys cannot approach the maximum values of the RCo$_5$-compounds, cf. table 5. Precipitation hardening can evidently only be achieved with Sm or Ce. In the system PrCo$_{5-x}$Cu$_x$, $_JH_c$-values of 30 kA/m were found at best [32]. It has not been possible to make useful compact magnets from MMCo$_{5-x}$Cu$_x$ either.

Figure 1 shows the demagnetization curves at room temperature of some RCo$_5$-magnets and some commonly used permanent-magnet materials.

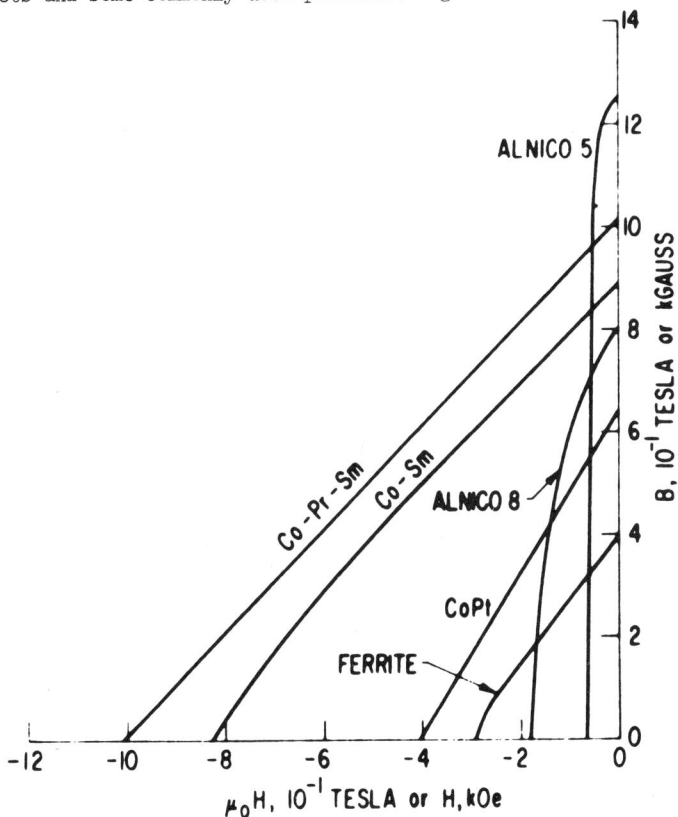

FIGURE 1

Comparison of the demagnetization curves of two cobalt-rare earth alloys with those of several commercial magnets [55].

6. INFLUENCE OF TEMPERATURE UPON THE MAGNETIC PROPERTIES

A damaging effect on the microstructure and, consequently, a loss mainly of $_JH_c$ was found in a $SmCo_5$-magnet after it had been annealed for 30 min at $650^{\circ}C$ [33]. This is qualitatively comparable to the behaviour of Alnico. A damaging effect on the grain structure must be anticipated at temperatures above about $400^{\circ}C$, which is probably due to the eutectoid decomposition mentioned above [75]. Additional damage occurs in the case of open pores by a reaction with O_2.

Demagnetization curves, measured at temperatures between 77 and $475^{\circ}K$, are shown in figure 2 in respect of a specimen having very high characteristics [34]. Conspicuous is the considerable effect of temperature upon $_JH_c$.

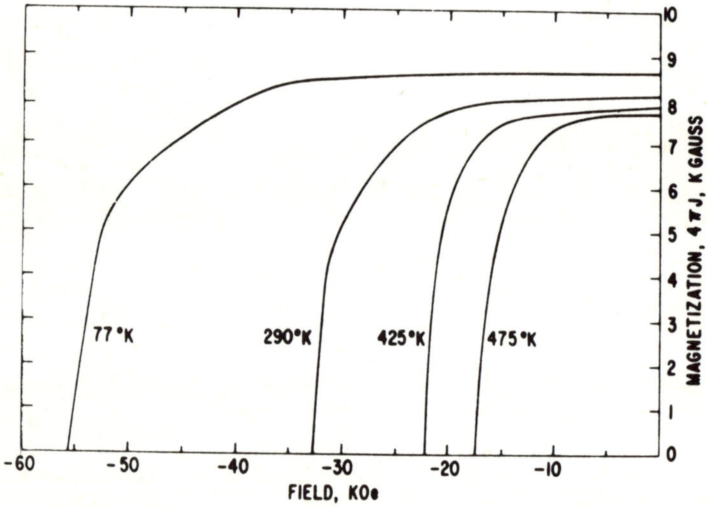

FIGURE 2

Demagnetization curves for a sintered Co-Sm sample
measured at temperatures between 77 K and $475^{\circ}K$ [34]

From data published in [33,34], the temperature coefficients (TK) can be calculated for B_r and $_JH_c$, which are compared in table 6 with the corresponding values of commonly used materials [35,36]. The TK's of $SmCo_5$ are roughly comparable to those of PtCo, while there are characteristic differences in relation to those of ferrite and Alnico. A similarly strong temperature dependence of $_JH_c$ was also found in $SmCo_5$-powder [37].

TABLE 6

Temperature coefficients TK in $\%/°C$ of remanence B_r and intrinsic coercivity $_JH_c$

$$TK = \frac{1}{X_r} \frac{dX}{dT} \; ; \quad X_r = \text{value at room temperature}$$

Material	TK (B_r) (20 - 200°C)	TK $(_JH_c)$ (20 - 200°C)	TK $(_JH_c)$ (-196 - 20°C)	
$SmCo_5$	- 0.024	- 0.23	- 0.34	[33,34]
PtCo	- 0.02	- 0.35	-	[35]
$BaO·6 \, Fe_2O_3$	- 0.19	+ 0.40	-	[35]
Alnico	- 0.02	- 0.06 to + 0.03	-	[35]

As far as actual use is concerned, the above described change of the demagnetization curve is of secondary interest only, while the irreversible and reversible changes of the magnetic flux, however, are of immediate interest.

Some data on irreversible losses are given in table 7. Specimens of $SmCo_5$ with $B/\mu_o H = - 0.5$ exhibited very high scattering losses [31,32]. They depend decisively on the form of the $J(H)$-demagnetization curve: The losses are the smaller the more rectangular the curve is when $_JH_c$ is as large as possible. The fact that such considerable losses may occur at working points close to $_JH_c$ is a consequence of the high negative TK $(_JH_c)$ described above. The higher the working points, the smaller are, therefore, the losses. Qualitatively, this is quite similar to the behaviour of Alnico-magnets, which may be better or worse depending on grade and conditions [38]. Up to at least $300^{\circ}C$, hard ferrites, however, show practically no losses at all as a result of their high <u>positive</u> $TK(_JH_c)$.

TABLE 7

Irreversible loss after temporary heating to $300^{\circ}C$

Material	$B/\mu_o H$	Irreversible loss in %	
$SmCo_5$	- 0.5	3 to 41	[33]
"	- 20	2.7	[33]
Alcomax III	- 9.5	1.1	[38]
Alnico V	- 2.4	10.5	[38]
$BaO \cdot 6 Fe_2 O_3$	all	0	[70]

TABLE 8

Temperature coefficients $TK (B) = \frac{1}{B} \frac{dB}{dT}$ in %/$^{\circ}C$ of open circuit remanence above and below room temperature (RT), resp.

Material	TK (B)		
	below RT	above RT	
$SmCo_5$	-0.013 to -0.037	-0.036 to -0.049	[33]
$(Sm,MM)Co_5$	?	- 0.04	[31]
PtCo	-0.015	- 0.0025	[70]
$BaO \cdot 6 Fe_2 O_3$	-0.19	- 0.19	[70]
Alnico	-0.045 to +0.046	-0.010 to -0.024	[38,70]
$SmCo_5$ with binder	-0.026	- 0.026	[71]

Some values of the reversible temperature coefficient of the open circuit remanence are compiled in table 8. The TK of $SmCo_5$ is at the upper limit of the values known for metallic permanent magnets, but still substantially smaller than that of ferrites.

7. COMPARISON OF RCo_5-MAGNETS WITH OTHER PERMANENT MAGNETS

Let us first consider the magnetic properties. According to figure 1, the RCo_5-alloys are clearly superior to the hitherto known permanent magnet materials in respect to their resistance to demagnetization. This furnishes advantages in applications with low load lines and/or large demagnetizing fields, as for instance in travelling-wave tubes with spatially periodic focussing [39-41], klystrons, magnetrons, electrical watches, in motors and generators [42], and in permanent magnet suspension systems for high-speed inter-urban transport or materials handling systems [43-46]. In the case of small demagnetizing fields, however, the highly remanent Alnico grades are still superior owing to their higher saturation polarization.

This picture does not change even if the behaviour of $SmCo_5$ below room temperature is considered. Owing to the high increase of $_JH_c$, the magnets then become even more stable. Things are different above room temperature. Under such conditions, substantial irreversible losses may occur under strong demagnetizing influences owing to the considerable decrease of $_JH_c$. It remains to be seen, whether this drawback can be eliminated by better control of the manufacturing conditions.

RCo_5-alloys suffer damage to their structure at temperatures around $400^\circ C$ and up, while high-grade Alnico-alloys, e.g., Alnico 500 or 450, show such effects only at temperatures above about $550^\circ C$.

Finally, the need of using very high fields for magnetization, for instance, of more than 2.5 MA/m, should be emphasized.

At present, the main obstacle is the price of the RE-metals plus the absence of mass-production facilities. At the present stage, $SmCo_5$-type magnets can compete only with the PtCo magnets which are even more expensive. This situation might change once we have learnt to use natural mixtures instead of the expensive, pure RE metals and suitable mass-production facilities have been established.

8. DIRECTIONS OF FUTURE DEVELOPMENT

In recent years, RCo_5-compounds have mainly been developed and studied for their suitability as permanent magnets. Other compounds composed of RE and 3d-transition metals $R_m T_n$ appeared to possess less optimum combinations of high J_s and a high easy axis H_A and were therefore somewhat neglected. In more recent times, however, it was found that suitable combinations of these properties exist in systems of the type $R_2(Co_{1-x}T_x)_{17}$.

TABLE 9

Saturation polarization J_s, Curie temperature T_c and type of anisotropy A of R_2Fe_{17}- and R_2Co_{17}-compounds (EP = easy plane, EA = easy axis) [47-49]

Composition	J_s (20°C) (T)	T_c (°C)	A
R_2Fe_{17} R = Y, Pr ··· Lu	0.65 ··· 1.12	-50 ··· +180	EP
R_2Co_{17} R = Y, Ce ··· Lu	0.65 ··· 1.40	+780 ··· +950	-
R = Sm, Er, Tm	-	-	EA
R = all others	-	-	EP

Table 9 gives J_s, T_c and type of anisotropy of the marginal compounds R_2Fe_{17} and R_2Co_{17}. They crystallize in the hexagonal and/or rhombohedral lattice. The J_s's are higher than in the neighbouring phases richer in R. T_c of the Fe-compounds are very low in comparison, while those of the Co-compounds are even above those of the RCo_5-compounds. As far as the type of anisotropy is concerned, however, only R_2Co_{17} with R = Sm, Er and Tm show an easy axis, all others easy planes.

It is amazing that it is possible in the case of a fairly large number of compositions to change at room temperature with fairly simple means the easy plane into a desired easy axis [47-49]: In $Sm_2(Co_{1-x}Fe_x)_{17}$, the easy axis is retained up to about x = 0.5. In the cases with R = Y, Ce, MM and Pr, an easy axis exists, depending on the element, above x_1 = 0.05 to 0.2

and below $x_2 = 0.45$ to 0.6. With the light RE's only in the case of R = Nd an easy plane exists for all x. Also, the magnitude of the anisotropy appears to be attractive. For Sm_2Co_{17}, H_A was determined at 4.8 $\cdots$ 8.0 MA/m, and for $Sm_2(Co_{0,5}Fe_{0,5})_{17}$ H_A was determined at $\approx$ 3.2 MA/m [48]. The formation of solid solutions exerts a particularly favourable influence upon J_s; it attains higher values than in the marginal compounds as for instance, 1.53 T in the case of $Sm_2(Co_{0,5}Fe_{0,5})_{17}$ and $Y_2(Co_{0,4}Fe_{0,6})_{17}$ [48]. While it is true that T_c decreases as x increases, it proceeds at first much less than proportionally, so that still T_c's of 600 to 750°C are obtained for x = 0.5, for instance [49].

Similar changes from an easy plane to an easy axis were found in compounds of the type $R_2(Co_{1-x}Mn_x)_{17}$. The anisotropy fields of these and the analogous Fe-compounds range between 0.95 and 1.9 MA/m [50].

Another possibility to change the type of anisotropy is to utilize the homogeneity range of Y_2Co_{17} towards the Y-rich side. In compounds of the composition $Y_{2+y}Co_{17-2y}$, an easy axis is likewise obtained at y = 0.13 and up [51,52].

Further development of the alloys of the above type should be aimed at producing high $_JH_c$-values in powders and in compact specimens. A start has already been made: In Sm_2Co_{17}-containing specimens, a $_JH_c$-value of 760 kA/m was achieved [48]. If this can be accomplished to an adequate extent, then $(BH)_{max}$-values should be attainable which are more than twice as high as those of the best $(Sm,Pr)Co_5$-magnets available today, cf. figure 3 [48]. Even the costs of the raw materials might be substantially reduced; on the one hand, because about 30% less R would be required, and on the other hand, because almost 50% of the cobalt could be replaced by iron. The future will show whether these fantastic potentialities can be realized in the development of permanent magnet materials and whether $(BH)_{max}$-values at room temperature can be produced of magnitudes so far only obtained at extremely low temperatures: A $(BH)_{max}$ = 580 kJ/m^3 with B_r = 1.72 T, $_BH_c$ = 1370 kA/m and $_JH_c$ = 1670 kA/m was measured in the case of a single crystal of Dy_3Al_2 [53].

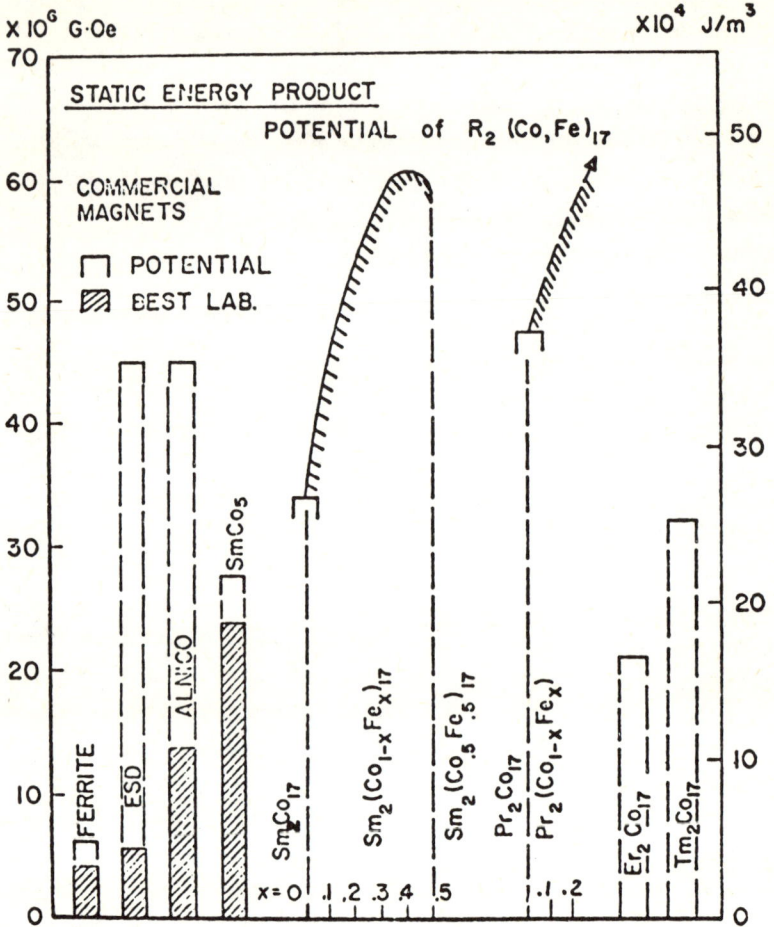

FIGURE 3

Theoretically possible energy products for magnets made from selected $R_2(Co,Fe)_{17}$ phases compared to best experimental and theoretical values for other magnet types [48].

References

[1] K. Strnat, IEEE Trans. Magn. MAG-6, 182-190 (1970).

[2] W.A.J.J. Velge, K.H.J. Buschow, J. Appl. Phys. 39, 1717 (1968).

[3] E.A. Nesbitt, J. Appl. Phys. 40, 1259 (1969).

[4] K.H.J. Buschow, W.A.J.J. Velge, Z. angew. Phys. 26, 157 (1969).

[5] T. Shibata, T. Katayama, T. Mizuhara, Jap. J. Appl. Phys. 10, 1479
     (1971).

[6] E. Tatsumoto, T. Okamoto, H. Fujii, C. Inoue, J. de Phys. 32, C 1,
     550 (1971).

[7] T. Okamoto, H. Fujii, C. Inoue, E. Tatsumoto, Intermag-Konferenze
     Kyoto 1972, Vortrag 33.4.

[8] E.A. Nesbitt, H.J. Williams, J.H. Wernick, R.C. Sherwood, J. Appl.
     Phys. 32, 342 S, (1961).

[9] J.J. Becker, AIP Conf. Proc. No. 5, Part 2, 1067 (1972).

[10] H. Zijlstra, J. Appl. Phys. 42, 1510 (1971).

[11] K. Backmann, F, Hofer, Z. angew. Phys. 32, 41 (1971).

[12] F.F. Westendorp, J. Appl. Phys. 42 5727 (1971).

[13] J.J. Becker, J. Appl. Phys. 41, 1055 (1970).

[14] K. Strnat, J. Tsui, Proc. 8th Rare Earth Res. Conf. Vol. 1, 3 (1970).

[15] K. Strnat, A.E. Ray, C. Herget, Journal de Phys. 32, C1, 552 (1971).

[16] D.K. Das, IEEE Trans. Magn. MAG-7, 432 (1971).

[17] R.E. Cech, J. Appl. Phys. 41, 5247 (1970).

[18] J.W. Walkiewicz,J.S. Winston, M.M. Wong, Proc. 9th Rare Earth Res.
     Conf., 242 (1971).

[19] T. Shibata, T. Katayama, Intermag-Konferenz Kyoto, 1972, Vortrag 37.2.

[20] J.B.Y. Tsui, K.J. Strnat, IEEE Trans. Magn. MAG-7, 427 (1971).

[21] M.G. Benz, D.L. Martin, Appl. Phys. Lett. 17, 176 (1970)

[22] F.F. Westendorp, Sol. State Com. 8, 139 (1970).

[23] J. Schweizer, K.J. Strnat, J.B.Y. Tsui, IEEE Trans. Magn. MAG-7,
     429 (1971).

[24] K.H.J. Buschow, P.A. Naastepad, F.F. Westendorp, J. Appl. Phys. 40,
     4029 (1969).

[25] Y. Kimura, K. Kamino, Trans. Jap. Inst. Met. 11, 132 (1970).

[26] G.Y. Chin, M.L. Green, E.A. Nesbitt, R.C. Sherwood, J.H. Wernick,

IEEE Trans. Magn. MAG-8, 29 (1972)

[27] Y. Tawara, H. Senno, Intermag-Konferenz Kyoto 1972, Vortrag 37.5.

[28] A. Hoffmann, H. Stäblein, Techn. Mitt. Krupp, Forsch.- Ber. 24, 113 (1966).

[29] A. Hoffmann, P. Pant, Techn. Mitt, Krupp, Forsch.- Ber. 28, 117 (1970).

[30] H. Stäblein, AIP Conference Proc. No. 5, Part 2, 950 (1972).

[31] M.G. Benz, D.L. Martin, J. Appl. Phys. 42, 2786 (1971).

[32] B.A. Samarin, Appl. Phys. Lett. 17, 196 (1970).

[33] D.L. Martin, M. G. Benz, IEEE Trans. Magn. MAG-8, 35 (1972).

[34] D.L. Martin, M.G. Benz, Intermag-Konferenz Kyoto 1972, Vortrag 37.6.

[35] H. Dietrich, Cobalt No. 35, 78 (1967).

[36] H. Stäblein, Techn. Mitt. Krupp, Forsch.-Ber. 29, 101 (1971).

[37] R.A. McCurrie, G.P. Carswell, Phil. Mag. 23, 333 (1971).

[38] M. McCaig, Cobalt No. 5, 26 (1959).

[39] D.L. Martin, M.G. Benz, IEEE Trans. Magn. MAG-7, 291 (1971).

[40] D.K. Das, W.J. Harrold, IEEE Trans. Magn. MAG-7, 281 (1971).

[41] W.J. Harrold, AIP Conf. Proc. No. 5, Part 2, 1006 (1972).

[42] E. Richter, AIP Conf. Proc. No. 5, Part 2, 991 (1972).

[43] D.J. Iden, C.E. Ehrenfried, H.J. Garrett, AIP Conf. Proc. No. 5, Part 2, 1026 (1972).

[44] J.B.Y. Tsui, D.J. Iden, K.J. Strnat, A.J. Evers, IEEE Trans. Magn. MAG-8, 188 (1972).

[45] W. Baran, Z. angew. Phys. 32, 216 (1971).

[46] W. Baran, Intern. J. Magnetism, 1972, in press.

[47] K.J. Strnat, AIP Conf. Proc. No. 5, Part 2, 1047 (1972).

[48] K.J. Strnat, Intermag-Konferenz Kyoto 1972, Vortrag 33.1.

[49] A.E. Ray, K.J. Strnat, Intermag-Konferenz Kyoto 1972, Vortrag 33.2.

[50] H.J. Schaller, R.S. Craig, W.E. Wallace, J. Appl. Phys. 43, 3161 (1972).

[51] S. Yajima, M. Hamano, H. Umebayashi, J. Phys. Soc. Jap. 32, 861 (1972).

[52] M. Hamano, S. Yajima, H. Umebayashi, Intermag-Konferenz Kyoto 1972, Vortrag 33.3.

[53] B. Barbara, C. Bécle, R. Lemaire, D. Paccard, IEEE Trans. Magn. MAG-7, 654 (1971).

[54] K. Schüler, K. Brinkmann, Dauarmagnete, Werkstoffe und Anwendungen, Springer 1970.

[55] D.L. Martin, M.G. Benz, AIP Conf. Proc. No. 5, Part 2, 970 (1972).

[56] E.A. Nesbitt, G.Y. Chin, P.K. Gallagher, R.C. Sherwood, J.H. Wernick, J. Appl. Phys. 41, 1107 (1970).

[57] Y. Iwama, M. Takeuchi, Intermag-Konferenz Kyoto 1972, Vortrag 61.3.

[58] K.J. Strnat, J.C. Olson, G. Hoffer, unpublished.

[59] J.J. Becker, J. Appl. Phys. 42, 1537 (1971).

[60] D.L. Martin, M.G. Benz, GE-Rep. 70-C-261, Aug. 1970.

[61] R.J. Charles, D.L. Martin, L. Valentine, R.E. Cech, AIP Conf. Proc. No. 5, Part 2, 1072 (1972).

[62] K.J. Strnat, J.B.Y. Tsui, J. Schweizer, Proc. 9th Rare Earth Res. Conf. 1971, Vol. 1, 252.

[63] W.A.J.J. Velge, K.H.J. Buschow, Conf. Magn. Mat. and Appl., London 1967.

[64] J.B.Y. Tsui, K.J. Strnat, R.S. Harmer, J. Appl. Phys. 42, 1539 (1971).

[65] H. Schuchert, Techn. Mitt. Krupp, Forsch.-Ber. 28, 121 (1970).

[66] R.E. Johnson, C.J. Fellows, Kobalt No. 53, 159 (1971).

[67] Y. Tawara, H. Senno, Japan J. Appl. Phys. 7, 966 (1968).

[68] T.J. Cullen, J. Appl. Phys. 42, 1535 (1971).

[69] E.A. Nesbitt, G.Y. Chin, R.C. Sherwood, J.H. Wernick, Appl. Phys. Lett. 16, 312 (1970).

[70] R.J. Parker, R.J. Studders, Permanent Magnets and Their Application, John Wiley & Sons Inc., 1962.

[71] H. Umebayashi, U. Fujimura, Jap. J. Appl. Phys. 10, 1585 (1971).

[72] M.G. Benz, D.L. Martin, AIP Conf. Proc. No. 5, Part 2, 1082 (1972).

[73] S. Eoner, E.J. McNiff Jr, D.L. Martin, M.G. Benz, Appl. Phys. Lett. 20, 447 (1972).

[74] R.C. Sherwood, E.A. Nesbitt, G.Y. Chin, M.L. Green, Mat. Res. Bull. 7, 489 (1972).

[75] F.J.A. den Broeder, K.H.J. Buschow, J. Less-Com. Mat. 29, 65 (1972).

[76] M.G. Benz, D.L. Martin, J. Appl. Phys. 43, 3165 (1972).

THE PRODUCTION, QUALITY CONTROL, AND APPLICATIONS OF
PURE RE METALS AND ALLOYS

by

Karl A. Gschneidner, Jr.

Ames Laboratory USAEC and Department of Metallurgy

Iowa State University, Ames, Iowa 50010, U.S.A.

ABSTRACT

The light lanthanide metals are usually prepared by the metallothermic reduction method using Ca or Li, or by the electrolytic method. Yttrium and the heavy lanthanides are prepared only by the Ca or Li reduction method. The volatile lanthanides, Sm, Eu, Tm and Yb are prepared by the La reduction of the oxide. Mischmetal, primarily because of economic reasons, is prepared electrolytically. The quality control with respect to the production of high purity rare earth metals is discussed. The important metallurgical uses of the rare earths are examined, and the reasons for the utilization are noted.

1. PRODUCTION

There are two basic methods used to prepare the light trivalent lanthanides (La, Ce, Pr, Nd). One of these methods is used to prepare the heavy trivalent lanthanides (Gd, Tb, Dy, Ho, Er, Lu) and yttrium. The divalent lanthanides (Eu and Yb) and the adjacent trivalent lanthanides which are one lower in atomic number (Sm and Tm) are prepared by a different method. These are described below along with the production of mischmetal.

1.1 Light Lanthanides - Metallothermic Method [1-6]

The light lanthanide metals, La, Ce, Pr and Nd, may be prepared by the reduction of the trifluoride using Ca metal as the reductant. The fluoride is prepared by reacting the oxide with HF gas or NH_4HF_2 at low temperatures

(200 - 500°C). The former reactant is preferred for high-purity metals, since there is less residual oxygen in the resultant fluoride and since the NH_4HF_2 introduces some Fe into the fluoride which is carried over into the metal. For the lowest oxygen contents the RF_3 should be melted under HF gas before the reduction step. Platinum ware is used in all steps to form the fluoride.

After fluorination the RF_3 is mixed with 10% excess Ca metal and placed in a Ta crucible, which is then heated inductively to about 1400°C or to a temperature about 100°C above the lanthanide metal's melting point, whichever is higher. The denser lanthanide metal sinks to the bottom, and the CaF_2 floats on top of it. The slag is easily removed at room temperature, and the metal is then vacuum cast at 1000 -1200°C to remove the excess Ca. For the highest purity metals, freshly prepared triply distilled Ca metal is used for the reductant, and the vacuum casting is carried out to much higher temperatures, ∼ 1800°C. The high vacuum casting temperature, however, causes the lanthanide metals to dissolve significant amounts of Ta, but if the sample is slowly cooled and held just above its melting point, most of the Ta precipitates out along the Ta crucible wall. Thus, the high-Ta region at the edge of the ingot is machined off leaving behind the essentially Ta-free lanthanide metal.

Other variations of this process are to use the chloride instead of the fluoride and to substitute Li for Ca as the reductant. The chloride has the disadvantage of being hygroscopic while the fluoride is not. Other variations have appeared in the literature, but the fluoride-Ca reduction method is the one most universally used. The metallothermic process is used to prepare metals 99 to 99.99% pure with respect to all impurities depending upon the care one takes in the metal preparation steps.

1.2 Light Lanthanides - Electrolytic Method [7-11]

The lanthanide metals are prepared by the reduction of the chloride or fluoride or oxide using an electrolytic cell. For the reduction of the oxide a complex fluoride solvent, RF_3-LiF-BaF_2, which dissolves about 5% of the oxide at the operating temperature, is used. The cell walls can be made of a ceramic (e.g., firebrick) or graphite or molybdenum; the anode of graphite

and the cathode of iron or molybdenum or tungsten. The choice of material depends upon the purity desired for the final product. The operating conditions are: temperatures of $800 - 1000°C$, high currents 200 to 2500 amp, low voltages 4 to 15 V and current densities of 2 to 10 amp/cm^2. The advantage of this method over the metallothermic one is that the lanthanide metals can be produced at a much lower cost. In general, the purities of the electrolytically prepared metals are less than those obtained by metallothermic reduction, although in principle one should be able to prepare as pure a metal electrolytically. One limitation of this method is that it cannot be (or at least to date has not been) used to prepare the heavy lanthanides in pure form, primarily because the cell and electrode materials cannot withstand the high temperatures required because of the heavy lanthanides' high melting points.

Mischmetal is prepared using the electrolytic method. Since the price is so important the cheapest workable materials are used in the construction of the cells and electrodes. In general, the quality of the mischmetal produced is 98% or less pure. In part this is due to the addition of Mg to reduce the oxidation of mischmetal at room temperature. The mischmetal also contains appreciable quantities of other metallic impurities such as Fe, Al, Si and Pb. Some of these impurities come from the original chloride used as the starting material, and the remainder is introduced in the metal reduction process. Since mischmetal was primarily used as an alloying agent or as lighter flints (60 - 90% mischmetal/40 - 10% Fe alloy) the impurity contents were not critical. However, with the recent development of mischmetal ordnance devices (projectile liners) [12], the effect of impurities is much more critical at least for this application.

1.3 Heavy Lanthanides and Yttrium [1-6]

These metals (Gd, Tb, Dy, Ho, Er, Lu and Y) are prepared metallothermically in the same manner as described above for the light lanthanides, with one important exception. The exception is that the high temperature vacuum casting operation is replaced by vacuum distillation or sublimation of high-purity metals. Because the heavy lanthanides dissolve appreciable amounts of tantalum, even at the melting point [13,14], it is important that these metals

are distilled. Furthermore, the slower the sublimation or distillation rate
the purer the final rare earth metal.

1.4 Lanthanothermic Reduction Method [1-6]

The low vapour pressure of La is utilized in the preparation of Sm, Eu,
Tm and Yb metal. The oxide of Sm (Eu, Tm or Yb) is mixed with La chips and
placed into a Ta distillation vessel. Upon heating, the La reacts with the
oxide to form the other rare earth metal (Sm, Eu, Tm and Yb), but because
they have very high vapour pressures they evaporate and thus the reaction is
driven to completion, leaving behind La_2O_3. The volatile rare earth metal
is usually redistilled for further purification. Other reactive metals with
high boiling points, such as Ce, Zr and Th, can be used to prepare these
metals.

2. QUALITY CONTROL

The quality control used in the production of the rare earth metals depends on the nature of the product. Most of the following remarks will be
concerned with the production of very high purity rare earth metals, 99.9%
pure or greater.

For routine metal production the starting materials and the final product are generally analyzed by emission spectroscopy, spark mass spectroscopy, vacuum fusion and wet chemical methods. Thus, there is a complete
analysis for rare earth impurities, non-rare earth metallic impurities and
the non-metallic elements in a given rare earth metal. For our high-purity
production the starting oxide must contain less than 30 ppm by weight total
other rare earths and less than 10 ppm Fe. For those rare earth metals which
are prepared by Ca reduction of the fluoride, the Ca and Si concentrations
in the oxide are not too critical, since Si impurity is removed during the
second fluorination step. The Ca concentration in the oxide for the La reduction of Sm, Eu, Tm and Yb is important since it will be distilled or sub-
limed along with the desired final product. The Si concentration is not too
critical since it will remain behind as $LaSi_x$ compound(s).

When difficulties are being experienced or when improvements in the

metal production are being attempted, analyses are made at each step of the
process and the results inter-compared. For example, in the fluorination
process using HF gas, it was found that the fluoride after the first step
contained about 200 ppm oxygen, which is subsequently reduced to less than
20 ppm in the second step - the melting of the fluoride under HF.

For less pure metals it is not necessary to have as complete analyses.
It is probably adequate to have knowledge of a few of the more important impurities which are likely to be found in the metal production steps, or for
those impurities which are of special interest in relation to the end use of
the metal.

3. APPLICATIONS

The major uses of the rare earth metals are as additives to ductile
iron, steels, non-ferrous metals, such as Mg, and superalloys, as lighter
flints, in ordnance and in research. In most of the above applications the
metals are used as mixed rare earth metals in their natural distribution -
this material is known as mischmetal. The separated pure metals are used
only for additions to superalloys and in research.

The ductile iron market is the largest outlet for rare earth metals.
The rare earths are added ($\sim$ 0.2 wt.%) to spheroidize the graphite, counter-
act the influence of deleterious impurities and increase liquid fluidity.
The acceptance of mischmetal over Mg, even though mischmetal costs more, is
due to the fact that Mg volatilizes from the melt making it difficult to control the quality of the ductile iron, but the rare earths do not volatilize,
and thus the quality of the final product is easier to control. The addition
is generally made as a ferro-mischmetal silicide or ferro-cerium silicide.

The addition of rare earths to steels is continually gaining acceptance.
The rare earths are added ($\sim$ 0.2 wt.%) to control the sulphur concentration
and thus improve primarily the workability and the transverse impact values.
The major use of rare earths in he steel market has been their addition to
plate and pipeline steels. The rare earth sulphides are essentially the most
stable sulphides known; only CaS has a comparable free energy of formation
[15]. The rare earths are generally added in the form of a ferro-mischmetal

(or cerium) silicide.

The addition of rare earths to magnesium improves the high temperature strength, creep resistance and fatigue properties, especially at elevated temperatures. Some Zn-R-Zr magnesium-base alloys have been introduced over the past three or four years as magnesium casting alloys because of their improved castability, low rejection rate and good weldability. Minor non-ferrous uses include the addition to copper-base and aluminium-base alloys and bronzes.

Pure rare earth metals are added to a number of chromium-base, cobalt-base and nickel-base superalloys [16-20]. Essentially about 0.1 wt.% of Y, La or Ce is added to improve the high temperature oxidation and/or resistance to corrosion in various corrosive environments, e.g., salt water, combustion atmospheres containing sulphur. The rare earths are thought to form a complex oxide which provides a protective coating on the superalloy. In the case of sulphur environments the protective mechanism is provided by the formation of a stable rare earth oxysulphide, R_2O_2S [18,19].

The pyrophoric nature of the light rare earths, especially cerium, accounts for two important uses - lighter flints and ordnance. A variety of mischmetal-iron combinations (60 to 80% mischmetal with the remainder being mostly iron plus additional minor alloying agents) are used as lighter flints and industrial sparking tools.

The use of mischmetal in munitions as projectile linings is a recent development [12]. A 95 wt.% mischmetal 5 wt.% Mg alloy is investment-cast into a close-end sleeve 2.5 cm in diameter by 7.5 cm long. This ordnance device can be ignited by either firing the projectiles at high velocity against steel or by an explosive charge within the projectile.

A major use of pure separated rare earth metals is in fundamental and applied research. Some of the research is being orientated toward determining more information about the fundamental properties and nature of these metals and their alloys. It is from these results that new applications will be generated. The remaining research is directed towards making some of the more promising ideas and developments into commercial realities. It is apparent to anyone following the literature that research on the rare earths, especially the applied work, has greatly expanded, even in countries which have experienced cutbacks in funding science. Furthermore, the dominance of rare earth research by the United States about five years ago (about 50% of papers published were from the U.S.) is no longer the case today. Both of these trends are healthy and should maintain keen world-wide competition in the science, technology and industry involved with the rare earths.

References

[1] A.H. Daane, p. 102 in The Rare Earths, F.H. Spedding and A.H. Daane, eds, Wiley, New York (1961).

[2] C.E. Habermann, A.H. Daane and P.E. Palmer, Trans. Met. Soc. AIME 233, 1038 (1965).

[3] K.A. Gschneidner, Jr., p. 99 in Trans. Vacuum Metallurgy Conf., 1965, L.M. Bianchi, ed., Am. Vac. Soc., Boston, Mass. (1966).

[4] F.H. Spedding, B.J. Beaudry, J.J. Croat and P.E. Palmer, p. 151 in Materials Technology - An Interamerican Approach, Am. Soc. Mech. Eng., New York (1968).

[5] F.H. Spedding, B.J. Beaudry, J.J. Croat and P.E. Palmer, p. 25 in Les Eléments des Terres Rares, Vol. 1, Centre National de la Recherche Scientifique, Paris (1970).

[6] J.L. Moriarty, J. Metals 20 [11], 41 (1968).

[7] E. Morrice and R.G. Knickerbocker, p. 126 in The Rare Earths, F.H. Spedding and A.H. Daane, eds, Wiley, New York (1961).

[8] E.S. Shedd, J.D. Marchant and T.A. Henrie, U.S. Bur. Mines, Rept. Invest. No. 6362 (1964).

[9] E. Morrice, E.S. Shedd and T.A. Henrie, U.S. Bur. Mines, Rept. Invest. No. 7146 (1968).

[10] I.S. Hirschhorn, J. Metals 20, [3], 19 (1968).

[11] E.S. Shedd, J.D. Marchant and M.M. Wong, U.S. Bur. Mines, Rept. Invest. No. 7398 (1970).

[12] RIC News 6, [4], 4 (December 1971); published quarterly by the Rare Earth Information Center, Institute for Atomic Research, Iowa State University, Ames, Iowa 50010, U.S.A.

[13] D.H. Dennison, M.J. Tschetter and K.A. Gschneidner, Jr., J. Less-Common Metals 10, 108 (1966).

[14] D.H. Dennison, M.J. Tschetter and K.A. Gschneidner, Jr., J. Less-Common Metals 11, 423 (1966).

[15] K.A. Gschneidner, Jr. and N. Kippenhan, IS-RIC-5 (August 1971); this report is available free from the Rare Earth Information Center, Institute for Atomic Research, Iowa State University, Ames, Iowa 50010, U.S.A.

[16] G.C. Wood, Werkstof. Korr. 22, 491 (1971).

[17] J.E. Antill, Werkstof. Korr. 22, 513 (1971).

[18] P. Elliott and T.K. Ross, Werkstof. Korr 22, 531 (1971).

[19] A.U. Seybolt, Corr. Sci. 11, 751 (1971).

[20] J.K. Tien and F.S. Pettit, Met. Trans. 3, 1587 (1972).

PANEL DISCUSSION

RARE EARTH APPLICATIONS IN LUMINESCENCE

Melvin Tecotzky, Chairman,
United States Radium Corporation, P.O. Box 409,
Hackettstown, New Jersey 07840, U.S.A.

RARE EARTH APPLICATIONS IN LUMINESCENCE

Rare earth luminescence has been known and studied since before the turn of the century. Early investigators such as Crooks and Urbain reported on Rare Earth luminescence [1]. However, it is just within the past ten years that efficient rare earth luminescence has become commercially significant. With the discovery of the solid state laser, the search was on for host materials for trivalent rare earth ions.

The first large scale commercial use of rare earth luminescence was in the cathode-ray tube used in colour television.

Ballman, Linares, and Van Uitert of Bell Laboratories reported on the solid state red emitting laser yttrium orthovanadate activated with europium [2].

The first commercial rare earth phosphor used in colour television was europium activated yttrium orthovanadate reported by Levine and Palilla [3], and introduced by Sylvania. This rare earth phosphor soon became the standard of the industry. This inspired additional research for brighter red phosphors and led to the introduction of yttrium oxide and gadolinium oxide, both activated by europium, and yttrium oxysul. activated with europium which was introduced by RCA [4]. These latter three phosphors were brighter than the vanadate and soon replaced it in the television tube as the red phosphor. These three phosphors are currently being used throughout the world as the red emitting phosphor for colour TV. I would guess that yttrium oxysulphide is used to the greatest extent at the current time.

This brief introduction brings us to our topic - rare earth applications in luminescence. I will also discuss aspects of rare earth luminescence which have not been covered by the previous speakers.

Rare earth phosphors in cathode-ray tubes is what I would like to cover first in this round table discussion. Commercial colour television is the big volume application for cathode-ray tubes.

As we have said, the red emitting phosphor in the present colour television set is a europium activated rare earth phosphor (table 1). Does anyone here see a new rare earth phosphor replacing the present red phosphor used?

TABLE 1

Red Emitting Rare Earth Phosphors for Colour Television

YVO_4:Eu

YVO_4:Eu:Bi

Y_2O_3:Eu

Gd_2O_3:Eu

Y_2O_2S:Eu

DISCUSSION

Yocom:

Lehmann of Westinghouse has reported on an efficient CaS:Eu:Ce phosphor which is red emitting. However, it is a band emitter and some useful light is lost in the far red portion of the spectrum. It is not as suitable as the red phosphors currently being used.

Tecotzky:

Actually the most urgent phosphor requirement in colour television is not a new red, but a new green. The cathode-ray tube people like to operate their three guns in the tube at unity gun ratio. This is difficult if one of the phosphors is more efficient than the others. Currently the tube manufacturers are green limited so the greatest need is for a brighter green emitting phosphor.

Let us continue then and see if there is an opportunity for introducing a green emitting rare earth phosphor and blue emitting rare earth phosphor in our current television tube.

Willi Lehmann of Westinghouse has reported on cerium activated calcium sulphide [5]. The activator is a rare earth but the host is not. He reports it to be comparable in performance to the currently used zinc cadmium sulphide activated with copper. To my knowledge, no one has introduced this in a commercial TV tube. One reason may be lack of stability during screening; another may be that if it is only comparable to the present green phosphor, why go to the trouble of making a change. Are there any comments from the group about this?

Yocom:

What is needed is a new host for a rare earth green. There is none currently known that has the required brightness.

Mathers:

CaS:Ce could be a problem on screening.

Blasse:

I think it is a waste of time to look for small advances. Your work should be directed at new concepts that completely change the state of the art. I don't consider a 10 or 20% brightness gain in green that sort of advance. The introduction of a rare earth red emitting phosphor was a significant change. A photoluminescent process where 1 photon absorbed produces 2 photons emitted would also be a significant advance and worthy of your efforts.

Tecotzky:

   Your point is well taken, but in the commercial world, television sets
are sold and advertised on the basis of improved brightness.

Mathers:

   Overall cost also is a very important factor. Research in luminescence is now more applied than basic.

   As far as a blue emitting phosphor is concerned, Shrader, Larach, and
Yocom have reported that zinc sulphide activated with thulium and charge
compensated with lithium is a highly efficient phosphor [6]. The emission,
however, is at too high a wavelength - about 4750 Å instead of a more
favourable 4500 Å, and thus the visual colour is not deep enough blue.
Do you think it may be possible to obtain an efficient rare earth blue of
suitable colour?

Yocom:

   The ZnS:Tm phosphor emission consists of three manifolds. The 4750 Å
is about 60% of the overall efficiency. I believe this phosphor has the
highest overall radiant efficiency of any rare earth containing phosphor
known. Once again, as with the green, we need a new phosphor host.

Tecotzky:

   Can anyone answer the question why thulium is so efficient in zinc
sulphide and shows poor efficiency in other hosts?

   No answer was given by the group.

   Thus far, we have limited out discussion to the current shadow mask
television tube.
   Several years ago Meyer and Palilla described an all rare earth phosphor screen for a high current density cathode-ray tube [7].

The presently used shadow mask tube is operated at low current density and only about 16% of the electrons reach the screen. With another type of cathode-ray tube design in which post deflection is used, the focussing action occurs closer to the screen and higher current densities are employed. It has been suggested that if a tube of this type were used, rare earth phosphors for green and blue as well as red, could be employed replacing the current green and blue sulphide phosphors. The green and blue sulphide phosphors would show non-linear behaviour at increased current densities. Since the rare earth phosphors show greater linearity, their performance would be excellent in this type of post deflection tube operating at higher current densities. Terbium activated indium borate, yttrium silicate, and yttrium phosphate were mentioned as potential rare earth green emitting phosphors for this type of tube.

Divalent europium strontium halophosphate was also mentioned as a potential rare earth blue emitting phosphor. Thulium, of course, is another potential blue activator (table 2).

TABLE 2

Potential Green Emitting Phosphors for a Post Deflection Cahode-Ray Tube

$$InBO_3 : Tb$$
$$Y_2 SiO_5 : Tb$$
$$YPO_4 : Tb$$
$$Ln_2 O_2 S : Tb^*$$
$$Ln_2 O_3 : Tb^*$$

Potential Blue Emitting Phosphors for a Post Deflection Cathode-Ray Tube

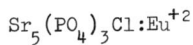

$$Sr_5 (PO_4)_3 Cl : Eu^{+2}$$
$$ABC : Tm^{**}$$

* $Ln = La^{+3}$, Gd^{+3}, or Y^{+3}

** ABC = Unknown host

The basic question is: Will the tube manufacturers consider introducing this type of tube and forget about their large investment in the shadow mask tube? I would presume they would only do it if it offers cost savings or perhaps a quality advantage.

May we have some comments on these points?

DISCUSSION

Mathers:

Cost is a very important factor in television manufacturing today. Perhaps a significant quality improvement or cost savings would be a reason to introduce this type of tube, but I don't know if the whole industry would go to it unless someone like RCA introduced it.

Tecotzky:

As you remember, the whole industry followed Sylvania when they introduced the yttrium vanadate red phosphor.

In addition to cathode-ray tubes for commercial television, we also have specialty types of cathode-ray tubes which are important in instrumentation and industry for displays. In this area many rare earth phosphors have replaced the classical sulphide type phosphors. Applications include business machine and computer displays, optical character recognition machines, avionics displays for aircraft cockpit instrument panels, oscilloscopes, monitors, projection, and special purpose applications [8].

The rare earth phosphors introduced in the specialty tubes show good linearity and can be driven at high current densities to obtain high brightness displays. This is important when one wants to view a cathode-ray tube in high ambient, such as an aircraft cockpit. With a high brightness display the picture will not be washed out in bright sunlight. The rare earth phosphors which are narrow line emitters can be viewed through narrow band pass filters which improves screen contrast.

Recently, Thomas Electronics has registered three new rare earth phosphors with JEDEC - P43, P44, and P45.

P43 is terbium activated gadolinium oxysulphide. The primary emission peaks are in the yellow green; the secondary peaks in the blue and red result in an overall unsaturated pale green colour. It is an efficient phosphor used in high brightness displays.

P44 is terbium activated lanthanum oxysulphide. It is also quite efficient with most of its light emission coming at about 5440 Å. The phosphor has been found suitable for CRT screens with a band pass filter.

P45 is a terbium activated yttrium gadolinium oxysulphide which exhibits an emission colour similar to P4 - the phosphor system used in black and white television. It has found use in projection type tubes.

Terbium activated indium borate is also used in specialty tube applications. It is a narrow line green emitting phosphor.

Colour display without using the shadow mask tube has long been a goal in the specialty tube area.

Two approaches that have recently been used to achieve colour displays make use of rare earth or rare earth activated phosphors.

One approach is a current sensitive phosphor system. Through the control of current density to the phosphor screen, it is possible to achieve multicolour displays in cathode-ray tubes by using sublinear phosphors of different emitting colours. At low current density the colour of the sublinear phosphor, e.g. $NaTaO_3$:Tb (green) is dominant; while at high current density, the colour of the superlinear phosphor (ZnCd)S:AgNi (orange) is dominant. As the current density increases the brightness of the sublinear phosphor saturates or shows no increased brightness while the brightness of the superlinear phosphor increases with increased current. The colour range achieved depends on the emitted colours of the individual phosphors, as well as the degree of sub- and super-linearity (table 3).

Other rare earth systems that have been investigated are sublinear $NaTaO_3$:Tb plus linear YVO_4:Eu. Selective optical filters have been used to improve contrast of these screens [9].

The other approach to multicolour displays is the voltage sensitive phosphor system. Through the control of electron beam energy or voltage, it is possible to produce a multicolour cathode-ray tube display through the use of a layered phosphor screen. Phosphors which emit in different colours are separated by barrier or dielectric layers. A colour display

TABLE 3

Current Sensitive Phosphor Systems

$NaTaO_3$:Tb	Green
(ZnCd)S:Ag:Ni	Orange
$NaTaO_3$:Tb	Green
YVO_4 :Eu	Orange
Zn_2SiO_4 :Mn	Green
Gd_2O_3 :Pr	Red

is achieved by electron penetration of the phosphor by voltage control of the electron beam. At low voltage the electrons excite the first phosphor layer which emits light of a particular colour. By increasing the voltage, the electrons pass through the first phosphor layer, through the barrier or dielectric layer, which is inert to electron excitation, and excite the second phosphor which emits light of a different colour. It is thus possible to obtain a colour display through the control of voltage in a cathode-ray tube [10].

Recently two different rare earth phosphor systems have been reported to obtain a voltage sensitive screen without the tedious job of making a layered screen.

In the first system europium activated yttrium oxide is deposited on an unactivated yttrium oxide core. This phosphor is mixed with a zinc cadmium sulphide:Ag screen which has had its surface poisoned with cobalt. The resulting two phosphors are settled in a CRT. At low voltage the electrons activate the europium activated yttrium oxide and one obtains a red display. The green phosphor is inactive at low voltage since its surface has been poisoned. As the voltage is increased the electrons penetrate further into the phosphor particles. The unactivated yttrium oxide core is inert and the green emitting zinc cadmium sulphide starts to emit. One can obtain a range of colours from red to green [11].

The second rare earth approach to voltage sensitive tubes is through the use of a composite phosphor particle containing gadolinium oxide and gadolinium oxysulphide activated with praseodymium. The core of the particle contains praseodymium activated gadolinium oxide, which is a red emitting phosphor. The outside of the particle is praseodymium activated gadolinium oxysulphide, which is a green emitting phosphor. In a cathode-ray tube at low voltage, this phosphor is green and at high voltage, this phosphor is red. At intermediate voltages the colours obtained are yellow green and yellow orange. There are other rare earth combinations that would work, such as, yttrium oxyfluoride with an yttrium oxide core, both activated with praseodymium. There may still be other combinations that would yield voltage sensitive screens [12] (figure 1).

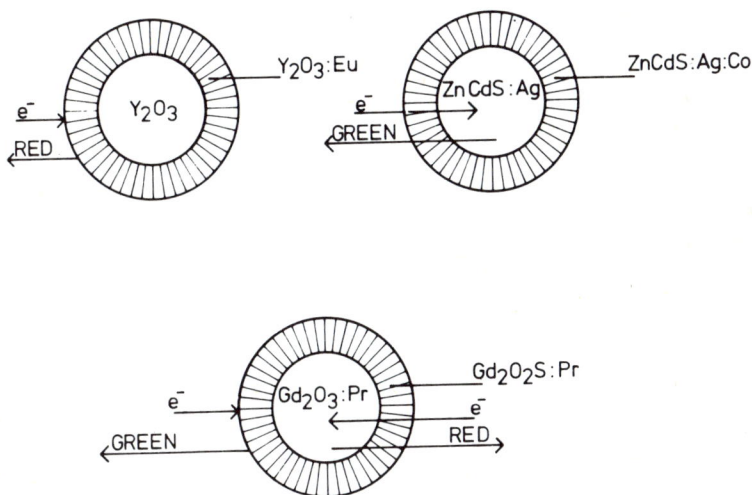

FIGURE 1

Voltage sensitive phosphor systems

Rare earth phosphors are also important as rapid decay phosphors. Decay times of less than 100 ns are achieved with yttrium aluminium garnet and yttrium silicate, both activated with cerium. Tb activated $Sr_3(PO_4)_2$ has been reported useful in reducing flicker because of its longer decay time [13].

There are rare earth phosphors capable of emission in the ultraviolet (Y_2O_3:Gd) as well as the infrared (Ln_2O_2S:Yb, Er, Pr, Tm, Ho).

There has been work on rare earth ions as activators in the II:VI hosts, such as ZnS:Tm which has been reported to have an efficiency of over 20% in ZnS [6]. A number of other rare earth activated sulphide phosphors have been studied [14].

Recently there have been a large number of papers written on rare earth up conversion phosphors [15]. These phosphors emit visible light when excited by infrared radiation coming from a Si doped GaAs diode. This is accomplished by co-doping the host lattice with ytterbium which absorbs and transfers energy to a fluorescing ion such as erbium, holmium, and thulium to produce green, red, and blue emission. A large number of rare earth hosts has been found to be useful, such as rare earth halides, oxyhalides, oxides, $BaYF_5$, $MLnF_4$, etc. Applications people speak of are as indicator lights, improved efficiency is, however, needed.

The larger applications for rare earth phosphors in photoluminescence involve fluorescent lamps. YVO_4:Eu and Y_2O_3:Eu and $YPVO_4$:Eu are used in high pressure mercury lamps and show excellent temperature stability. Other rare earth activated phosphors useful in fluorescent lamps are $Mg(PO_3)_2$:Ce (yellow orange warm colours) and $SrAl_2O_4$:Eu^{+2} (green).

Rare earth activated phosphors have also been found useful in photocopy lamps, e.g. $Sr_3(PO_4)_2$:Eu^{+2} and $Sr_2P_2O_7$:Eu^{+2}.

I would think that costs would prevent the rare earths from being a real major phosphor in the fluorescent lighting field. The halo phosphates are quite inexpensive.

The rare earths hold quite a prominent position in the solid state laser field. Neodymium activated yttrium aluminium garnet is one of the standard solid state lasers used today.

Recently, work has been reported on neodymium activated lanthanum oxysulphide as a laser material by a group at Lockheed [16]. Based on their measurements, they predict that this material would be more efficient than YAG.Nd. To my knowledge, no one has yet grown good quality laser rods of Nd:La$_2$O$_2$S. Many laboratories in the United Stated, however, are working toward this end. I would expect that other efficient rare earth solid state lasers will be developed.

Liquid lasers with rare earth elements are also known. Europium and terbium with organic chelates and Nd in SeOCl$_2$ or POCl$_3$ solvents containing SnCl$_4$ or SbCl$_5$. I am not aware of any large scale applications for these liquid lasers [17].

Europium activated LiF and CaF$_2$ have been reported to be useful scintillator materials.

Electroluminescence (EL) has also been achieved with rare earths. Zinc sulphides activated with rare earth fluorides is deposited in thin film form on conducting substrates. EL has also been achieved by coevaporating ZnS and a rare earth metal. Fifty ft. Lamberts has been reported for ZnS:TbF$_3$ [18].

Direct current EL has also been reported for rare earth activated ZnS [19]. EL of rare earth ions has also been reported in non-aqueous solvents by forming electronically excited ions by electrode processes [20].

Do you think there is a future for rare earth EL as a commercial product?

DISCUSSION
Brown:

What about flat screen television?

Yocom:

This is possible in the future; it is not around the corner. Matsushita is working on tubes using DC Electroluminescence. They have achieved about 10 ft. Lamberts brightness. Aron Vecht's work on DC rare earth electroluminescence offers the potential of a range of colours and increased life. Life of a phosphor screen is one of the unknowns.

The last application of rare earth phosphors I wish to mention is their use as X-ray phosphors. A good X-ray phosphor should have good stopping power for X-rays and good luminescent efficiency. X-ray image intensifier tubes in the past have used ZnCdS:Ag as the input phosphor. New tubes on the market now are using Gd_2O_2S:Tb or CsI:Tl. Reports indicate that La_2O_2S:Tb and Gd_2O_2S:Tb and La_2O_3:Tb are promising materials for this application [21]. These materials should be useful in medical as well as industrial radiography.

In medical radiography an intensifying screen is used with film to decrease X-ray dosage to patients. Calcium tungstate has been used for decades for this application. Recent reports indicate that there are some rare earth phosphors that could be considered as tungstate substitutes; these are barium phosphate:Eu, La_2O_5:Tb, La_2O_2S:Tb, Gd_2O_2S:Tb, and $YGdPO_4$:Tb [22].

Tecotzky:

Barium sulphate : europium is being used as an X-ray phosphor.

I have tried to cover the various areas of luminescence where rare earths are making a contribution. Are there areas that the group knows about that I have missed?

DISCUSSION

No additional comments.

References

[1] G. Urbain, Ann. Chim. Phys., 8^e Serie, 18, 293 (1909).

[2] A.A. Ballman, R.C. Linares and L.G. Van Uitert, U.S. Patent 3,152,085.

[3] A.K. Levine and F.C. Palilla, Appl. Phys. Lett. 5, 118 (1964).

[4] M.R. Royce and A.L. Smith, Electrochemical Society Meeting, Boston, Mass., May 1968.

R.E. Shrader and P.N. Yocom, Electrochemical Society Meeting, Boston, Mass., May 1968.

M. Tecotzky, S.A. Ring, K.A. Wickersheim, and R.A. Buchanan, Electrochemical Society Meeting, Boston, Mass., May 1968.

K.A. Wickersheim, M. Tecotzky, L.E. Sobon, S.A. Ring, and R.A. Buchanan, Electrochemical Society Meeting, Boston, Mass., May 1968.

[5] W. Lehmann, J. Electrochem. Soc. 118, 1164 (1971).

[6] R.E. Shrader, S. Larach, and P.N. Yocom, J. Appl. Phys. 42, 4529 (1971).

[7] V.D. Meyer and F.C. Palilla, J. Electrochem. Soc. 116, 535 (1969).

[8] P. Seats, IEEE Conference on Display Devices, December 1970.

[9] T. Sisneros, P. Wacher, and F. Avella, Electrochemical Society Meeting, Houston, Texas, May 1972.

[10] D.H. Pritchard, U.S. Patent 3,204,143.

R.D. Kell, U.S. Patent 3,523,905.

[11] J.S. Prener and J.D. Kingsley, Electrochemical Society Meeting, Cleveland, Ohio, October 1971.

[12] M. Tecotzky and J.J. Mattis, Electrochemical Society Meeting, Cleveland, Ohio, October 1971.

[13] A. Bril, W.L. Wanmaker, and J.W. terVrugt, J. Electrochem. Soc. 115, 776 (1968).

[14] S. Larach, R.E. Shrader, and P.N. Yocom, J. Electrochem. Soc. 116, 471 (1969).

J.D. Kingsley, J.S. Prener, and M. Aven, Phys. Rev. Lett. 5, 136 (1963).

[15] R.A. Hewes and J.F. Sarver, Bull. Am. Phys. Soc. 13, 687 (1968).

L.G. Van Uitert et al., Appl. Phys. Lett. 15, 53 (1969).

H.J. Guggenheim and L.F. Johnson, App. Phys. Lett. 15, 51 (1969).

L.F. Johnson et al., Appl. Phys. Lett. 15, 48 (1969).

J.P. Wittke, I. Landany, and P.N. Yocom, Proc. IEEE 58, 1283 (1970).

[16] R.V. Alves, R.A. Buchanan, K.A. Wickersheim and E.A.C. Yates, J. Appl. Phys. 42, 3043 (1971).

L.E. Sobon, K.A. Wickersheim, R.A. Buchanan, and R.V. Alves, J. Appl. Phys. 42, 3049 (1971).

[17] A. Heller, Physics Today 20, 34 (1967).

C. Brecher and K.W. French, J. Phys. Chem. 73, 1785 (1969).

[18] E.W. Chase, R.T. Hepplewhite, D.C. Krupka, and D. Kahng, J. Appl. Phys. 40, 2512 (1969).

D.C. Krupka and D.M. Mahoney, J. Appl. Phys. <u>43</u>, 2314 (1972).

D.C. Krupka, J. Appl. Phys. <u>43</u>, 476 (1972).

[19] M.S. Waite and A. Vecht, Appl. Phys. Lett. <u>19</u>, 472 (1971).

[20] A. Heller, K.W. French, and P.O. Haugsjaa, J. Chem. Phys. <u>56</u>, 2368 (1972).

[21] S.P. Wang et al., IEEE Trans. Nucl. Sci. N. S. 17, 49 (1970).

K.A. Wickersheim, R.V. Alves, and R.A. Buchanan, IEEE Trans. Nucl. Sci. N.S. 17, 57 (1970).

J.G. Rabatin and E. Bradshaw, U.S. Patent 3,666,676.

[22] S.Z. Toma, Electrochemical Society Meeting, Washington D.C. May 1971.

R.A. Buchanan, S.I. Finklestein, and K.A. Wickersheim, Radiological Society Meeting, Chicago, Ill., November 1971.

A.P. D'Silva and V. Fassel, Electrochemical Society Meeting, Washington D.C., May 1971.

-/jet
7.12.1972.

LIST OF PARTICIPANTS

Alstad, J.	Chemistry Department, University of Oslo, Norway.
Bjune, A.	Institutt for Atomenergi, Kjeller, Norway.
Blasse, G.	Fysisch Laboratorium, Rijksuniversiteit, Utrecht, The Netherlands.
Bonnevie-Svendsen, M.	Institutt for Atomenergi, Kjeller, Norway.
Brown, C.S.	The General Electric Company Ltd., Hirst Research Centre, Wembley, England.
Brunfelt, A.O.	Mineralogical-Geological Museum, Oslo, Norway.
Braaten, G.	A/S Megon & Co., Oslo, Norway.
Cabral, J.M.P.	Laboratorio de Física de Engenharia Nucleares, Sacavém, Portugal.
Carlson, O.N.	Ames Laboratory USAEC, Iowa State University, Ames, Iowa 50010, USA.
Carvalho, R.A.G. de	Faculty of Engineering, University of Porto, Portugal.
Cetincelik, M.	Nuclear Energy Forum of Turkey, P.K. 37-Bakanliklar, Ankara, Turkey.
Conzemius, R.	Ames Laboratory USAEC, Iowa State University, Ames, Iowa 50010, USA.
Danielsen, B.E.	Institutt for Atomenergi, Kjeller, Norway.
Danielssen,T.	Chemistry Department, University of Oslo, Norway.
Das, H.A.	Reactor Centrum Nederland, Petten (NH), The Netherlands.
Davies, K.E.	Rare Earth Products Ltd., Widnes, Lancs., England.
DeKalb, E.L.	Ames Laboratory USAEC, Iowa State University, Ames, Iowa 50010, USA.
Dugan, M.	Institutt for Atomenergi, Kjeller, Norway.
Fassel, V.O.	Ames Laboratory USAEC, Iowa State University, Ames, Iowa 50010, USA.
Faye, G. Chr.	Geological Survey of Norway, Trondheim, Norway.

Follo, A.	Institutt for Atomenergi, Kjeller, Norway.
Gjelsvik, N.	Institutt for Atomenergi, Kjeller, Norway.
Grade, M.R.	Centro des Estudos de Radioquimica, Lisboa, Portugal.
Griffin, W.L.	Geological Institute, University of Oslo, Norway.
Grøttum, B.	Institutt for Atomenergi, Kjeller, Norway.
Gschneidner, K.A., Jr.	Rare Earth Information Center, Iowa State University, Ames, Iowa 50010, USA.
Hannestad, G.	Institutt for Atomenergi, Kjeller, Norway.
Haskin, L.A.	Department of Chemistry, University of Wisconsin, Madison, Wisconsin 53706, USA.
Herrmann, E.	Technische Universität Dresden, Sektion Chemie, 8027 Dresden, Mommsenstr., DDR.
Hobbs, D.J.	Johnson Matthey Chemicals Ltd., Royston, Herts., England.
Hukin, D.A.	Clarendon Laboratory, Oxford University, Oxford, England.
Hundere, I.	Institutt for Atomenergi, Kjeller, Norway.
Haaland, J.	Institutt for Atomenergi, Kjeller, Norway.
Jones, D.W.	Centre for Materials Science, University of Birmingham, Birmingham, England.
Khaladji, J.	Rhone -Progil, 17010 La Rochelle, France.
Khan, Y.	Institut für Werkstoffe der Elektrotechnik, University of Bochum, 463 Bochum, Germany.
Kvalheim, A.	Geological Survey of Norway, Trondheim, Norway.
Larach, S.	RCA/David Sarnoff Research Center, Princeton, New Jersey 08540, USA.
Lausch, J.	Sektor für Kernchemie, Hahn-Meitner Institut, 1 Berlin 39, Germany.
Lipp, S.	RCA/David Sarnoff Research Center, Princeton, New Jersey 08540, USA.

Mathers, J.E.	Chemical and Metallurgical Division, GTE Sylvania, Towanda, Penna. 18848, USA.
Michelsen, O.B.	Institutt for Atomenergi, Kjeller, Norway.
Molnár, F.	Joint Institute of Nuclear Research, Moscow, USSR.
Müller, P.	Sektor Kernchemie, Hahn-Meitner Institut, 1 Berlin 39, Germany.
Ooyen, J. van	Reactor Centrum Nederland, Petten (NH), The Netherlands.
Pappas, A.C.	Chemistry Department, University of Oslo, Norway.
Rannestad, A.	Scientific Affairs Division, NATO, 1110 Brussels, Belgium.
Skarestad, M.	The University of Manchester, Manchester, England.
Shrader, R.E.	The University of Princeton, New Jersey 08540, USA.
Stäblein, H.	Krupp Research Center, D43 Essen, Germany.
Steinberg, M.	Department of Inorganic and Analytical Chemistry, The Hebrew University of Jerusalem, Israel.
Steinnes, E.	Institutt for Atomenergi, Kjeller, Norway.
Stijfhoorn, D.E.	Institutt for Atomenergi, Kjeller, Norway.
Tecotzky, M.	United States Radium Corporation, Hackettstown, New Jersey 07840, USA.
Tjølsen, P.	A/S Megon & Co, Oslo, Norway.
Whitley, J.E.	Scottish Research Reactor Centre, East Kilbride, Glascow, Scotland.
Yocom, P.N.	RCA/David Sarnoff Research Center, Princeton, New Jersey 08540, USA.
Ødegård, M.	Geological Survey of Norway, Trondheim, Norway.
Østgaard, E.	Norges Lærerhøgskole, Trondheim, Norway
Åmli, R.	Mineralogical-Geological Museum, Oslo, Norway.